## Praise for *Gemba Kaizen*

It's exciting to see an updated version of a classic book, *Gemba Kaizen*, which shares a wealth of new healthcare examples and case studies from around the world. A true *sensei* and master of *kaizen*, Mr. Imai shares sage and timeless advice on engaging all team members in process improvements and radical redesigns which are deeply meaningful to all stakeholders. The methods in this book will help you improve quality and safety, reduce waiting times, and improve the long-term financial position of your organization. Highly recommended!

—Mark Graban, author of *Lean Hospitals*
and co-author of *Healthcare Kaizen*

Every business faces the iron triangle of quality, cost, and delivery. Conventional thinking claims you cannot have all three. Not only does Mr. Imai turn that thinking on its head, but he shows you in *Gemba Kaizen* exactly how to do it.

—Matthew E. May, author of *The Elegant Solution*
and *The Laws of Subtraction*

Masaaki Imai has done it again. The second edition of his famous book *Gemba Kaizen* not only describes all the tools necessary for any type of business to implement a lean strategy but also includes a large number of excellent case studies. These show how *kaizen* can be used to improve hospitals, supermarkets, airport management, a bus line, and even software development. This is a must-read for the leadership of any business.

My first exposure to lean [the term hadn't been invented yet, we called it Just-in-Time or the Toyota Production System] was at the beginning of 1982, during my first General Manager job at the General Electric Company. We created a simple *kanban* system between one of my plants and one of my suppliers. We dropped raw material inventory from 40 days to 3 days and got a lot of unexpected side benefits in the areas of productivity, quality, freed up space, 5S improvements, etc. Professor Schoenburger later did a story on this where he said that this was the first real lean activity at The General Electric Company. In late 1985 I joined the Danaher Company as one of two Group Executives. One of my company presidents, George Koenigsaecker, and I began introducing lean to Danaher in 1986. One of the things that really helped us improve our knowledge of lean at the time was Masaaki Imai's first book, *Kaizen*. This was the most definitive work on the subject and was a great help. Imai helped us even more in early 1987 when he ran a seminar in the Hartford, CT area [just down the street from Jake Brake]. Imai used a Japanese consulting firm, Shingijutsu, to help run his seminar and be responsible for the hands-on factory *kaizen* part of the week. The three principals of Shingijutsu all had spent years working for Taiichi Ohno, the father of the Toyota production system. Koenigsaecker and I agreed that getting Shingijutsu to help us at Danaher would be a home run for us and George worked diligently the rest of the week convincing them. We became their first, and for four years, *only* American client and our lean knowledge increased dramatically.

In 1991, I left Danaher to become CEO of The Wiremold Company, also in the area. I, of course, brought Shingijutsu along with me and by 1996 Masaaki Imai was back in my life as he included a chapter on Wiremold and what we had done in his new book, *Gemba Kaizen*. We have stayed in touch over the years and Imai has become a true leader in the lean movement throughout the world through his Kaizen institute. He clearly understands that lean is a strategy, not just "some manufacturing thing" and that it can apply to any business. He and I have discussed why is it so difficult for most business leaders to understand this and to embrace lean. Unfortunately there is no simple answer to this other than the fact that most people just don't like to change and implementing lean is massive change (everything has to change) if you are to be successful. This latest edition of *Gemba Kaizen* goes a long way to helping to solve this problem. First of all, it lays out the lean philosophy and tools in a very simple way so that executives should not only understand them but more importantly, not be afraid to try them. More importantly however, Imai makes the case that lean is a *strategy* and that it can be applied to any business. His case studies of non-manufacturing companies where lean has had a dramatic impact really help to make the point. Every leader of any type of organization should read this book and follow what it says.

—Art Byrne, Operating Partner at J W Childs Associates, LP
and author of *The Lean Turnaround*

# Gemba Kaizen

## About the Author

More than any other business authority in the world, **Masaaki Imai** has championed the concept of *kaizen* over the past three decades in thought, word, and action. Mr. Imai is considered one of the leaders of the quality movement and a pioneer of modern business operational excellence. Mr. Imai is an international lecturer, consultant, and founder of the Kaizen Institute, a leading continuous improvement consultancy with offices worldwide. Mr. Imai's first book, *Kaizen*—translated into 14 languages—is the reference on the subject. *Gemba Kaizen* picks up where *Kaizen* left off, introducing real-world application of continuous process improvement methods in production and service businesses. The second edition is fully revised with brand-new case studies, updated chapters, and current references. In 2010 Mr. Imai was honored for his lifetime of achievement with the first ever Fellowship of the Quality Council of India, the apex quality body of the government of India.

# Gemba Kaizen

## A Commonsense Approach to a Continuous Improvement Strategy

**Masaaki Imai**

**Second Edition**

New York Chicago San Francisco
Lisbon London Madrid Mexico City
Milan New Delhi San Juan
Seoul Singapore Sydney Toronto

The McGraw-Hill Companies

Cataloging-in-Publication Data is on file with the Library of Congress.

McGraw-Hill books are available at special quantity discounts to use as premiums and sales promotions, or for use in corporate training programs. To contact a representative please e-mail us at bulksales@mcgraw-hill.com.

**Gemba Kaizen: A Commonsense Approach to a Continuous Improvement Strategy, Second Edition**



12 13 14 15 QFR 21 20

ISBN 978-0-07-179035-2
MHID 0-07-179035-7



The pages within this book were printed on acid-free paper.

**Sponsoring Editor**
Judy Bass

**Acquisitions Coordinator**
Bridget Thoreson

**Editorial Supervisor**
David E. Fogarty

**Project Manager**
Patricia Wallenburg

**Copy Editor**
James Madru

**Proofreader**
Paul Tyler

**Indexer**
Judy Davis

**Production Supervisor**
Pamela A. Pelton

**Composition**
TypeWriting

**Art Director, Cover**
Jeff Weeks

# CONTENTS

# FOREWORD

The middle of the last century saw the dawning of a new industrial revolution, founded on the work and teachings of process innovators such as Deming, Juran, Ishikawa, Taguchi, and Crosby. By embracing these principles in both practice and spirit, Japanese companies led the world in this deep transformation.

In less than thirty years, these practices led to new performance standards for not only quality but also productivity in a transformation that has spread to all areas of industry, including the service sectors. The benefits to companies and society have been tremendous—companies perform better financially, employees work in safer and more pleasant environments, and customers enjoy higher quality products and services at affordable prices.

I find it particularly fascinating that this extraordinary change can be traced to two basic ideas. The first is that extremely simple tools and concepts—listening to the customer, SPC (statistical process control), PDCA (plan, do, check, act), cause-effect diagrams—can hold the key to substantial improvements in large, complex operations. The other, no less important, is the application of discipline and determination throughout the organization in applying these principles and concepts in order to make them a reality.

It is in this latter category that Japanese companies stood out from their counterparts. By instilling the spirit of *kaizen*—continuous improvement—throughout their workplaces, companies in Japan not only improved their processes, but became extremely adept in developing and adopting new ideas.

From the 1980s on, the results obtained by the Japanese companies were so superior to those of their competitors in almost all sectors that there emerged a significant movement to understand the Japanese "secrets" and how to apply them.

My company, Embraco, began to manufacture refrigeration equipment in Brazil in the early 1970s, originally as a technological licensee of a European company. By the end of the 1980s, Embraco had initiated TQM (total quality management) using the Japanese model as a benchmark.

Beginning in the 1990s, the company began to expand globally, and at the same time, adopt global manufacturing standards, which were applied in facilities in Europe and China. Through these efforts, combined with significant investments in product innovation, Embraco became, by the end of the decade, a world leader in the manufacture of hermetic compressors.

In 2004, the company decided to take additional steps to improve operationally. The following year, the company established a partnership with Kaizen Institute to establish a new learning cycle aimed at educating employees and putting lean manufacturing concepts into practice.

Since its initiation, the program has contributed significantly to the enhancement of operational performance and, in many cases, has transformed Embraco's practices and routines. Two methods that are now firmly integrated into our processes are worth noting. The first, VSM (value stream mapping), serves as the benchmark for our annual improvement plans. This organizes how we deliver value to our customers.

The other method is the practice of implementing improvements with *gemba kaizen*, which means continuous improvement in the *gemba*—the workplace. It is through *gemba kaizen* that we have learned to work with multifunctional teams, acting in an intense and proficient way to create change for the better.

Our global environment, however, is also changing. As we evolve, new challenges confront us—challenges of understanding and responding even more quickly to the opportunities and demands of the market with differentiated products and services, of giving an attractive return to stockholders so that they continue to invest in our business, and last but not least, of contributing to a more sustainable world.

In order to achieve these aims, it is necessary to maintain an efficient operation that minimizes waste and cultivates a mindset of continuous improvement. This is a task for everyone in the organization, to be executed with discipline and determination, following what the Japanese showed us fifty years ago—an example that never ceases to be useful, even in our current days.

This book will certainly bring more inspiration and knowledge to help us move forward in this endless journey towards operational excellence.

MARIO SERGIO USSYK
*Vice President, Corporate Management,*
*Quality & EHS*
*Embraco*

# PREFACE

My two books, *Kaizen: The Key to Japan's Competitive Success* (McGraw-Hill, 1986) and *Gemba Kaizen: A Commonsense, Low-Cost Approach to Management* (McGraw-Hill, 1997) laid the foundation for exploration of *kaizen* as both a personal philosophy and business improvement system for people outside of Japan. Initially grasped as a set of methods such as total quality control, total productive maintenance, just-in-time management, quality circles, and suggestion systems, the West is ever closer to understanding *kaizen* for what it truly is: a strategy to win by developing people into problem solvers.

The second edition of *Gemba Kaizen* reveals how *kaizen* has spread to every continent and culture, met with various unique challenges and demonstrated its success. *Gemba* means "actual place" or "workplace" in Japanese, and this book gives you a look into more than thirty actual places where *kaizen* was successfully made a part of the culture. The book explains how to use a commonsense, low-cost approach to managing the workplace—the place where value is added—whether that place be the production line, hospital, government department, shopping center, airport, or engineering firm. This is not a book of theory, but a book of action. Its ultimate message is that no matter how much knowledge the reader may gain, it is of no use if it is not put into practice daily. *Gemba Kaizen* provides not more theoretical knowledge, but a simple frame of reference to use in solving problems. To that purpose, it provides many checklists, examples, and case studies.

## The Commonsense, Low-Cost Approach to a Continuous Improvement Strategy

Today's managers often try to apply sophisticated tools and technologies to deal with problems that can be solved with a commonsense, low-cost approach. They need to unlearn the habit of trying ever more sophisticated technologies to solve everyday problems. Furthermore, leaders must embrace *kaizen* and business excellence not as a tool or technique but as a never-finished pillar of their strategy.

Putting common sense into practice is the subject of this book. It is for everybody: managers, engineers, supervisors, and rank and file employees. Along with putting common sense into practice, *Gemba Kaizen* deals with the roles of managers and the need to develop a learning organization. I believe that one of the roles of top management should be to challenge all managers to attain ever higher goals. In turn, first-line supervisors need to challenge workers to do a better job all the time. Unfortunately, many managers today have long ceased to play such a role.

Another problem besetting most companies today is the tendency to place too much emphasis on *teaching* knowledge, while disregarding group *learning* of fundamental values derived from common sense, self-discipline, order, and economy. Good management should strive to lead the company to learn these values while achieving "lean management."

There are two approaches to problem solving. The first involves innovation—applying the latest high-cost technology, such as state-of-the-art computers and other tools, and investing a great deal of money. The second uses commonsense tools, checklists, and techniques that do not cost much money. This approach is called *kaizen*. *Kaizen* involves *everybody*—starting with the CEO in the organization—planning and working together for success. This book will show how *kaizen* can achieve significant improvement as an essential building block that prepares the company for truly rewarding accomplishments.

## Back to Basics: Housekeeping, *Muda* Elimination, and Standardization

During the past 27 years since *Kaizen* was first published, many have looked for and asked "what is next?" but many times they are overlooking what is directly in front of them. We must go back to the basics and ask how well we have kept a steady, long-term focus on *kaizen*. Everyone in the company must work together to follow three ground rules for practicing *kaizen* in the *gemba*:

- Housekeeping
- *Muda* elimination

▲ Standardization

Housekeeping is an indispensable ingredient of good management. Through good housekeeping, employees acquire and practice self-discipline. Employees without self-discipline make it impossible to provide products or services of good quality to the customer.

In Japanese, the word *muda* means waste. Any activity that does not add value is *muda*. People in the *gemba* either add value or do not add value. This is also true for other resources, such as machines and materials. Suppose a company's employees are adding nine parts *muda* for every one part value. Their productivity can be doubled by reducing *muda* to eight parts and increasing the added value to two parts. *Muda* elimination can be the most cost-effective way to improve productivity and reduce operating costs. *Kaizen* emphasizes the elimination of *muda* in the *gemba* rather than the increasing of investment in the hope of adding value.

A simple example illustrates the cost benefits of *kaizen*. Suppose that operators assembling a household appliance are standing in front of their workstations to put certain parts into the main unit. The parts for assembly are kept in a large container behind the operators. The action of turning around to pick up a part takes an operator five seconds, while actual assembly time is only two seconds.

Now let's assume the parts are placed in front of the operator. The operator simply extends his or her arms forward to pick up a part—an action that takes only a second. The operators can use the time saved to concentrate on the (value-adding) assembly. A simple change in the location of the parts—eliminating the *muda* involved in the action of reaching behind—has yielded a four-second time gain that translates into a threefold increase in productivity!

Such small improvements in many processes gradually accumulate, leading to significant quality improvement, cost benefits, and productivity improvements. Applying such an approach throughout all management activities, especially at top management levels, gradually achieves a just-in-time, lean management system by teaching people the skills to see their work in a new way and by teaching them the skills to change how they work. By contrast, management primarily focused on innovation and breakthroughs might be inclined to buy software, equipment or capabilities that would enable the organization to perform their work much faster. But this would not eliminate the *muda* inherent in the current system. Furthermore, investing in the new device or capability costs money, while eliminating *muda* costs nothing. We must innovate, but on a foundation of *kaizen*. The case study from Densho Engineering and others in this book reveal how this is done.

The third ground rule of *kaizen* practices in the *gemba* is standardization. Standards may be defined as the best way to do the job. For products or services created as a result of a series of processes, a certain standard must be maintained at each process in order to assure quality. Maintaining standards is a way of assuring quality at each process and preventing the recurrence of errors.

As a general rule of thumb, introducing good housekeeping in the *gemba* reduces the failure rate by 50 percent, and standardization further reduces the failure rate by 50 percent of the new figure. Yet many managers elect to introduce statistical process control and control charts in the *gemba* without making efforts to clean house, eliminate *muda*, or standardize.

Supporting these rules of *kaizen* is the foundation of the house of *gemba*—namely, the use of such human-centered activities as learning together, teamwork, morale enhancement, self-discipline, quality circles, and suggestions. These are all methods not only for generating improvements in safety, quality and cost, but positive means to *kaizen* and develop our people.

Management (especially Western management) must regain the power of common sense and start applying it in the *gemba*. These low-cost practices will provide management with the opportunity for a future phase of rapid growth via innovation—something Western management excels at. When Western management combines *kaizen* with its innovative ingenuity, it will greatly improve its competitive strength.

MASAAKI IMAI
*Tokyo*

# ACKNOWLEDGMENTS

The first edition of *Gemba Kaizen* was born out of 10 years of teaching *kaizen*, following the publication of my book *Kaizen: The Key to Japan's Competitive Success* in 1986. The second edition of *Gemba Kaizen* comes 15 years after the original publication, and much has changed in the world. I have been fortunate to see the transforming effect of kaizen on people and organizations worldwide over the past three decades.

I wish to recognize and thank everyone who has taken up *kaizen*. The many cases and explanations of *kaizen* which are documented in this second edition are the fruits of the many workers, engineers, administrators, nurses, officials, managers and professionals who practice continuous improvement and were engaged in *gemba kaizen* at our clients' sites around the world. This book is truly a result of teamwork, collaboration and the *kaizen* spirit at work.

I would like to thank those people who assisted in writing the cases in the first edition of this book. They include Kevin Meyer of Specialty Silicone, and Arthur Byrne of J.W. Childs, Inc., Iwao Sumoge of Densho Engineering, Joao-Paulo Oliveira of Bosch, Natacha Muro and Fernando Coletti of La Buenos Aires, Nestor Herrerra of Molinos Rio de la Plata, Gary Buchanan and Valerie Oberle of Disney University, Darla Hastings of Quality Inc., Shoji Shiratori of Aisin Seiki, and Yutaka Mori of Toyoda Automatic Loom Works as well as Yoshikazu Sano and Katsuo Inoue of Toyoda Machine Works.

Besides those whose names appear in the book, I am particularly indebted to Professor Zenjiro Sawada at Kurume University, who gave me the inspiration for the House of Gemba Management through his book *Visual Control of Factory Management* (published in English in 1991 from Nikkan Kogyo Shinbun); Ichiro Majima, Dean of Faculty of Business Administration of Miyazaki Sangyo Keiei University, who provided much valuable information in writing this book; Kaizen consultants Kenji Takahashi, Yukio Kakiuchi, and many others who worked together with us in giving many *gemba kaizen* sessions at the clients' sites around the world.

Once again I am deeply indebted to my colleagues at the Kaizen Institute since their work has made it possible to advance the ideas from my books into actual practice. I would like to thank those who led the efforts to document their experiences and the stories of our customers, including Antonio Costa, Daniel Simoes, Vinod Grover, Sebastian Reimer, Jayanth Murthy, Bruno Fabiano, David Lu, Mike Wroblewski, Julien Bratu, Jefferson Escobar, Aakash Borse, Chris Schrandt, John Verhees, Wijbrand Medendorp, Brad Schmidt, Ruy Cortez, Alexandra Caramalho, and Euclides Coimbra. I would like to thank Jon Miller for managing this book project.

I must thank Jacob Stoller of StollerStrategies for bringing his editorial, journalistic, and creative skills, which were essential to the successful completion of this project.

I wish to thank my wife Noriko for her patience and for accompanying me on my travels around the world; also to those who assisted in the making of this book, in particular Patty Wallenburg of TypeWriting for the composition, and from McGraw-Hill, Pamela Pelton and David Fogarty for the production, Judy Bass, who provided the spark for making the second edition of this book a reality, and Philip Ruppel, President of McGraw-Hill Professional and McGraw-Hill Education International, without whom the original book would never have come to fruition.

# ABOUT KAIZEN INSTITUTE

Founded by Masaaki Imai in 1985, Kaizen Institute is the pioneer and global leader in promoting the spirit and practice of *kaizen*. Its global team of professionals is dedicated to building a world where it is possible for everyone, everywhere, every day is able to "*kaizen* it."

Kaizen Institute guides organizations (public and private) to achieve higher levels of performance in the global marketplace—easier, faster, better, and with lower costs. Kaizen Institute experts challenge clients to help develop leaders capable of sustaining continuous improvement in all aspects of their enterprise, which ultimately leads into Kaizen Institute's vision of a worldwide community of practice in *kaizen*.

Major services of Kaizen Institute, include, but are not limited to:

- Consulting and Implementation
  - Partnering with clients for long-term *kaizen* implementation
  - Operating system design and deployment
  - Breakthrough projects and turnarounds
- Education, Training, and Events
  - Business training, academic, and online training curriculum design
  - Kaizen practitioner, coach, and manager level certification
  - On/off-site training, workshops, seminars, corporate events, and leadership sessions
- Tours and Benchmarking
  - "Kaikaku" benchmark to best-in-class organizations in Japan and around the world
  - Building peer-to-peer learning and tour exchange network

Visit www.kaizen.com to learn more about *kaizen* and the world-changing purpose of Kaizen Institute.

## CHAPTER ONE

# An Introduction to *Kaizen*

Since 1986 when the book *Kaizen: The Key to Japan's Competitive Success* was published, the term *kaizen* has come to be accepted as one of the key concepts of management. In the first decade of the twenty-first century as the Toyota Motor Company surpassed General Motors to become the top automotive manufacturer in the world, awareness of the vital difference played by *kaizen* in Toyota's success also increased.

Today, organizations worldwide from manufacturers, to hospitals, to banks, to software developers, to governments are making a difference by adopting *kaizen* philosophies, mind-sets, and methodologies. Even though the names of these strategies may change over the decades from continuous quality improvement and total quality management, to just-in-time and operational excellence, to six sigma and lean manufacturing, the most successful of these strategies are customer-focused, *gemba*-oriented, and *kaizen*-driven.

The 1993 edition of the *New Shorter Oxford English Dictionary* recognized the word *kaizen** as an English word. The dictionary defines *kaizen* as "continuous improvement of working practices, personal efficiency, etc., as a business philosophy." Readers who are unfamiliar with *kaizen* may find it helpful to begin with a brief summary of the concepts of *kaizen*. For those who are already familiar with *kaizen*, this chapter may serve as a review.

In Japanese, *kaizen* means "continuous improvement." The word implies improvement that involves everyone—both managers and workers—and entails relatively little expense. The *kaizen* philosophy assumes that our way

---

*Kaizen Institute AG has exclusive right to the use of *kaizen*®, as well as *gemba kaizen*®, as trademarks registered in major countries of the world.

of life—be it our working life, our social life, or our home life—should focus on constant improvement efforts. This concept is so natural and obvious to many Japanese that they don't even realize they possess it! In my opinion, *kaizen* has contributed greatly to Japan's competitive success.

Although improvements under *kaizen* are small and incremental, the *kaizen* process brings about dramatic results over time. The *kaizen* concept explains why companies cannot remain static for long in Japan. Western management, meanwhile, worships innovation: major changes in the wake of technological breakthroughs and the latest management concepts or production techniques. Innovation is dramatic, a real attention-getter. *Kaizen*, on the other hand, is often undramatic and subtle. But innovation is one-shot, and its results are often problematic, whereas the *kaizen* process, based on commonsense and low-cost approaches, ensures incremental progress that pays off in the long run. *Kaizen* is also a low-risk approach. Managers always can go back to the old way without incurring large costs.

Most "uniquely Japanese" management practices, such as total quality control (TQC) or companywide quality control and quality circles, and our style of labor relations can be reduced to one word: *kaizen*. Using the term *kaizen* in place of such buzzwords as *productivity*, *total quality control* (TQC), *zero defects* (ZDs), *just-in-time* (JIT), and the *suggestion system* paints a clearer picture of what has been going on in Japanese industry. *Kaizen* is an umbrella concept for all these practices. However, I hasten to add that these practices are not necessarily confined to Japanese management but rather should be regarded as sound principles to be applied by managers everywhere. By following the right steps and applying the processes properly, any company, no matter what its nationality, can benefit from *kaizen*. The widespread acceptance of *kaizen* into management thinking, including the successes of Kaizen Institute clients in more than 50 countries, bears this out.

## Major *Kaizen* Concepts

Management must learn to implement certain basic concepts and systems in order to realize *kaizen* strategy:

- *Kaizen* and management
- Process versus result

- Following the plan-do-check-act (PDCA)/standardize-do-check-act (SDCA) cycles
- Putting quality first
- Speak with data.
- The next process is the customer.

By way of introduction, top management must put forth a very careful and very clear policy statement. It then must establish an implementation schedule and demonstrate leadership by practicing a *kaizen* procedure within its own ranks.

## Kaizen *and Management*

In the context of *kaizen*, management has two major functions: maintenance and improvement (see Figure 1.1). *Maintenance* refers to activities directed toward maintaining current technological, managerial, and operating standards and upholding such standards through training and discipline. Under its maintenance function, management performs its assigned tasks so that everybody can follow standard operating procedures (SOPs). *Improvement*, meanwhile, refers to activities directed toward elevating current standards. The Japanese view of management thus boils down to one precept: Maintain and improve standards.

As Figure 1.2 shows, improvement can be classified as either *kaizen* or innovation. *Kaizen* signifies small improvements as a result of ongoing efforts. Innovation involves a drastic improvement as a result of a large investment of resources in new technology or equipment. (Whenever money is a key factor, innovation is expensive.) Because of their fascination

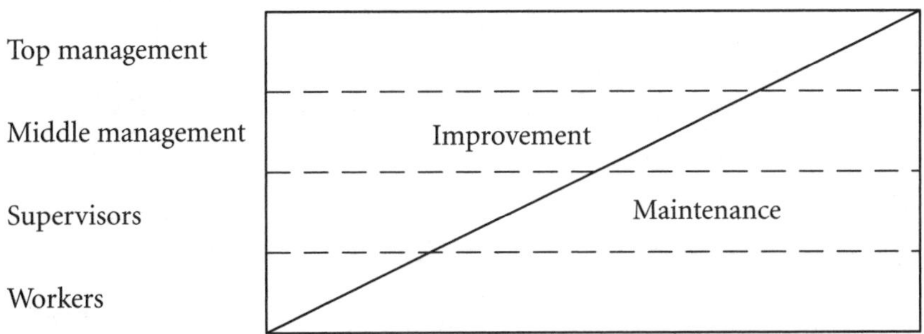


**Figure 1.1** Japanese perceptions of job functions.

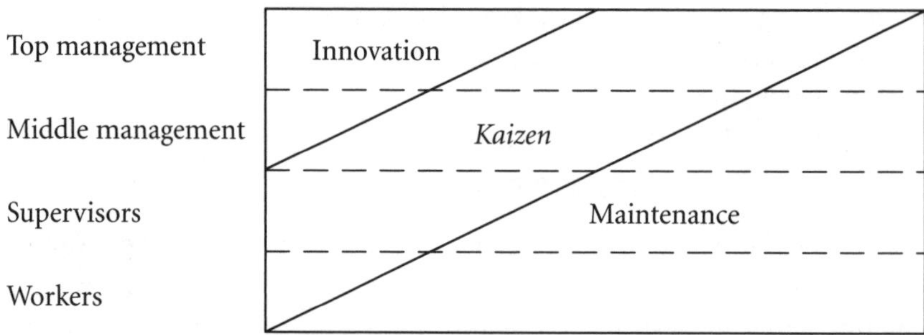


**Figure 1.2** Improvement broken down into innovation and *kaizen*.

with innovation, Western managers tend to be impatient and overlook the long-term benefits *kaizen* can bring to a company. *Kaizen*, on the other hand, emphasizes human efforts, morale, communication, training, teamwork, involvement, and self-discipline—a commonsense, low-cost approach to improvement.

## *Process versus Result*

*Kaizen* fosters process-oriented thinking because processes must be improved for results to improve. Failure to achieve planned results indicates a failure in the process. Management must identify and correct such process-based errors. *Kaizen* focuses on human efforts—an orientation that contrasts sharply with the results-based thinking in the West.

A process-oriented approach also should be applied in the introduction of the various *kaizen* strategies: the plan-do-check-act (PDCA) cycle; the standardize-do-check-act (SDCA) cycle; quality, cost, and delivery (QCD); total quality management (TQM); just-in-time (JIT); and total productive maintenance (TPM). *Kaizen* strategies have failed many companies simply because they ignored process. The most crucial element in the *kaizen* process is the commitment and involvement of top management. It must be demonstrated immediately and consistently to ensure success in the *kaizen* process.

## *Following the PDCA/SDCA Cycles*

The first step in the *kaizen* process establishes the *plan-do-check-act (PDCA) cycle* as a vehicle that ensures the continuity of *kaizen* in pursuing a policy

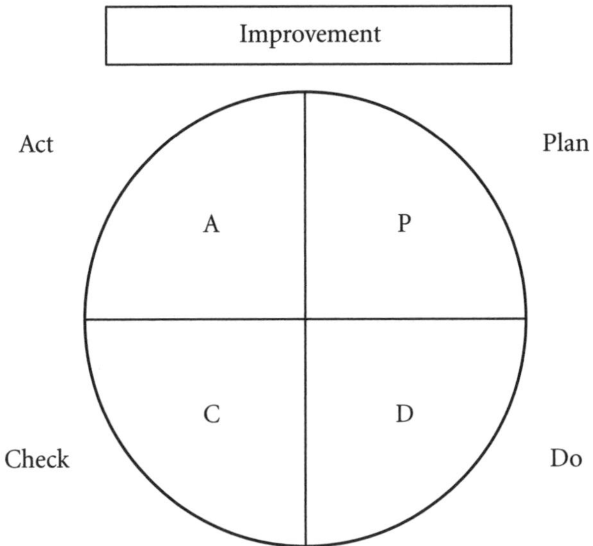


**Figure 1.3** The plan-do-check-act (PDCA) cycle.

of maintaining and improving standards. It is one of the most important concepts of the process (see Figure 1.3).

Plan refers to establishing a target for improvement (since *kaizen* is a way of life, there always should be a target for improvement in any area) and devising action plans to achieve that target. *Do* refers to implementing the plan. *Check* refers to determining whether the implementation remains on track and has brought about the planned improvement. *Act* refers to performing and standardizing the new procedures to prevent recurrence of the original problem or to set goals for the new improvements. The PDCA cycle revolves continuously; no sooner is an improvement made than the resulting status quo becomes the target for further improvement. PDCA means never being satisfied with the status quo. Because employees prefer the status quo and frequently do not have initiative to improve conditions, management must initiate PDCA by establishing continuously challenging goals.

In the beginning, any new work process is unstable. Before one starts working on PDCA, any current process must be stabilized in a process often referred to as the *standardize-do-check-act (SDCA) cycle* (see Figure 1.4).

Every time an abnormality occurs in the current process, the following questions must be asked: Did it happen because we did not have a standard?

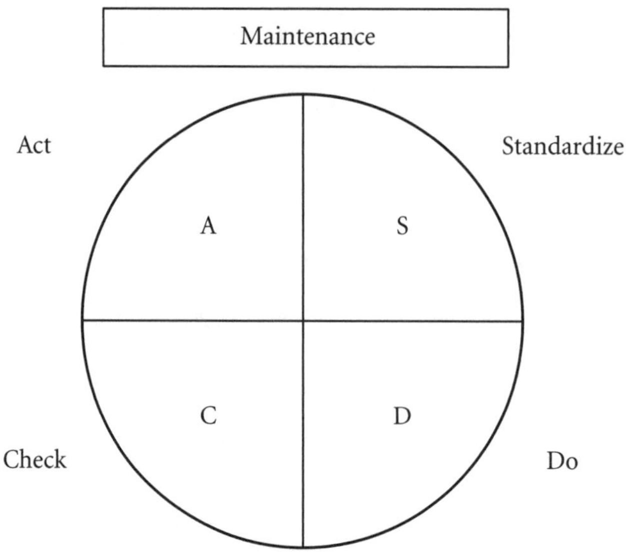


**Figure 1.4** The standardize-do-check-act (SDCA) cycle.

Did it happen because the standard was not followed? Or did it happen because the standard was not adequate? Only after a standard has been established and followed, stabilizing the current process, should one move on to the PDCA cycle.

Thus the SDCA cycle standardizes and stabilizes the current processes, whereas the PDCA cycle improves them. SDCA refers to maintenance, and PDCA refers to improvement; these become the two major responsibilities of management.

## *Putting Quality First*

Of the primary goals of quality, cost, and delivery (QCD), quality always should have the highest priority. No matter how attractive the price and delivery terms offered to a customer, the company will not be able to compete if the product or service lacks quality. Practicing a quality-first credo requires management commitment because managers often face the temptation to make compromises in meeting delivery requirements or cutting costs. In so doing, they risk sacrificing not only quality but also the life of the business.

### *Speak with Data*

*Kaizen* is a problem-solving process. In order for a problem to be correctly understood and solved, the problem must be recognized and the relevant data gathered and analyzed. Trying to solve a problem without hard data is akin to resorting to hunches and feelings—not a very scientific or objective approach. Collecting data on the current status helps you to understand where you are now focusing; this serves as a starting point for improvement. Collecting, verifying, and analyzing data for improvement is a theme that recurs throughout this book.

### *The Next Process Is the Customer*

All work is a series of processes, and each process has its supplier as well as its customer. A material or a piece of information provided by process A (supplier) is worked on and improved in process B and then sent on to process C. The next process always should be regarded as a customer. The axiom "the next process is the customer" refers to two types of customers: internal (within the company) and external (out in the market).

Most people working in an organization deal with internal customers. This realization should lead to a commitment never to pass on defective parts or inaccurate pieces of information to those in the next process. When everybody in the organization practices this axiom, the external customer in the market receives a high-quality product or service as a result. A real quality-assurance system means that everybody in the organization subscribes to and practices this axiom.

## Major *Kaizen* Systems

The following are major systems that should be in place in order to successfully achieve a *kaizen* strategy:

- Total quality control (TQC)/total quality management (TQM)
- A just-in-time (JIT) production system (Toyota Production System)
- Total productive maintenance (TPM)
- Policy deployment
- A suggestion system
- Small-group activities

## *Total Quality Control/Total Quality Management*

One of the principles of Japanese management has been total quality control (TQC), which, in its early development, emphasized control of the quality process. This has evolved into a system encompassing all aspects of management and is now referred to as *total quality management* (TQM), a term used internationally.

Regarding the TQC/TQM movement as a part of *kaizen* strategy gives us a clearer understanding of the Japanese approach. Japanese TQC/TQM should not be regarded strictly as a quality-control activity; TQC/TQM has been developed as a *strategy* to aid management in becoming more competitive and profitable by helping it to improve in all aspects of business. In TQC/TQM, *Q*, meaning "quality," has priority, but there are other goals, too—namely, cost and delivery.

The *T* in TQC/TQM signifies "total," meaning that it involves everybody in the organization, from top management through middle managers, supervisors, and shop-floor workers. It further extends to suppliers, dealers, and wholesalers. The *T* also refers to top management's leadership and performance—so essential for successful implementation of TQC/TQM.

The *C* refers to "control" or "process control." In TQC/TQM, key processes must be identified, controlled, and improved on continuously in order to improve results. Management's role in TQC/TQM is to set up a plan to check the process against the result in order to improve the process, not to criticize the process on the basis of the result.

TQC/TQM in Japan encompasses such activities as policy deployment, building quality-assurance systems, standardization, training and education, cost management, and quality circles.

## *The Just-in-Time Production System*

Originating at Toyota Motor Company under the leadership of Taiichi Ohno, the just-in-time (JIT) production system aims at eliminating non-value-adding activities of all kinds and achieving a lean production system that is flexible enough to accommodate fluctuations in customer orders. This production system is supported by such concepts as *takt* time (the time it takes to produce one unit) versus cycle time, one-piece flow, pull production, *jidoka* ("autonomation"), U-shaped cells, and setup reduction.

To realize the ideal JIT production system, a series of *kaizen* activities must be carried out continuously to eliminate non-value-adding work in *gemba*. JIT dramatically reduces cost, delivers the product in time, and greatly enhances company profits.

## *Total Productive Maintenance*

An increasing number of manufacturing companies now practice *total productive maintenance* (TPM) within as well as outside of Japan. Whereas TQM emphasizes improving overall management performance and quality, TPM focuses on improving equipment quality. TPM seeks to maximize equipment efficiency through a total system of preventive maintenance spanning the lifetime of the equipment.

Just as TQM involves everybody in the company, TPM involves everybody at the plant. The five S of housekeeping (discussed in Chapter 5), another pivotal activity in *gemba*, may be regarded as a prelude to TPM. However, 5S activities have registered remarkable achievements in many cases even when carried out separately from TPM.

## *Policy Deployment*

Although *kaizen* strategy aims at making improvements, its impact may be limited if everybody is engaged in *kaizen* for *kaizen*'s sake without any aim. Management should establish clear targets to guide everyone and make certain to provide leadership for all *kaizen* activities directed toward achieving the targets. Real *kaizen* strategy at work requires closely supervised implementation. This process is called Policy Deployment, or in Japanese, *hoshin kanri*.

First, top management must devise a long-term strategy, broken down into medium-term and annual strategies. Top management must have a plan-to-deploy strategy, passing it down through subsequent levels of management until it reaches the shop floor. As the strategy cascades down to the lower echelons, the plan should include increasingly specific action plans and activities. For instance, a policy statement along the lines of "We must reduce our cost by 10 percent to stay competitive" may be translated on the shop floor to such activities as increasing productivity, reducing inventory and rejects, and improving line configurations.

*Kaizen* without a target would resemble a trip without a destination. *Kaizen* is most effective when everybody works to achieve a target, and management should set that target.

## The Suggestion System

The *suggestion system* functions as an integral part of individual-oriented *kaizen* and emphasizes the morale-boosting benefits of positive employee participation. Japanese managers see its primary role as that of sparking employee interest in *kaizen* by encouraging them to provide many suggestions, no matter how small. Japanese employees are often encouraged to discuss their suggestions verbally with supervisors and put them into action right away, even before submitting suggestion forms. They do not expect to reap great economic benefits from each suggestion. Developing *kaizen*-minded and self-disciplined employees is the primary goal. This outlook contrasts sharply with that of Western management's emphasis on the economic benefits and financial incentives of suggestion systems.

## Small-Group Activities

A *kaizen* strategy includes *small-group activities*—informal, voluntary, intracompany groups organized to carry out specific tasks in a workshop environment. The most popular type of small-group activity is *quality circles.* Designed to address not only quality issues but also such issues as cost, safety, and productivity, quality circles may be regarded as group-oriented *kaizen* activities. Quality circles have played an important part in improving product quality and productivity in Japan. However, their role often has been blown out of proportion by overseas observers, who believe that these groups are the mainstay of quality activities in Japan. Management plays a leading role in realizing quality—in ways that include building quality-assurance systems, providing employee training, establishing and deploying policies, and building cross-functional systems for QCD. Successful quality-circle activities indicate that management plays an invisible but vital role in supporting such activities.

## The Ultimate Goal of *Kaizen* Strategy

Since *kaizen* deals with improvement, we must know which aspects of business activities need to be improved most. And the answer to this question is quality, cost, and delivery (QCD). My previous book, *Kaizen: The Key to Japan's Competitive Success,* used the term *quality, cost,* and *scheduling* (QCS). Since that time, QCD has replaced QCS as the commonly accepted terminology.

*Quality* refers not only to the quality of finished products or services but also to the quality of the processes that go into those products or services. *Cost* refers to the overall cost of designing, producing, selling, and servicing the product or service. *Delivery* means delivering the requested volume on time. When the three conditions defined by the term QCD are met, customers are satisfied.

QCD activities bridge such functional and departmental lines as research and development, engineering, production, sales, and after-sales service. Therefore, cross-functional collaborations are necessary, as are collaborations with suppliers and dealers. It is top management's responsibility to review the current position of the company's QCD in the marketplace and to establish priorities for its QCD improvement policy.

Following the chapters of this book, I have assembled a number of cases that illustrate how various companies from both manufacturing and service sectors have implemented the concepts and systems of *gemba kaizen.*

## CHAPTER TWO

# *Gemba Kaizen*

Adoption of the word *gemba* has lagged behind adoption of the *kaizen* concept in the world. This is unfortunate but understandable; being present on the *gemba* can be a greater mind-set and behavior change than simply doing *kaizen.*

The *Cambridge Business English Dictionary* is one of a few sources, as of November 2011, to give the definition of *gemba* as an English word:

> gemba
> /ˈgembə/ noun
> in Japanese business theory, the place where things happen in manufacturing, used to say that people whose job is to manufacture products are in a good place to make improvements in the manufacturing process

This definition captures the spirit of *gemba* as it pertains to *kaizen,* but we must first understand *gemba* in its broader context beyond manufacturing.

In Japanese, *gemba* means "real place"—the place where real action occurs. Japanese use the word *gemba* in their daily speech. Whenever an earthquake shakes Japan, the TV reporters at the scene refer to themselves as "reporting from the *gemba.*" The *gemba* may be any workplace, crime scene, filming location, or even an archaeological excavation site. The *gemba* is where the action is and where the facts may be found. In business, the value-adding activities that satisfy the customer happen in the *gemba.*

Within Japanese industry, the word *gemba* is almost as popular as *kaizen.* Joop Bokern, one of the first *kaizen* consultants in Europe, had worked at Philips Electronics N.V. in Europe as production manager, as plant director, and finally as corporate quality manager. Bokern said that whenever he visited a Japanese company, he had a rule of thumb to determine whether the company was a good one or not. If, in his conversation with the Japanese manager, he heard the word *kaizen* within the first 5 minutes and the word *gemba* within the first 10 minutes, he concluded that it must be a good company. Bokern's example shows that *kaizen* and *gemba* are subjects close to managers' hearts and that they often make decisions based on their understanding of their *gemba.*

All businesses practice three major activities directly related to earning profit: developing, producing, and selling. Without these activities, a company cannot exist. Therefore, in a broad sense, *gemba* means the sites of these three major activities.

In a narrower context, however, *gemba* means the place where the products or services are formed. This book will use the word in this narrower context because these sites have been one of the business arenas most neglected by management. Managers seem to overlook the workplace as a means to generate revenue, and they usually place far more emphasis on such sectors as financial management, marketing and sales, and product development. When management focuses on the *gemba*, or work sites, they discover opportunities for making the company far more successful and profitable.

In many service sectors, the *gemba* is where the customers come into contact with the services offered. In the hotel business, for instance, *gemba* is everywhere: in the lobby, the dining room, guest rooms, the reception desk, the check-in counters, and the concierge station. At banks, the tellers are working in the *gemba*, as are the loan officers receiving applicants. The same goes for employees working at desks in offices and for telephone operators sitting in front of switchboards. Thus *gemba* spans a multitude of office and administrative functions. Most departments in these service companies have internal customers with whom they have interdepartmental activity, which also represents the *gemba.* A telephone call to a general manager, production manager, or quality manager at a Japanese plant is likely to get a response from the manager's assistant to the effect that "He is out at the *gemba.*"

## *Gemba* and Management

In or at the *gemba*, customer-satisfying value is added to the product or service that enables the company to survive and prosper. Figure 2.1 places *gemba* at the top of the organization, showing its importance to the company.

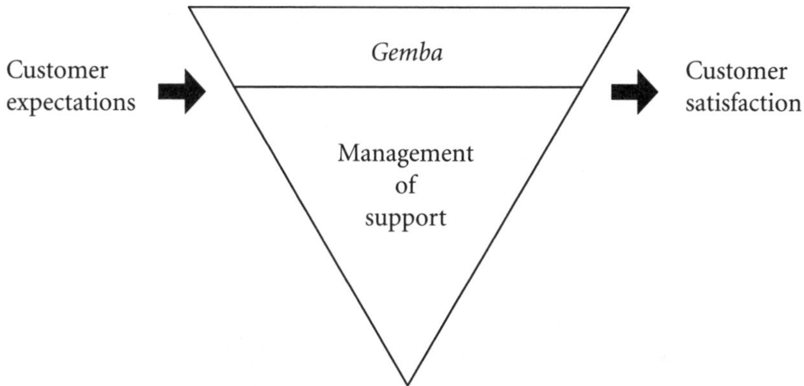


**Figure 2.1** In this view of *gemba*-management relations, management's role is to provide support to the *gemba*, which is seen as being at the top of management structure.

The regular management layers—top management, middle management, engineering staff, and supervisors—exist to provide the necessary support to the work site. For that matter, *gemba* should be the site of all improvements and the source of all information. Therefore, management must maintain close contact with the realities of the *gemba* in order to solve whatever problems arise there. To put it differently, whatever assistance management provides should start from the specific needs of the work site. When management does not respect and appreciate the *gemba*, it tends to "dump" its instructions, designs, and other supporting services, instructions, designs, other supporting services—often in complete disregard of actual requirements.

Management exists to help the *gemba* do a better job by reducing constraints as much as possible. In reality, however, I wonder how many managers correctly understand their role. More often than not, managers regard the *gemba* as a failure source, where things always go wrong, and they neglect their responsibility for those problems.

In some Western companies where the influence of strong unions practically controls the *gemba*, management avoids involvement in *gemba* affairs. Sometimes management even appears afraid of the plant and seems almost lost or helpless. Even in places where the union does not exercise a firm grip, *gemba* work is left to veteran supervisors who are allowed by management to run the show as they please. In such cases, management has lost control of the workplace.

Subsequent chapters will discuss in depth what management of the *gemba* really means. Supervisors should play a key role in *gemba* management, and yet they often lack the basic training to manage or to do their most important job: maintaining and improving the standards and achieving quality, cost, and delivery.

Eric Machiels, who came to Japan from Europe as a young student to learn about Japanese management practices, was placed in a Japanese automotive assembly plant as an operator: Comparing his experience there with his previous experience in European *gemba*, Machiels observed much more intense communication between management and operators in Japan, resulting in a much more effective two-way information flow between them. Workers had a much clearer understanding of management expectations and of their own responsibilities in the whole *kaizen* process. The resulting constructive tension on the work floor made the work much more challenging in terms of meeting management expectations and giving workers a higher sense of pride in their work.

Maintaining *gemba* at the top of the management structure requires committed employees. Workers must be inspired to fulfill their roles, to feel proud of their jobs, and to appreciate the contribution they make to their company and society. Instilling a sense of mission and pride is an integral part of management's responsibility for their *gemba.*

This approach contrasts sharply with perceptions of *gemba* (Figure 2.2) that regard it as a place where things always go wrong, a source of failure and customer complaints. In Japan, production-related work is sometimes referred to as *3K*, signifying the Japanese words for "dangerous" (*kiken*), "dirty" (*kitanai*), and "difficult" (*kitsui*). Once upon a time, the *gemba* was a place that good managers avoided. Being assigned a position at or close to the *gemba* amounted to a career dead end. Today, in contrast, the presidents of some better-known Japanese companies have rich backgrounds in *gemba* areas. They possess a good understanding of what goes on in the *gemba* and provide support accordingly.

The two opposite views of the *gemba*—as sitting on top of the management structure (inverted triangle) and as sitting at the bottom of the management structure (normal triangle)—are equally valid in terms of *gemba*-management relations. *Gemba* and management share an equally important place—the *gemba* by providing the product or service that satisfies the customer and management by setting strategy and deploying policy to

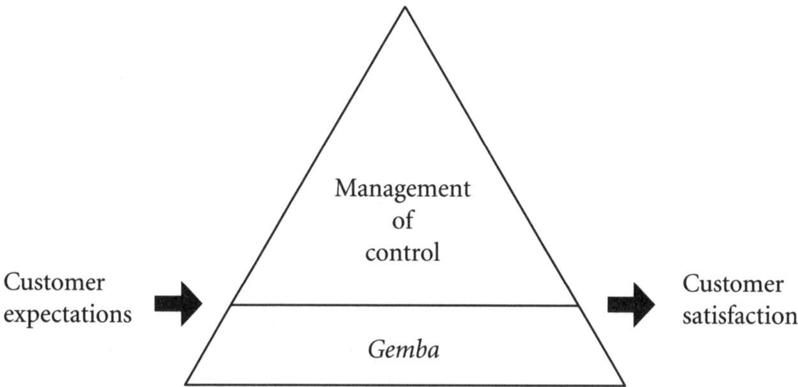


**Figure 2.2** In this view of *gemba*-management relations, management's role is to manage *gemba* by providing policies and resources.

achieve that goal in the *gemba.* Thus the thrust for improvement should be both bottom-up and top-down. In Figure 2.2, management stays on top of the organization. It takes the initiative in establishing policies, targets, and priorities and in allocating resources such as manpower and money. In this model, management must exercise leadership and determine the kind of *kaizen* most urgently needed. This process of achieving corporate objectives is called *policy deployment.* Because of their attachment to the *gemba*-management relationship as shown in the regular triangle (Figure 2.2), many managers tend to believe that their job is always to tell the *gemba* what to do. However, by looking at the inverted triangle (Figure 2.1) showing *gemba* at the top, managers can see that they should listen to and learn from employees at the *gemba* in order to provide appropriate help. *Gemba* becomes the source for achieving commonsense, low-cost improvements.

The respective roles of management and *gemba* in these two models never should be confused. Assistant Professor Takeshi Kawase of Keio University writes, in *Solving Industrial Engineering Problems* (published by Nikkan Kogyo Shinbun in Japanese, 1995):

> People within a company can be divided into two groups: those who earn money and those who don't. Only those frontline people who develop, produce, and sell products are earning money for the company. The ideal company would have only one person who does not earn money—the president—leaving the rest of the employees directly involved in revenue-generating activity.

The people who do not earn money are those who sit on top of the money earners—all employees with titles such as *chief, head,* or *manager,* including the president and all staff, and spanning areas that include personnel, finance, advertising, quality, and industrial engineering. No matter how hard these people may work, they do not directly earn money for the company. For this reason, they might be better referred to as *dependents.* If money earners stop work for one second, the company's chances of making money will be lost by one second.

The trouble is that non–money earners often think that they know better and are better qualified than money earners because they are better educated. They often make the job of the latter more difficult. Non–money earners may think, "Without us, they cannot survive," when they should be thinking, "What can we do to help them do their job better without us?"

If we say "the customer is king," we should say "the *gemba* is Buddha." Historically, the corporate staff played a leading role in regard to the *gemba*; the staff was accountable for achieving greater efficiency by providing guidance for *gemba* people to follow. The shortcoming of this system is the separation between those who pass down directives and those who carry them out. The new approach should be what we might call a *gemba*-centered approach, where *gemba* is accountable not only for production but also for quality and cost and personnel assist from the sidelines. The following are the conditions for successful implementation of a *gemba*-centered approach:

- *Gemba* management must accept accountability for achieving quality, cost, and delivery (QCD).
- *Gemba* must be allowed enough elbow room for *kaizen.*
- Management should provide the target for the *gemba* to achieve but should be accountable for the outcome. (Also, management should assist the *gemba* in achieving the target.)
- Needs of the *gemba* are more easily identified by the people working there.
- Somebody on the line is always thinking about all kinds of problems and solutions.
- Resistance to change is minimized.
- Continual adjustment becomes possible.
- Solutions grounded in reality can be obtained.

- Solutions emphasize commonsense and low-cost approaches rather than expensive and method-oriented approaches.
- People begin to enjoy *kaizen* and are readily inspired.
- *Kaizen* awareness and work efficiency can be enhanced simultaneously.
- Workers can think about *kaizen* while working.
- It is not always necessary to gain upper management's approval to make changes.

The benefits of a *gemba*-centered approach are many.

## The House of *Gemba*

Two major activities take place in the *gemba* on a daily basis as regards resource management—namely, maintenance and *kaizen*. The former relates to following existing standards and maintaining the status quo, and the latter relates to improving such standards. *Gemba* managers engage in one or the other of these two functions, and quality, cost, and delivery (QCD) are the outcome.

Figure 2.3 shows a bird's-eye view of activities taking place in the *gemba* that achieve QCD. A company that produces quality products or services at a reasonable price and delivers them on time satisfies its customers, and they, in turn, remain loyal. (For a more detailed explanation of QCD, see Chapter 3.)

## Standardization

In order to realize QCD, the company must manage various resources properly on a daily basis. These resources include personnel, information, equipment, and materials. Efficient daily management of resources requires standards. Every time problems or irregularities arise, the manager must investigate, identify the root cause, and revise the existing standards or implement new ones to prevent recurrence. Standards become an integral part of *gemba kaizen* and provide the basis for daily improvement.

Properly applied, *kaizen* can improve quality, reduce cost considerably, and meet customers' delivery requirements without any significant investment or introduction of new technology. Three major *kaizen* activities—standardization, 5S (which cover various housekeeping tasks), and the

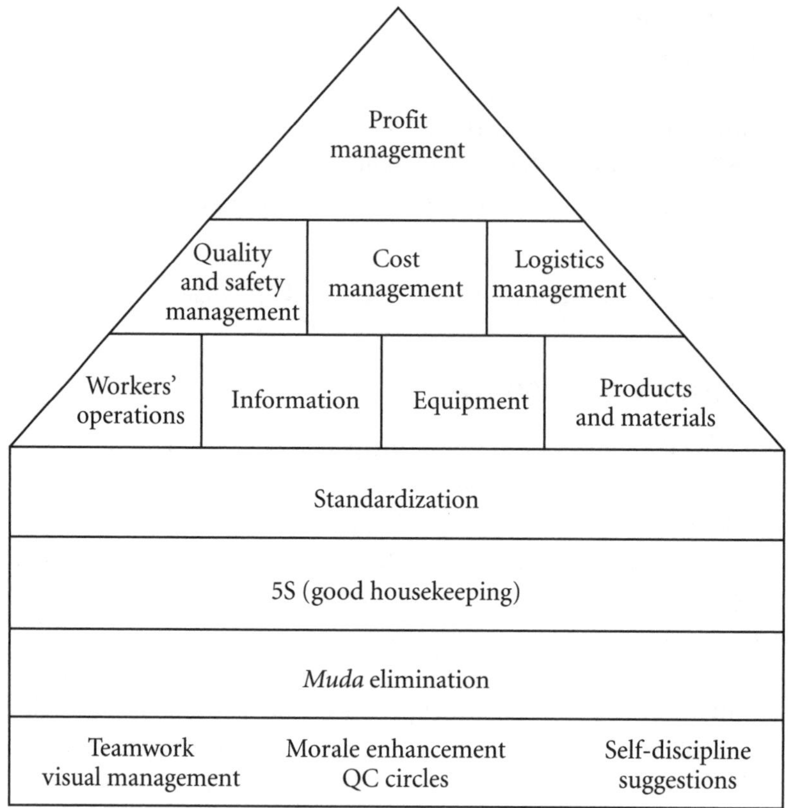


**Figure 2.3** House of *gemba* management.

elimination of *muda* (waste)—contribute to successful QCD. These three activities are indispensable in building lean, efficient, and successful QCD. Standardization, *muda* elimination, and 5S are easy to understand and implement and do not require sophisticated knowledge or technology. Anybody—any manager, any supervisor, or any employee—can readily introduce these commonsense, low-cost activities. The difficult part is building the self-discipline necessary to maintain them.

Standardization in the *gemba* often means the translation of technological and engineering requirements specified by engineers into workers' day-to-day operational standards. Such a translating process does not require technology or sophistication. It does require a clear plan from management deployed in logical phases. (For details of standards, refer to Chapter 4.)

## The Five S (5S) of Good Housekeeping

The *kaizen* principle of 5S stands for five Japanese words that constitute good workplace organization. Today, practicing 5S has become almost a must for any company engaged in manufacturing. An observant *gemba* management expert can determine the caliber of a company in five minutes by visiting the plant and taking a good look at what goes on there, especially in regard to *muda* elimination and 5S. A lack of 5S in the *gemba* should be considered a visual indicator of inefficiency, *muda*, insufficient self-discipline, low morale, poor quality, high costs, and an inability to meet delivery terms. Suppliers not practicing 5S will not be taken seriously by prospective customers. These five points of housekeeping represent a starting point for any company that seeks to be recognized as a responsible manufacturer eligible for world-class status. (The implications of 5S will be explained in detail in Chapter 5.)

Recently, before starting assembly operations in Europe, a Japanese automobile manufacturer sent purchasing managers to visit several prospective European suppliers. Eagerly anticipating new business, one of the suppliers prepared a detailed schedule for receiving the potential customers, starting with an hour-long presentation, complete with graphs and charts, on the company's efforts to improve quality. Next, the visitors would receive a tour of the *gemba*. On arrival, the purchasing managers were shown into the conference room. However, they insisted on being taken to the *gemba* right away, skipping the conference agenda. Once at the plant, they stayed only a few minutes before preparing to leave.

Bewildered, the general manager of the plant implored, "Please tell us about your findings!" The purchasing group replied, "We saw a very low level of housekeeping and found the plant very disorderly. Even worse, we saw some workers smoking while working on the line. If management allows these things to happen in the *gemba*, it cannot be serious enough about making components vital for automotive safety, and we do not want to deal with management that is not serious enough."

## *Muda* Elimination

*Muda* means "waste" in Japanese, but the implications of the word include anything or any activity that does not add value. At the *gemba*, only two types of activities go on: value-adding or non-value-adding activities. A worker looking at an automatic machine while the machine processes a piece does not add any value. The machine does the only value-adding work, no matter how attentively and affectionately the worker may look at it. When a maintenance engineer walks a long distance with a tool in his hand, he is not adding any value. The value is added by using the tool to fix, maintain, or set up the machine.

Customers do not pay for non-value-adding activities. Why, then, do so many people in the *gemba* engage in non-value-adding activities?

A manager of one factory once checked how far a worker walked in the *gemba* in the course of a year and found that the worker walked a distance of 400 kilometers. Jogging for health should be done in the gym, not in the *gemba*! Ironically, some factories are equipped with gyms that have running tracks, but the workers spend more time jogging in the *gemba* during working hours than in the gym during off-hours.

Once, when I was at Dallas–Fort Worth Airport in Texas, I needed to have my ticket endorsed in order to switch to another airline. After I had stood in line at the ticket counter for several minutes, my turn came, whereupon I was told that I had to go to another desk in another terminal to get the endorsement. I had to take a tram to the other terminal because the terminals at Dallas–Fort Worth are so far apart (a big *muda* in *kaizen* terms!). At the counter there, I waited in line again for several minutes. When my turn came, the airline employee stamped my ticket with a bang and said, "Here you are, sir!" I asked myself, "Did I deserve to wait almost half an hour for this?" At what moment did I get my value? Bang! That was the moment of truth, as far as I was concerned. When a company in the service industry conducts its business inefficiently, the company is not only wasting its own resources but also stealing the valued customer's time.

Any work that takes place in the *gemba* is actually a series of processes. Assuming 100 processes from receipt of raw materials and components until final assembly and shipment, the value-adding time at each process is just like that bang! Just think about how little time it takes to press a sheet of metal, shape a piece of work on a lathe, process a sheet of paper, or give a signature for approval. These value-adding activities take only seconds. Even supposing that each process takes one minute, value-adding activity for 100 processes should take no more than a total of 100 minutes. Why is it, then, that in most companies, days or weeks pass from the time raw materials and parts are brought in to when finished products emerge or for a document to go through the production process? There is far too much *muda* between the value-adding moments. We should seek to realize a series of processes in which we can concentrate on each value-adding process. We should seek to realize a series of processes in which we can concentrate on each value. There is far too much *muda* between the value-adding moments. We should seek to realize a series of processes in which we can concentrate on each value-adding process—Bang! Bang! Bang!—and eliminate intervening downtime. (Chapter 6 offers a more detailed explanation of *muda.*)

*Muda* elimination and good housekeeping often go hand in hand. Facilities where *muda* has been eliminated are orderly and show a high level of 5S discipline.

Good housekeeping indicates good employee morale and self-discipline. Any company can achieve a high level of self-discipline among employees temporarily. Sustaining that level, however, is a very challenging job. And the moment it disappears, its absence shows up in the form of a disorderly *gemba.* Increased morale and self-discipline within the *gemba* require involvement, participation, and information sharing with employees. Certain activities expedite the process of *kaizen* and maintain its momentum, eventually bringing change to the culture. These include teamwork, such as quality-circle and other small-group activities and employee suggestion schemes, in which workers remain continuously on the lookout for potential *kaizen* targets. When *gemba* employees participate in *kaizen* activities and notice the dramatic changes that have taken place as a result, they grow much more enthusiastic and self-disciplined.

More positive communication on policy deployment at the *gemba,* as well as in a company's offices, worker participation in setting up goals for *kaizen,* and the use of various kinds of visual management also play a vital role in sustaining the momentum of *kaizen* in the *gemba.* (Chapter 7 addresses employee empowerment, involvement, and participation.)

## The Golden Rules of *Gemba* Management

Most managers prefer their desk as their workplace and wish to distance themselves from the events taking place in the *gemba.* Most managers come into contact with reality only through their daily, weekly, or even monthly reports and meetings.

Staying in close contact with and understanding the *gemba* are the first steps in managing a production site effectively. Hence the five golden rules of *gemba* management:

1. When a problem (abnormality) arises, go to the *gemba* first.
2. Check the *gembutsu* ("relevant objects").
3. Take temporary countermeasures on the spot.
4. Find the root cause.
5. Standardize to prevent recurrence.

## *Go to the* Gemba *First*

Management responsibilities include hiring and training workers, setting the standards for their work, and designing the product and processes. Management sets the conditions in the *gemba*, and whatever happens there reflects on management. Managers must know firsthand the conditions in the *gemba*—thus the axiom, "Go to the *gemba* first." As a matter of routine, managers and supervisors should immediately go to the site and stand in one spot attentively observing what goes on. After developing the habit of going to the *gemba*, a manager will develop the confidence to use the habit to solve specific problems.

Kristianto Jahja, a *kaizen* consultant who worked for the joint venture in Indonesia between the Astra Group and Toyota Motor Company, recalls the first time he was sent to Toyota's plant in Japan for training. On the first day, a supervisor who was assigned as his mentor took him to a corner of the plant, drew a small circle on the floor with chalk, and told him to stay within the circle all morning and keep his eyes on what was happening.

Thus Kristianto watched and watched. Half an hour and then an hour went by. As time passed, he became bored because he was simply watching routine and repetitive work. Eventually, he became angry and said to himself, "What is my supervisor trying to do? I'm supposed to learn something here, but he doesn't teach me anything. Does he want to show his power? What kind of training is this?" Before he became too frustrated, though, the supervisor came back and took him to the meeting room.

There, Kristianto was asked to describe what he had observed. He was asked specific questions, such as, "What did you see there?" and "What did you think about that process?" Kristianto could not answer most of the questions. He realized that he had missed many vital points in his observations.

The supervisor patiently explained to Kristianto the points he had failed to answer, using drawings and sketches on a sheet of paper so that he could describe the processes more clearly and accurately. At this point, Kristianto understood his mentor's deep understanding of the process and realized his own ignorance.

Slowly but steadily, it became clear to Kristianto: The *gemba* is the source of all information. Then his mentor said that to qualify as a Toyota worker, one must love the *gemba* and that every Toyota employee believes that the *gemba* is the most important place in the company.

Says Kristianto, "Definitely, this was the best training I ever had because it helped me to truly become a *gemba* man, and this *gemba* thinking has influenced me throughout my career. Even now, every time I see a problem, my mind immediately shouts out loud and clear: Go to the *gemba* first and have a look!"

This is a common training method in Japanese *gemba*. Taiichi Ohno is credited with having developed the Toyota Production System. When Ohno noticed a supervisor out of touch with the realities of the *gemba*, he would take the supervisor to the plant, draw a circle, and have the supervisor stand in it until he gained awareness. Ohno urged managers, too, to visit the *gemba*. He would say, "Go to the *gemba* every day. And when you go, don't wear out the soles of your shoes in vain. You should come back with at least one idea for *kaizen*."

When he first began introducing just-in-time concepts at Toyota, Ohno encountered resistance from all quarters. One source of strong opposition was the company's financial people, who only believed written financial reports and often did not support allocating resources to *gemba*-related *kaizen* because doing so did not always yield immediate bottom-line results. To soften this opposition, Ohno urged accountants to go to the plant. He told them to wear out two pairs of shoes per year just walking around the site observing how inventory, efficiency, quality, and so on were improved and how the improvements contributed to cost reductions that ultimately produced higher profits.

In his later years, Ohno made public speeches sharing his experiences. He is reported to have opened one such speech by asking, "Are there any financial people in the audience?" When several people raised their hands, he told them, "You are not going to understand what I am going to say. Even if you understand, you are not going to be able to implement it, since you live far away from the *gemba*. Knowing how busy you are, I believe your time will be better spent if you go back to your desk to work." He said this facetiously, knowing that the support of financial managers is crucial for *gemba kaizen*.

Fuji Xerox President Akira Miyahara started his career at the Fuji Photo Film Company as a cost accountant. Knowing that the *gemba* was the source of the real data, he would go to the *gemba* to ascertain the information he needed. When he received the data about rejects for his financial report, he felt compelled to go to the *gemba* and observe the reason for the rejects because he believed that an accountant's job was not simply to deal with numbers but also to understand the process behind the numbers. Because Miyahara was seen in the *gemba* so often, the supervisor finally had to prepare a special desk for his use near the production line.

Miyahara's attachment to the *gemba* remained with him even after he was transferred to Fuji Xerox Company and was promoted to other management positions. When he was general manager of the sales division, for instance, the *gemba* was where his sales and service people were—that is, at the customers' sites. He accompanied service reps and visited customers, which gave him a far better understanding of the customers' needs than did reading the reports.

I once traveled to Central America and visited a branch of Yaohan, a Japanese supermarket chain headquartered in Hong Kong, whose stores span the globe. I asked the general manager, who had his office in the corner of a warehouse, how often he went to the *gemba* (at a supermarket, the *gemba* is the sales floor, warehouse, and checkout counter). The manager answered very apologetically, "You know, I have an assistant who is in charge of the *gemba*, so I don't go there as often as I should." When I pressed him to tell me exactly how often, he said, "Well, I must go there about thirty times a day." This manager felt apologetic about "only" going to the *gemba* 30 times a day!

"As I walk through the *gemba*," he told me, "I not only look around me to see how many customers we have, whether merchandise is properly displayed, which items are popular, and so on, but I also look up at the ceiling and down at the floor to see if there is any abnormality. Going through the *gemba* and looking straight ahead is something any manager can do, you know?"

One place that is definitely not the *gemba* is the manager's desk. When a manager makes a decision at her desk based on data, the manager is not in the *gemba*, and the source of the original information must be questioned carefully.

An example will illustrate. Because of its volcanic terrain, Japan has many hot-spring resorts. A key attraction of the spas is the open-air bath (*rotemburo*), where guests can soak while enjoying a view of river or mountains. I recently spent several days at a large hot-spring hotel that had both an indoor and an outdoor bath. Most guests would bathe in the indoor bath first and then walk down the stairs to the *rotemburo*. I normally found about half the guests in each bath. One evening I found the indoor bath almost empty. When I went in, I found out why: The water was too hot. Consequently, there was a crowd in the *rotemburo*, where the temperature was fine.

Clearly, something was wrong with the indoor bath. A housekeeper who was bringing in additional towels and cleaning the bath had apparently not noticed anything amiss. When I brought the problem to her attention, she quickly made a telephone call, and the temperature was restored to normal.

Later, I discussed this incident with the hotel's general manager, a good friend of mine. He told me that the temperature of the indoor bath was set at 42.5 degrees Celsius and that of the outdoor bath at 43 degrees Celsius. The manager went on to explain, "We have a monitoring room whereby our engineer keeps close watch over the temperatures of the baths, along with room temperatures, fire alarms, and such. Whenever he sees an abnormality on the meters, he's supposed to take corrective action."

To this, I responded, "Wrong! The person who watches the meters is only relying on secondary information. The information on the baths is first collected by the thermometer submerged in the tub and then transferred to the monitoring room by the electromechanical device, which moves the dial on the chart. Anything could go wrong in this process. The reality in the *gemba* is that at that time on that day, there were very few people in the indoor bath, and if the housekeeper had been trained to be more attentive, she could have noticed the situation, stuck her hand in the water, and found that it was too hot.

"The information you get directly from the *gemba*," I told my friend, "is the most reliable. The feeling of the hot water you experience with your hand is the reality. Often, you don't even need substantiating data when you are in the *gemba* because what you feel and see are the original data! People in the *gemba* should be responsible for quality because they are in touch with reality all the time. They are better equipped to maintain quality than the person in the monitoring room!"

Dr. Kaoru Ishikawa, one of the gurus of quality management in Japan, used to say, "When you see data, doubt ... [them]! When you see measurements, doubt them!" He knew that many data are collected in the company to please the boss and that measurements are often made or recorded incorrectly by devices. At best, measurements are only secondary information that does not always reflect the actual conditions.

Many Western managers tend to choose not to visit the *gemba*. They may take pride in not going to the site and not knowing much about it. Recently, on learning from the president of one company that he never visited the plant, I suggested that he do so once in a while. He replied, "I am an engineer by background, and I know how to read and interpret data. So I can make a good decision based on the data. Why should I go to the plant?"

At another plant I visited, I was told that whenever the big shots came from corporate headquarters to visit, the senior managers had to spend hours on end in the conference room answering foolish questions by managers who did not understand what was going on in the *gemba* and who often left inappropriate and disturbing instructions. "We would be much better off without these meetings," the plant managers told me.

The plant managers' opinion of the meetings illustrate the tremendous gap between top management and the workplace, a condition that can make a company vulnerable to challenge from internal waste and external competition. This attitude at the management level usually fosters a similar disrespect from workers.

## *Check the* Gembutsu

*Gembutsu* in Japanese means "something physical or tangible." In the context of the *gemba*, the word can refer to a broken-down machine, a reject, a tool that has been destroyed, returned goods, or even a complaining customer. In the event of a problem or abnormality, managers should go to the *gemba* and check the *gembutsu*. By looking closely at the *gembutsu* in the *gemba*, repeatedly asking "Why?" and using a commonsense, low-cost approach, managers should be able to identify the root cause of a problem without applying sophisticated technology. If a reject is produced, for example, simply holding it in your hands, touching it, feeling it, closely examining it, and looking at the production method probably will reveal the cause.

Some executives believe that when one of the company's machines breaks down, the *gemba* for the managers is not where the machine is, but the conference room. There, the managers get together and discuss the problem without ever looking at the *gembutsu* (in this case, the machine), and then everybody disavows his or her culpability.

*Kaizen* starts with recognizing the problem. Once aware, we are already halfway to success. One of the supervisor's jobs should be to keep constant watch at the site of the action and identify problems based on *gemba* and *gembutsu* principles.

One supervisor recently remarked, "I walk through the *gemba* every day and try to look at the *gembutsu* to find something unusual so that I can take it back to my desk and start working on it. When I do not find any item for *kaizen*, I feel frustrated."

Soichiro Honda, the founder of Honda Motor Company, did not have a president's office; he was always found somewhere in the *gemba*. Being a mechanic by background and having worked close to the *gemba* all his life tuning and adjusting engines with screwdrivers and wrenches, he had many scars on his hands. Later in his life, when Honda visited nearby grade schools to talk with the children, he would proudly show them his hands and let them touch the scars there.

### *Take Temporary Countermeasures on the Spot*

Once I visited a plant where I found a small broom attached to a machine engaged in cutting operations. I noticed that the machine kept stopping because metal chips were falling on the belt that was driving the machine, thus clogging the belt's movement. At this point, the operator would pick up the broom and sweep the chips off the belt to start the machine again. After a while, the machine would stop, and the operator would repeat the same process to get it started again.

If a machine goes down, it must be started promptly. The show must go on. Sometimes kicking the machine will do the job! However, temporary measures address only the symptoms, not the root cause, of machine stoppages. This is why you need to check the *gembutsu* and keep asking "Why?" until you identify the root causes of the problem.

Determination and self-discipline never stop the *kaizen* effort at the third stage (temporary countermeasures). They continue to the next stage, identifying the real cause of the problem and taking action.

## *Find the Root Cause*

Many problems can be solved quite readily using the *gemba-gembutsu* principles and common sense. With a good look at the *gembutsu* at the site of the problem and determination to identify root causes, many *gemba*-related problems can be solved on the spot and in real time. Other problems require substantial preparation and planning to solve; examples include some engineering difficulties or the introduction of new technologies or systems. In these cases, managers need to collect data from all angles and also may need to apply some sophisticated problem-solving tools.

For instance, if chips falling on a conveyor belt are causing stoppages, a temporary guide or cover can be fashioned from cardboard on the spot. Once the effectiveness of the new method has been confirmed, a permanent metal device can be installed. Such a change can be made within hours or certainly within a day or two. The opportunities for making such a change abound in the *gemba*, and one of the most popular axioms of *gemba kaizen* is, "Do it now. Do it right away!"

Unfortunately, many managers believe that one must make a detailed study of every situation before implementing any *kaizen*. In reality, about 90 percent of all problems in the *gemba* can be solved right away if managers see the problem and insist that it be addressed on the spot. Supervisors need training on how to employ *kaizen* and what role they should play.

One of the most useful tools for finding the root cause in the *gemba* is to keep asking "Why?" until the root cause is reached. This process is sometimes referred to as the *five whys* because chances are that asking "Why?" five times will uncover the root cause.

Suppose, for example, that you find a worker throwing sawdust on the floor in the corridor between machines.

**Your question:** "Why are you throwing sawdust on the floor?"

**His answer:** "Because the floor is slippery and unsafe."

**Your question:** "Why is it slippery and unsafe?"

**His answer:** "Because there's oil on it."

**Your question:** "Why is there oil on it?"

**His answer:** "Because the machine is dripping."

**Your question:** "Why is it dripping?"

**His answer:** "Because oil is leaking from the oil coupling."

**Your question:** "Why is it leaking?"

**His answer:** "Because the rubber lining inside the coupling is worn out."

Very often—as in this case—by asking "Why?" five times, we can identify the root cause and take a countermeasure, such as replacing the rubber lining with a metal lining to stop the oil leakage once and for all. Of course, depending on the complexity of the problem, the question "Why?" may need to be asked more or fewer times. However, I have noticed that people tend to look at a problem (in this case, oil on the floor) and jump to the conclusion that throwing sawdust on it will solve everything.

## *Standardize to Prevent Recurrence*

A manager's task at the *gemba* is to realize QCD. However, all manner of problems and abnormalities occur at plants every day; there are rejects, machines break down, production targets are missed, and people arrive late for work. Whenever a given problem arises, management must solve it and make sure that it will not recur for the same reason. Once a problem has been solved, therefore, the new procedure needs to be standardized and the standardize-do-check-act (SDCA) cycle invoked. Otherwise, people are always busy firefighting. Thus the fifth and last golden rule of *gemba* management is standardization. When a problem occurs in the *gemba*, whether a machine breaks down from metal chips clogging the conveyors or hotel guests complain about the way fax messages are handled, the problem first must be carefully observed in light of *gemba-gembutsu* principles. Next, the root causes must be sought out, and finally, after the effectiveness of the procedure devised to solve the problem has been confirmed, the new procedure must be standardized.

In this manner, every abnormality gives rise to a *kaizen* project, which eventually should lead either to introducing a new standard or to upgrading the current standard. Standardization ensures the continuity of the effects of *kaizen.*

One definition of a *standard* is "the best way to do a job." If *gemba* employees follow such a standard, they ensure that the customer is satisfied. If a standard means the best way, it follows that the employee should adhere to the same standard in the same way every time. If employees do not follow standards in repetitive work—which is often the case in manufacturing *gemba*—the outcome will vary, leading to fluctuations in quality. Management must clearly designate standards for employees as the only way to ensure customer-satisfying QCD. Managers who do not take the initiative to standardize the work procedure forfeit their job of managing the *gemba.*

At Giorgio Foods, Inc., in Temple, Pennsylvania, the administrative rooms once were located upstairs, whereas the *gemba* was downstairs. Upstairs, walls separated the rooms for each function: sales, marketing, engineering, research and development, and personnel.

But the company's chairman, Fred Giorgio, decided that everyone whose job was to support the *gemba* should move their desks to the *gemba.* He declared, "We are all going to move to the *gemba*, and we are going to work together in a big room without walls!" An uproar of protests followed: "It will be too noisy!" "We won't be able to concentrate on our work!" "Some subordinates will quit!" "We won't be able to keep the company's secrets!" Giorgio was undaunted. He said, "If a secret leaks out this way, then it can't be kept a secret anyway. If people don't like it, let them go!"

In the end, though, everybody moved, if not wholeheartedly. Today, a visitor to the company can see at a glance everyone working in one big room. If the visitor is attentive, she will find Fred Giorgio among them, inconspicuously sitting at a small desk flanked by two other desks, each occupied by an executive of the company. "Before," says Giorgio," "whenever I wanted to have a meeting with managers, I had to check who was in and who was out before calling such a meeting. Now, I look around and see who is present. Then, I yell out and say, 'Hey, let's go to the meeting room and discuss this matter!' No *muda*!"

This arrangement of the company's staff offers other advantages as well. At the entrance to the administrative floor are two small rooms: a telephone operator's room and the personnel department. In the wall of the former is a window allowing the operator to see at a glance who is in and who is out. And because employees must pass the personnel office whenever they have business on the administrative floor, it has become easier for them to approach personnel people to discuss matters of concern.

Says Tony Puglio, former department manager of the labeling department at Giorgio Foods, "Five years ago, I spent a lot of time in my office doing paperwork. I thought I had all the answers and I could do everything myself. Now I find that we can make a difference through our *kaizen* activities like the quality circle meetings and listening to workers' suggestions, going out to the workplace, spending more time there, looking at each and every problem and correcting them. I found out that my employees have great ability—artistic talent and practical skills—that I didn't realize they had. They were able to do all the *kaizen* work themselves and make a difference on the lines.

"I spend around 90 percent of my time on the shop floor, which enables me to see workers' problems," continues Puglio. "Before, when they would come into the office and tell me their problems, I would listen to them but not do much about it. I didn't realize that we'd been running like this for years and years, and I assumed everything was okay. But it wasn't. By going on the floor, I could really see what the workers were talking about.

"Now I notice that everybody is putting a little more effort in; they're excited, and they're proud of their department. They're keeping everything in order and in place, and they are keeping everything much cleaner. The workplace looks good, and when people come in, they want to be at work. They feel good about themselves. They look good, and they feel good. They see that these changes are working, and it's making a difference, making their jobs a little bit easier."

## Application of the Golden Rules

Let me explain how these golden rules have been applied in my own experience. Electronic communication has become indispensable for business. Anyone like myself, who spends more than half his time traveling around the world on business, cannot accomplish their business without e-mail, mobile devices, and fax machines. During a hotel stay lasting a few days, I had a series of problems with the way the hotel handled incoming faxes. I was supposed to have received an urgent fax from Tokyo. When I called my executive assistant there, I was assured that the transmission had gone through a few hours before. Because the document had not been delivered to me, I had to inquire at the front desk. The person at the desk was sure that no fax had arrived for me. Earlier, at this same hotel, I had received several faxes addressed to me, together with several meant for somebody else. I was so annoyed that I finally asked myself what I would do if I were the general manager of this hotel and received many complaints from customers on the way employees handled faxes. My conclusion: Apply the golden rules, by all means!

So I put myself in the shoes of a hotel manager interested in applying *gemba kaizen.* The first step was to go to the *gemba,* in this case, the lobby. I stood on an elevated platform in a corner of the lobby (but did not draw a chalk circle) and stayed there for a few minutes, watching attentively how people at the front office handled faxes. It did not take five minutes to find out that there were no special procedure! For instance, there was no fixed place to store the incoming documents (no standard). Some employees put them in the key boxes. Others left them on the desk. Still others put them wherever they found a space. Also, when the fax sheets (*gembutsu*) came out of the fax machine (another *gembutsu*) in the reverse order of pages, employees didn't even take the time to put them in the right order. This appeared to be the reason somebody else's faxes were delivered to me along with my own.

If I had actually been the hotel's general manager, after observing this situation, I would have called a meeting with the *gemba* people and asked them to work out procedures for handling faxes. We might have agreed that the documents' pages should be arranged in the right sequence and that all incoming faxes should be placed in the key boxes, for example. We also might have arranged to record the times that faxes were delivered to guests (standardization) to avoid any arguments over whether or not a guest received a fax. Discussing and agreeing on the new procedures probably would have taken no more than half an hour. (This is the essence of "Go to the *gemba*, and do it right away.") The agreed-on procedure then would be followed consistently. In response to problems or customer complaints, the procedure could be refined so that the hotel's fax-handling system could be continuously improved over time.

# CHAPTER THREE

# Quality, Cost, and Delivery at the *Gemba*

Many businesses that have moved their design, production, delivery, or service capability off-shore in the pursuit of ever-lower prices over the recent decades have learned this lesson the hard way: You get what you pay for. A sustainable competitive advantage must be built not on unit cost alone but on a total cost that reflects the interaction of quality, cost, and delivery (QCD).

Quality, cost, and delivery are not distinctly separate subjects but rather are closely interrelated. It is pointless to buy products or services lacking in quality, no matter how attractive their price. Conversely, it is meaningless to offer products or services of good quality and attractive price if those products cannot be delivered in time to meet the customer's need and in the quantity that the customer wants.

## Quality: More Than Just a Result

*Quality* in this context means the quality of products or services. In a broad sense, however, it also means the quality of the processes and of the work that yields those products or services. We may call the former *result quality* and the latter *process quality*. By this definition, quality runs through all phases of company activity—namely, throughout the processes of developing, designing, producing, selling, and servicing the products or services.

Figure 3.1, a quality-assurance (QA) system diagram of Toyoda Machine Works, shows how quality-assuring activities take place on an ongoing basis at a tool manufacturing company. One might say that this diagram shows all the key steps of process quality. Reading from top to bottom, the figure shows the flow of activities from the identification of

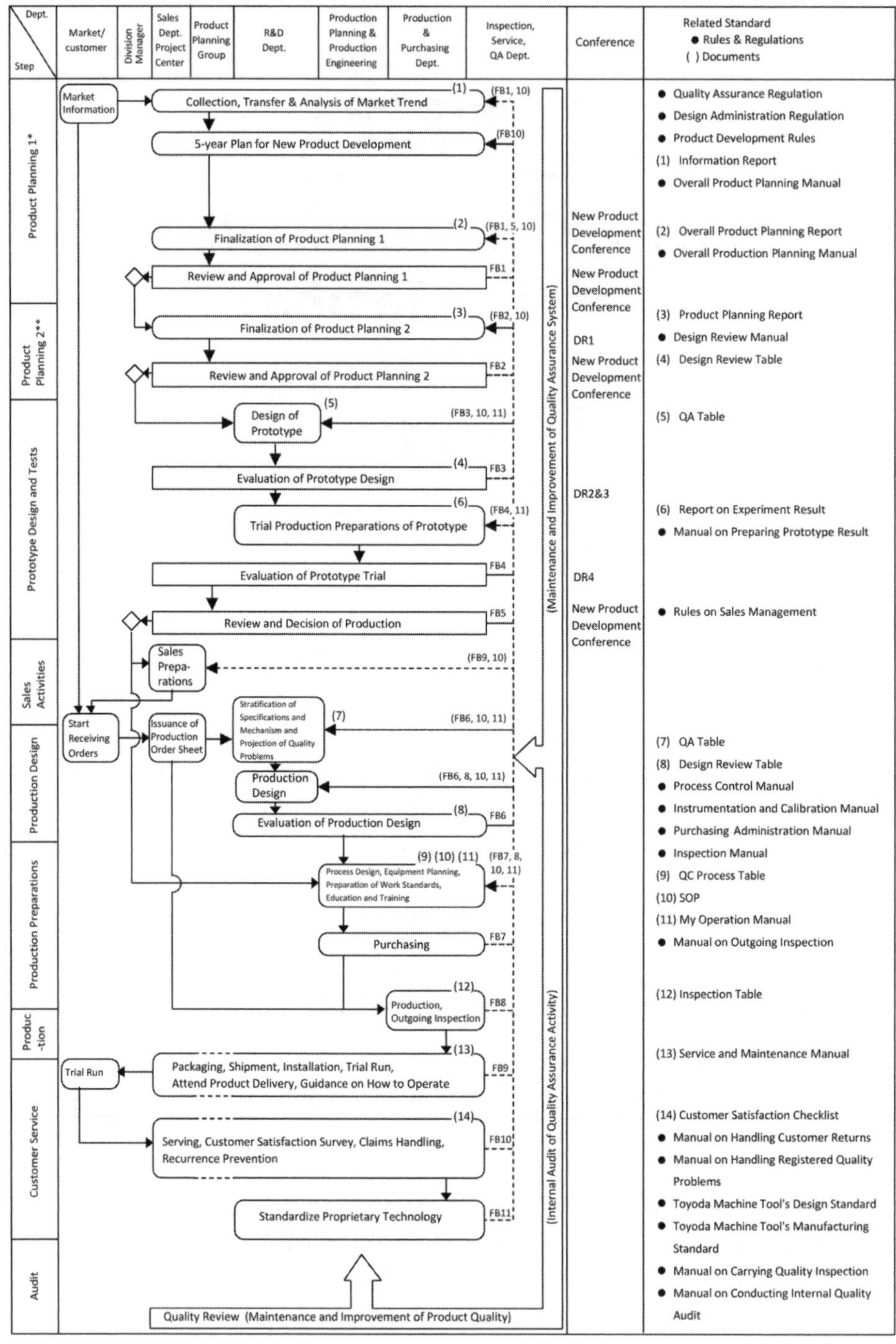


* Refers to product planning from the standpoint of meeting customer needs.

** Refers to product planning from the standpoint of making the product.

**Figure 3.1** QA system diagram.

customer requirements through such stages as product planning 1 (the customer's standpoint), product planning 2 (the manufacturer's standpoint), prototype design and test, sales activities, production design, production preparations, production, customer service, and audit.

Reading the diagram from left to right shows the involvement of people from various departments. The main body of the diagram shows activities that ensure quality at every process. The flow of quality-related information also appears here. For instance, below the "Division Manager" column appear four stages of design review (DR), meaning that the division manager is involved in all design review stages.

The "Conference" column shows cross-functional meetings and conferences in which the departments concerned must participate at each key stage before moving on to the next stage.

The last column on the right shows the related standards, regulations, or documents corresponding to each stage of QA. This diagram shows that before the *gemba* starts making the products, a long list of quality-assuring actions take place. For instance, items 8 through 12 of "Standards and Regulations" (including the process control manual, the instrumentation and calibration manual, the inspection manual, the quality-control process table, the standard operating procedures manual, my operation manual, and the manual on outgoing inspection) list typical procedures that ensure quality at the *gemba.* However, the diagram also shows that items 1 through 7 have been completed by the time the *gemba* work begins.

Activities that precede the *gemba* work (standards 1 through 8) are called *upstream management.* Traditionally, when quality was perceived primarily as a matter of workmanship, quality-related improvement efforts focused mainly on the *gemba.* While workmanship remains one of the most important pillars of quality, people increasingly recognize that quality in the area of design, product concepts, and understanding of customer requirements must precede *gemba* work.

Most activities in the *gemba* relate to workmanship and seldom reach upstream management, although *gemba*-based *kaizen* activities arise from management's policy deployment, which, in turn, identifies the need for *kaizen* upstream as well. Top management must establish standards for quality of planning.

Planning correctly the first time around—accurately understanding customer needs, translating this understanding into the engineering and designing requirements, and making advance preparations for a smooth startup—makes it possible to avoid many problems at the *gemba* during process stages as well as in after-sales service.

The job of developing a new product or designing a new process starts with paperwork. Bugs or malfunctions can be rectified with the stroke of a pen at no cost. Malfunctions identified later, in the production stage or—even worse—after the product has been delivered to the customer, necessitate very expensive corrections.

*Quality function deployment* (QFD) is a powerful tool that enables management to identify the customer's needs, convert those needs into engineering and designing requirements, and eventually deploy this information to develop components and processes, establish standards, and train workers.

The system diagram in Figure 3.1 shows that the company is using the tools of QFD in the daily QA listed in the right-hand column. These tools include QA tables, which are matrixes correlating between such items as customers' requirements and corresponding engineering parameters.

Upstream management plays an indispensable role in ensuring quality. On the other hand, if the *gemba* is not sufficiently robust, the company will not be able to enjoy the full benefits of even the most effective upstream management. Such a situation is analogous to making a sophisticated plan to climb Mount Everest only to find that one's legs are too weak to make the climb.

## Quality Management at the *Gemba*

The *gemba* confronts quality issues from a different angle than upstream management. While upstream management requires sophisticated tools, such as design reviews, design of experiments, value analysis, value engineering, and the various tools of QFD, many problems in the *gemba* relate to simple matters, such as workmanship and handling the difficulties and variations that come up every day (e.g., inadequate working standards and operators' careless mistakes).

In order to reduce variability, management must establish standards, build self-discipline among employees to maintain standards, and make certain that no defects are passed on to the next customer. Most quality problems can be solved using *gemba-gembutsu* principles, the common-sense, low-cost approach explained in Chapter 2.

Management must introduce teamwork among operators because the operators' involvement is a key issue. Statistical quality control (SQC) is often employed effectively in the *gemba*, but SQC is a tool to confine the variability of the processes and will work well only if everybody—particularly management—understands the concept of variability control and makes an effort to practice it.

I once visited a plant whose manager was proud of her achievement of SQC. I saw many control charts posted on the walls in her room. But once I stepped into the *gemba*, I realized that nobody understood variability. The operators had no standards, and they did their jobs differently with each piece they assembled. Sometimes they didn't even have a designated place for assembly work. During my visit, machines broke down repeatedly and many rejects were produced. Yet this manager was proud of her SQC!

Professor Hitoshi Kume of Tokyo University has said, "I think that while quality control in the West aims at 'controlling' the quality and conformance to standards and specifications, the feature of the Japanese approach centers around improving (*kaizen*) quality. In other words, the Japanese approach is to do such *kaizen* systematically and continually."

The landmark case of line-assembly quality improvement of the dip-soldering process at Yokogawa Hewlett-Packard (YHP), in which the company succeeded in reducing the failure rate from 4,000 to 3 parts per million (ppm) between 1978 and 1982, may well illustrate his point. YHP's history of quality improvement is divided into two periods—1978 to 1979 and 1980 to 1982.

Quality improvement activities differed considerably during the two periods. During the first phase, for example, YHP took such actions as improving working standards, collecting and analyzing data on defects, introducing jigs for better control of the process, providing worker training, encouraging quality-circle activities, and reducing careless mistakes by operators. To do this, YHP assembled a project team of *gemba* supervisors and production engineers to collect data, train quality-circle members, and provide technical assistance in such areas as jig construction. These actions helped to drive the failure rate down to 40 ppm from its previous level of 4,000 ppm (see Figure 3.2).

Once the 40 ppm level had been reached, YHP needed to step up and refine these activities if it wanted to continue its momentum and make further gains (see Figure 3.3). At the same time, it had to apply new

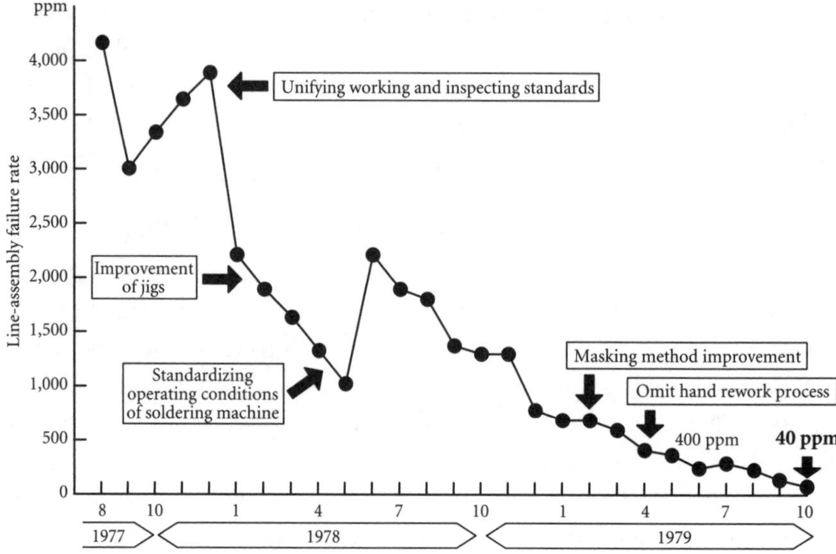


**Figure 3.2** Process quality improvement, phase 1.

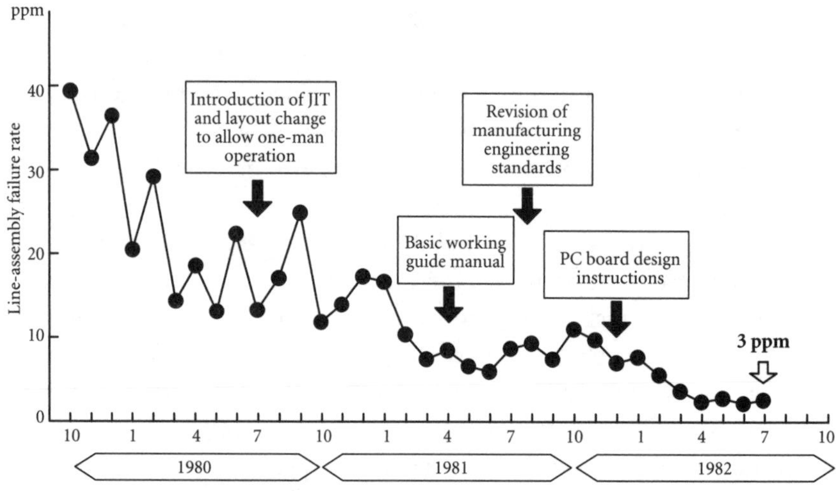


**Figure 3.3** Process quality improvement, phase 2.

technologies: revised engineering standards, improved PC board designs, and production layout. It also needed to redesign its equipment as well as its layout, incorporating the just-in-time concepts. All the while, YHP's quality circles maintained their activities to gain a better understanding of the process. They also contributed greatly to the continuous improvement of the process. As a result, YHP reached the level of 3 ppm in 1982.

Generally speaking, as long as the quality level remains in the percentile figures, companies can achieve dramatic improvement through such basic activities as reviewing the standards, housekeeping, collecting data on rejects, and conducting group activities for problem solving.

First, review existing procedures, asking such questions as

- ▲ Do we have standards?
- ▲ What about housekeeping (5S) in the *gemba*?
- ▲ How much *muda* exists in the *gemba*?

Then begin taking action. For example:

- ▲ Implement the five *gemba* principles.
- ▲ Train employees to be committed to never send rejects on to the next process.
- ▲ Encourage group activities and suggestions for problem solving.
- ▲ Start collecting data to gain a better understanding of the nature of the problems, and solve them.
- ▲ Start making simple jigs and tools to make the job easier and its results more reliable.

These down-to-earth activities alone should reduce reject rates to a tenth their original levels. When these fundamentals are lacking, the variables are so large that sophisticated technologies do little to improve the process. Only after the basic variables have been addressed are the more challenging applications of SQC and other sophisticated approaches cost-effective.

Quality begins when everybody in the organization commits to never sending rejects or imperfect information to the next process. Dr. Kaoru Ishikawa's axiom, "The next process is the customer," refers to the internal customer within the same company. One should never inconvenience the customers in the next process by sending rejects to them. At the *gemba*, such a state of mind is often referred to as, "Don't accept defects, don't make defects, and don't pass on defects." When everybody subscribes to and lives by this philosophy, a good QA system exists.

## Cost Reduction at the *Gemba*

In this context, the word *cost* does not mean cost cutting but cost management. Cost management oversees the processes of developing, producing, and selling products or services of good quality while striving to lower costs or hold them to target levels. Cost reduction at the *gemba* should come as a result of various activities carried out by management. Unfortunately, many managers try to reduce costs only by cutting corners; typical actions include firing employees, restructuring, and beating up suppliers. Such cost cutting invariably disrupts the process of quality and ends in quality deterioration. But today's customers are increasingly demanding; they want better quality at a lower price—complete with prompt delivery. When we respond to demand for lower prices simply by cost cutting, we soon find that quality and prompt delivery disappear. Cost management encompasses a wide spectrum of activities, including

1. Cost planning to maximize the margin between costs and revenues
2. Overall cost reduction at the *gemba*
3. Investment planning by top management

Opportunities for cost reduction onsite may be expressed in terms of *muda*. The best way to reduce costs in the *gemba* is to eliminate excess use of resources. To reduce costs, the following seven activities should be carried out simultaneously, with quality improvement being the most important. The other six major cost-reduction activities may be regarded as part of the process quality in a broader sense:

1. Improve quality.
2. Improve productivity.
3. Reduce inventory.
4. Shorten the production line.
5. Reduce machine downtime.
6. Reduce space.
7. Reduce lead time.

These efforts to eliminate *muda* will reduce the overall cost of operations.

## *Improve Quality*

Quality improvement actually initiates cost reduction. Quality here refers to the process quality of managers' and employees' work. Improving the quality of the work processes results in fewer mistakes, fewer rejects and less rework, shorter lead time, and reduced use of resources, therefore lowering the overall cost of operations. Quality improvement is synonymous with better yields as well.

*Process quality* includes the quality of work in developing, making, and selling products or services. At the *gemba*, the term specifically refers to the way products or services are made and delivered. It refers mainly to managing resources at the *gemba*; more specifically, it refers to managing man (worker activity), machine, material, method, and measurement—known collectively as the *five Ms* (5M).

## *Improve Productivity to Lower Costs*

Productivity improves when less input produces the same output or when output increases with the same input. *Input* here refers to such items as human resources, utilities, and material. *Output* means such items as products, services, yield, and added value. Reduce the number of people on the line; the fewer line employees, the better. This not only reduces cost but also, more important, reduces quality problems because fewer hands present fewer opportunities to make mistakes. I hasten to add that *kaizen* and productivity improvement must not result in firing of employees. There are many ways to use employees freed from a process where productivity has improved. Management must consider employees freed up by *kaizen* activities as resources for other value-adding activities and innovation efforts. When productivity goes up, cost goes down.

## *Reduce Inventory*

Inventory occupies space, prolongs production lead time, creates transport and storage needs, and eats up financial assets. Products and work-in-process sitting on the factory floor or in the warehouse do not yield any added value. On the contrary, they deteriorate in quality and even may become obsolete overnight when the market changes or competitors introduce a new product.

## *Shorten the Production Line*

In manufacturing, a longer production line requires more people, more work-in-process, and a longer lead time. More people on the line also means more mistakes, which lead to quality problems. One company's production line was 15 times longer than its competitor's line. The result—in terms of the number of people employed on the line, the quality level (more people producing more quality problems), the inventory (both work-in-process and finished products), and the much longer lead time—was an overall cost of operations that was much higher than it needed to be.

I once reviewed the layout of a production line that was to be introduced soon at a company that was manufacturing a new product. To my surprise, the new process was a carbon copy of the existing one, except that some of the existing machines were replaced with the latest models. The company had made no effort to shorten the line. Management had not included shortening the line as one of its targets, nor had the designers given it a thought.

In Japan, an engineer tasked with collecting catalogues from various machine makers and placing orders from among them to design a new layout is called a *catalogue engineer*—not a very prestigious title. Management should encourage such engineers to do a better production layout—to design ever-shorter assembly lines employing fewer and fewer people. It is an integral part of a *kaizen*-driven manager's work to constantly challenge employees to find better ways to do the job than the last time. The situation is exactly the same in nonmanufacturing activities.

## *Reduce Machine Downtime*

A machine that goes down interrupts production. Unreliable machinery necessitates batch production, extra work-in-process, extra inventory, and extra repair efforts. Quality also suffers. All these factors increase the cost of operations. Such problems are similar in the service sector. Downtime in the computer or communications system causes undue delay, greatly increasing the cost of machine operations. When a newly hired employee is assigned to a workstation without proper training to handle the equipment, the consequent delay in operation may be just as costly as if the equipment were down.

## *Reduce Space*

As a rule, manufacturing companies use four times as much space, twice as many people, and 10 times as much lead time as they really need. Despite the decades of information technology advancement and *kaizen* activity undertaken by many companies since 1985, this remains true for the majority of businesses today. Typically, *gemba kaizen* eliminates conveyor lines, shortens production lines, incorporates separate workstations into the main line of production, reduces inventory, and decreases transportation needs. All these improvements reduce space requirements. Extra space freed up by *gemba kaizen* may be used to add new lines or may be reserved for future expansion. A similar improvement can be introduced in a nonmanufacturing environment.

## *Reduce Lead Time (Throughput Time)*

*Lead time* begins when a company pays for raw materials and supplies and ends only when the company receives payment from its customer for products sold. Thus lead time represents the turnover of money. A shorter lead time means better use and turnover of resources, more flexibility in meeting customer needs, and a lower cost of operations. The lead time is the true measure of management's capability, and shortening this interval should be top management's paramount concern. *Muda* in the area of lead time presents a golden opportunity for *kaizen.*

Ways to cut lead time include improving and speeding feedback of customer orders and communicating better with suppliers; this reduces the inventory of raw materials and supplies. Streamlining and increasing the flexibility of *gemba* operations also can shorten production lead time. When everyone in an organization works toward this goal, there is a positive impact on cost-effectiveness.

### *Role of the* Gemba *in Overall Cost Reduction*

If the *gemba* cannot make its procedures very short, flexible, efficient, defect-free, and free of machine downtime, there is little hope either of reducing the inventory levels of supplies and parts or of becoming flexible enough to meet today's stringent customer demands for high quality, low cost, and prompt delivery. *Gemba kaizen* can be the starting point for improvements in all three categories.

Any *gemba* that is not sufficiently reliable or robust cannot sustain improvements made in other functional areas, such as product development and process designs, purchasing, marketing, and sales. *Kaizen* should start at the *gemba.* To put it another way, by carrying out *gemba kaizen* and identifying the problems manifested at the work site, we can identify the shortcomings of other supporting departments, such as research and development, design, quality control, industrial engineering, purchasing, sales, and marketing. In other words, *gemba kaizen* helps to identify shortcomings in upstream management. The *gemba* becomes a mirror that reflects the quality of the company's management systems and a window through which we see the *real* capabilities of management.

## Delivery

*Delivery* refers to the *timely* delivery of the *volume* of products or services. One of management's tasks is to deliver the required volume of products or services in time to meet customer needs. The challenge to management is how to live up to delivery commitments while meeting quality and cost targets. In line with the axiom, "Quality first," quality is the foundation on which cost and delivery are built.

A just-in-time (JIT) system addresses both cost and delivery issues, but it can be introduced only if a good QA system is in place. By eliminating all kinds of non-value-adding activities, JIT helps to reduce costs. Indeed, synchronizing the flow of goods and services using JIT is a practical way to drastically cut costs for companies that have never tried it before.

Equally important, JIT addresses delivery. The conventional approach has been to deliver products out of finished-goods inventory, with the customer paying for the added cost. In JIT, every effort is made to produce and deliver the product *just in time*—that is, to produce only as *many* as are needed and only *when* needed, thereby eliminating the cost of excessive inventory. Through various *kaizen* activities, JIT makes it possible to build such flexibility into the management system (see Chapter 11).

It is possible to realize improved quality, cost, and delivery simultaneously by employing various management systems that have been developed over the years and thus to make the company far more profitable than it has been in the past.

## Quality Improvement and Cost Reduction Are Compatible

The recurring theme of this chapter has been that improving quality and reducing cost are compatible objectives. In fact, quality is the foundation on which both cost and delivery can be built. Without creating a firm system to ensure quality, there can be no hope of building effective cost-management and delivery systems.

Not only is it *possible* both to improve quality and to reduce cost, but we *must* do both in order to meet today's customer requirements. Take, for example, international competition in the high-end consumer goods market. Suppose that one company subscribes to the old philosophy that better quality costs more money. The company's major means of ensuring quality has been to buy more expensive machines and testing equipment and hire more people to perform rework and inspections. The company has a reputation for world-class quality, but its prices are very high.

Suppose that a new company emerges as a competitor. This company believes that better quality and lower cost are compatible and has succeeded in building a product of equal or better quality to the first company, but at a lower price. How will the first company cope with its new rival? This is the real nature of the "clear and present danger" facing many of today's companies that continue to subscribe to the outdated notion that quality costs money.

The simultaneous realization of QCD is the task that the *kaizen*-minded manager must tackle in today's competitive environment. At a time when customers are demanding ever-better QCD, management must emphasize the proper priority to achieve all three: *Quality first!* Resist the temptation to cut costs at the expense of quality! And do not sacrifice quality for delivery!

# CHAPTER FOUR

# Standards

Often I am asked to reflect on my three decades of teaching *kaizen* and to offer advice as to where companies can do better. Many organizations have achieved great things through *gemba kaizen* and have made a difference in the lives of people. However, one area stands out as a consistent weakness of even the best organizations. It is the proper use of standards. This has become especially evident over the last decade as *kaizen* and variously named continuous improvement strategies grow in popularity worldwide, but sadly much of the gains are lost because they have built a weak foundation of standards.

Daily business activities function according to certain agreed-on formulas. These formulas, when written down explicitly, become *standards.* Successful management on a day-to-day level boils down to one precept: Maintain and improve standards. This means not only adhering to current technological, managerial, and operating standards but also improving current processes in order to elevate current standards to higher levels.

## Maintain and Improve Standards

Whenever things go wrong at the *gemba,* such as producing rejects or dissatisfying customers, management should seek out the root causes, take actions to remedy the situation, and change the work procedure to eliminate the problem. In *kaizen* terminology, managers should implement the standardize-do-check-act (SDCA) cycle.

With current standards in place and workers doing their jobs according to those standards with no abnormalities, the process is under control. The next step is to adjust the status quo and raise standards to a higher level. This entails the plan-do-check-act (PDCA) cycle.

In both cycles, the final stage of the cycle—act—refers to standardizing and stabilizing the job. Thus standardization becomes an inseparable part of everybody's job. As will be explained later, standards are the best way to ensure quality and the most cost-effective way to do the job.

Here, let's refer back to the anecdote in Chapter 2 about the way fax messages were improperly handled by the desk staff of a hotel at which I was staying. In a case such as this, each customer complaint gives rise to a need to review the existing standards. Depending on the level of sophistication involved, management might find that no standards existed at all to start with and that simply adding standards would make the system more robust. However, not every aspect of our work needs close scrutiny. For instance, if the hotel management had received no complaints from its guests, it might have concluded that its current way of handling fax messages was adequate. In such a case, one could look for *kaizen* in other areas rather than trying to improve fax-handling procedures. This would not mean, however, that one didn't need to look at the best benchmark practices of the industry and try to reach such a level even when no complaints had been received. An improved fax-handling procedure might have saved time and work for the staff, thus freeing them for other work.

We should establish priorities in reviewing standards based on such factors as quality, cost, delivery, safety, the urgency and the gravity of the consequences, and the severity of customer complaints. In daily routine work (what I call *maintenance*), workers either do the job the right way, causing no abnormalities, or encounter abnormalities, which should trigger a review of existing standards and perhaps lead to establishing new ones. The first requirement of management remains that of maintaining standards. The system is under control when standards exist that are followed by workers who produce no abnormalities. Once the system is under control, the next challenge is to improve the status quo.

Let's assume that a strong demand necessitates a production increase of 10 percent. In line with the *kaizen* spirit, making better use of the existing resources would be the best way to cope with such a demand. To meet the goal, operators must change their way of doing their jobs. The existing standards must be upgraded through *kaizen* activities. At this stage, we have left the maintenance stage and moved on to the *improvement stage.*

Once such improvement has started, new and upgraded standards can be installed and efforts made to stabilize the new procedures, initiating a

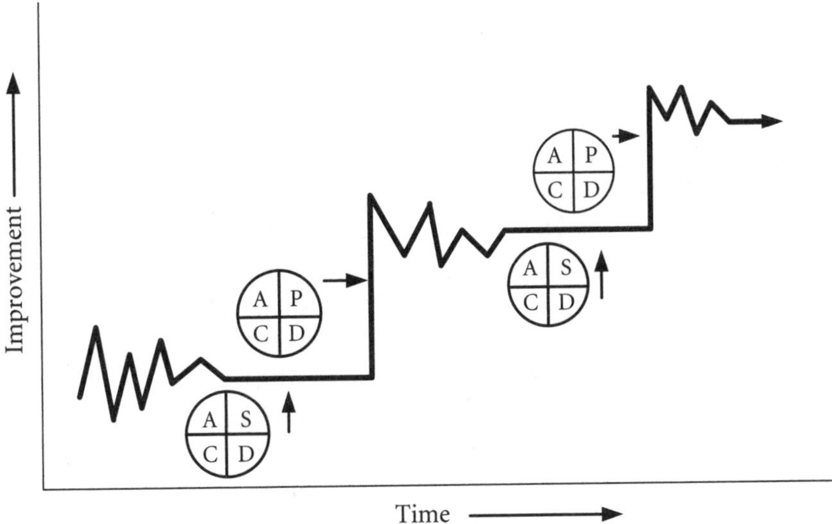


**Figure 4.1** How improvements are registered from SDCA cycles to PDCA cycles.

new maintenance stage. Figure 4.1 shows how improvements are registered within a company between the standardize-do-check-act (SDCA) cycle and the plan-do-check-act (PDCA) cycle.

## Operational Standards

Before we go any further, let's focus on the way we're using the word *standards*. In this context, there are two types of standards. One is *managerial standards*, which are necessary for managing employees for administrative purposes and which include administrative rules, personnel guidelines and policies, job descriptions, guidelines for preparing expense accounts, and so on. The other is called *operational standards*, which have to do with the way people do a job to realize quality, cost, and delivery (QCD). While managerial standards relate to the internal purpose of managing employees, operational standards relate to the external demand to achieve QCD to satisfy customers.

The standards referred to in this book are operational and point up a big disparity between Japanese and Western companies. Japan enthusiastically embraces the idea of establishing standards, whereas the West looks on standards with a certain degree of cynicism. In the West, the word *standards* is often misinterpreted to mean the imposition of unfair conditions on workers—the introduction of a wage system based on piecework, for instance. However, the word *standards* as used in this book means using the process that's the safest and easiest for workers and the most cost-effective and productive for the company to ensure quality for its customers.

In extreme cases, standards in the West are seen as something that goes against human nature. There is a feeling that people should not be bound by standards and that human beings should be given maximum freedom to do their job the way they want to. But it's important to distinguish between the ideas of *controlling* and *managing*. When management talks about *control*, it means control over the process, not the person. Management manages employees so that employees can control the process. Following standards is like driving a car. The driver must follow certain regulations, and yet, as a result, he or she gains the freedom to go where he or she wants to go. Likewise, when workers follow standards and do the job right, the customer is satisfied with the product or service, the company prospers, and the workers can look forward to job security.

## Key Features of Standards

Standards have the following key features:

1. *Standards represent the best, easiest, and safest way to do a job.* Standards reflect many years of wisdom and know-how on the part of employees in doing their jobs. When management maintains and improves a certain way of doing something, making sure that all the workers on different shifts follow the same procedures, those standards become the most efficient, safe, and cost-effective way of doing the job.
2. *Standards offer the best way to preserve know-how and expertise.* If an employee knows the best way to do the job and leaves without sharing that knowledge, his or her know-how also will leave. Only by standardizing and institutionalizing such know-how within the company does it stay in the company, regardless of the comings and goings of its individual workers.
3. *Standards provide a way to measure performance.* With established standards, managers can evaluate job performance. Without standards, there is no fair way to do this.

4. *Standards show the relationship between cause and effect.* Having no standards or not following standards invariably leads to abnormalities, variability, and waste. Let's apply this concept to the sport of skydiving, for example. When people first begin skydiving, they depend on their instructor to fold their parachutes. As they become more experienced, they begin folding their own parachutes with the help of the instructor. Before they can become full-fledged skydivers, they must learn how to fold their parachutes correctly by themselves. Suppose that a skydiver has folded her parachute for the first time in her life and is going to jump tomorrow. She goes to bed but cannot sleep and starts wondering, "Did I fold it right?" She gets out of bed, unfolds the parachute and starts all over again, goes back to bed, but still can't sleep. How many times does she need to fold it before she is convinced that everything is okay? The answer is that she should need to do it only once. The way to fold the parachute today is the best, easiest, and safest way, reflecting the experience of many thousands of parachutists—and the aftermath of various tragedies. Every time a parachute did not open, it gave rise to wrenching questions: "Where in our way of folding the parachute did we go wrong? How can we change and improve the process to prevent a recurrence?" What are the consequences of not following the folding standards? By the time you find it out, it may be too late.
5. *Standards provide a basis for both maintenance and improvement.* By definition, following standards means *maintenance*, and upgrading standards means *improvement*. Without standards, we have no way of knowing whether we have made improvements or not. Management's duty is, first and foremost, to maintain standards. When variability occurs owing to a lack of standards, one must introduce new standards. If variability occurs even with adherence to standards, management first must determine the cause and then either revise and upgrade the existing standards or train the operators to do the job as specified by the standards. Perhaps something about the existing standards is unclear, or perhaps the operators need more training to do the job properly.

Maintenance activities should constitute a majority of managers' tasks in their day-to-day activities in the *gemba*. Once maintenance stabilizes and controls the process, management can plan the next challenge: improve -

ment, or upgrading the existing standards. Where there is no standard, there can be no improvement. For these reasons, standards are the basis for both maintenance and improvement. For example:

- *Standards provide objectives and indicate training goals.* Standards can be described as a set of visual signs that show how to do the job. As such, standards should communicate in a simple, understandable manner. Normally, standards come in the form of written documents, but at times, pictures, sketches, and photos may facilitate understanding.
- *Standards provide a basis for training.* Once standards are established, the next step is to train operators to such an extent that it becomes second nature for them to do the job according to the standards.
- *Standards create a basis for audit or diagnosis.* In the *gemba*, work standards are often displayed, showing the vital steps and checkpoints of an operator's work. These standards no doubt serve as reminders to operators. Even more important, though, they help managers to check whether work is progressing normally. If maintaining and improving standards are the two major tasks of management, the primary job of *gemba* supervisors is to see whether standards are being maintained and, at the appropriate time, whether plans to upgrade current standards are being implemented.
- *Standards provide a means for preventing recurrence of errors and minimizing variability.* As already stated, standardization is the last step of the five *gemba* principles. It is also the next-to-the-last step in the *kaizen* stories, explained later in this chapter. Only when we standardize the effect of a *kaizen* project can we expect that the same problem does not recur. Quality control means variability control. Management's task is to identify, define, and standardize key control points at each process and make certain that such control points will be followed at all times.

Often, company A turns out to be better than company B in quality—not because A is superior in *all* aspects of the processes but because company A is making concerted efforts to ensure that *all* the processes are followed as specified in standards, whereas company B finds that one or two processes are not always followed.

Thus standardization is an integral part of quality assurance (QA), and without standards, there can be no way to build a viable quality system.

## Toyoda Machine Works

Yoshio Shima, director of Toyoda Machine Works, says that the benefit of building systems and standards for QA became apparent in the 1980s when the company introduced total quality management (TQM) with the aim of receiving the Deming Prize. Management's efforts to build a frame of reference for a QA system culminated in the company's receiving the Deming Prize in 1985.

Shima admits that the various standards instituted during those early days reflected not only essential steps for QA but also the management's wishful thinking, their vision of ideal procedure. "You might say that we started out by putting in the form of standardization first and later putting a soul into it," he says.

However, Shima found that after these standards were put into practice, they were not always usable. In order to remain practical, they had to be reviewed and upgraded constantly. Thus the journey to quality improvement at the company meant a never-ending review of existing standards.

Says Shima, "The difficult part of standardization is that standards are not unchangeable. If you believe that standards are writ in stone, you will fail. When the environment changes, standards change as well. You have to believe that standards are there to be changed." He goes on to say: "Once standards are in place and being followed, if you find a deviation, you know there is a problem. Then you review the standards and either correct the deviation from the standard or change the standard. It is a never-ending process of PDCA! That is the reason why you see numerous feedback routes on our QA system diagram" (see Figure 3.1).

Shima finds that the QA system diagram is very helpful because it affords him a bird's-eye view of the total system of quality assurance within the company. When quality problems arise with the customer, the company uses the system diagram to explain how it will solve the problem. The customer understands and appreciates management's efforts.

## The *Kaizen* Story*

The *kaizen story* is a standardized format to record *kaizen* activities conducted by such small-group activities as quality circles. The same

*In Japan the term *QC story* is commonly used. In view of the fact that the term *kaizen story* is used more often in other countries these days, the term *kaizen story* is used throughout this book.

standardized format is also employed to report *kaizen* activities conducted by staff and managers.

The *kaizen* story follows the plan-do-check-act (PDCA) cycle. Steps 1 through 4 relate to *P* (plan), step 5 relates to *D* (do), step 6 relates to *C* (check), and steps 7 and 8 relate to *A* (act). The *kaizen* story format helps anybody to solve problems based on data analysis. One of its merits is to help managers visualize and communicate the problem-solving process. It is also an effective way to keep records of *kaizen* activities.

Various problem-solving tools are often shown in the *kaizen* story to help the reader understand the process.

The *kaizen* story includes the following standardized steps:

1. *Selecting the theme.* The story begins with the reason why the particular theme was selected. Often the themes are determined in line with management policies or depend on the priority, importance, urgency, or economics of current circumstances.
2. *Understanding the current status and setting objectives.* Before starting the project, current conditions must be understood and reviewed. One way to do this is to go to the *gemba* and follow the five *gemba* principles. Another way is to collect data.
3. *Analyzing the data thus collected to identify root causes.*
4. *Establishing countermeasures based on the data analysis.*
5. *Implementing countermeasures.*
6. *Confirming the effects of the countermeasures.*
7. *Establishing or revising the standards to prevent recurrence.*
8. *Reviewing the preceding processes and working on the next steps.*

For an example of a *kaizen* story, see the case study "*Kaizen* Experience at Alpargatas."

## The Toyota Business Practice: The Standard Problem-Solving Story at Toyota

Many companies have their own so-called standard way of solving problems. Some have adopted the *kaizen* story approach just shown, some follow a nearly identical step-by-step approach called *kobetsu kaizen* most popular within total productive maintenance (TPM), and others have adopted the 8D approach from the automotive industry. I have found that many times a standards exists but is not used in the true meaning of a standard—a method to be followed, improved, and revised.

Today, the so-called A3 problem-solving method has become increasingly popular. The A3 refers only to the paper size that is the standard for summarizing the problem-solving story. The A3 problem-solving approach comes from Toyota, is based on the *kaizen* story, and follows an eight-step approach. In an effort to standardize and strengthen problem solving as Toyota operations became increasingly globalized, in the early 2000s, the *Toyota Business Practice* (TBP) was born.

The eight steps of the TBP problem-solving approach are

1. Clarify the problem.
2. Break down the problem.
3. Set a target to be achieved.
4. Analyze the root cause.
5. Develop countermeasures.
6. See countermeasures through.
7. Evaluate both results and process.
8. Standardize successful processes.

The TBP approach can seem very simple and quite similar to other eight-step approaches. As with many methodologies in *kaizen*, knowing a few simple key points and practicing them diligently makes the difference (see Figure 4.2).

Chris Schrandt is a senior consultant from the Kaizen Institute with over 30 years of experience in the field of quality. Ten of those years he spent at Toyota Motor Manufacturing Kentucky as quality engineering manager, during which he wrote "thousands" of A3 problem-solving documents. Chris shared some lessons learned from teaching the TBP approach to a wide variety of manufacturing and service companies after leaving Toyota.

About the typical approach to A3 and TBP problem solving, Chris shares:

> Not enough importance is placed on the problem statement itself. It must measurably define "the gap" between the current situation and the target condition. The problem statement must contain

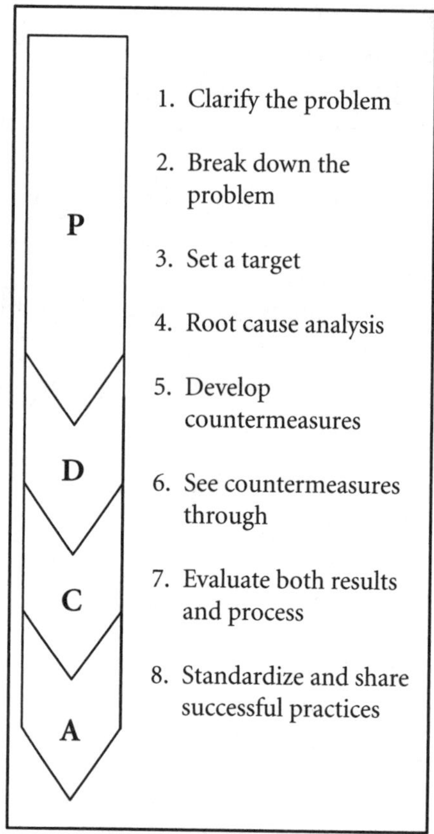


**Figure 4.2** The eight-step Toyota Business Practice (TBP) approach to problem solving.

neither a cause nor a countermeasure. The more time spent getting the problem statement right, the less time will be needed to actually address the problem.

There are many tools and methods for root cause analysis, but the one you must know and use is five-whys analysis. It is a simple and powerful tool, much like a chainsaw, and must be used properly and logically to arrive at potential root causes. The properly written problem-solving story connects the problem statement through the root cause analysis to the countermeasure action plan. The A3 should make sense read backwards or forwards.

It can't be repeated enough that the A3 is not a worksheet or a form. It is a logical story that contains thoughts about each phase of Plan, Do, Check, Act ... even if these are in advance of the investigation. People want a formula, a template to fill out to arrive at the answer. Resist the temptation to turn A3 thinking into a form-filling exercise.

The news is filled with so many companies that fail to follow their own standards or to adhere to their own problem-solving process. This is true even of Toyota! I asked Chris Schrandt why so many companies fail to consistently follow a standardized approach to problem solving. He replied:

> If a management team is too busy putting out many fires daily, this will distract from problem solving of any kind. Many times they are working on too many things, the wrong things. They have jumped to solutions rather than followed a problem-solving approach. Sometimes even after sorting and applying Pareto analysis to the issues, none clearly stand out as the vital few things that need to be done to put out many fires. So we throw money at all of them.
>
> Instead, the team that follows a standard problem-solving approach will be able to start with the recurring problems that they believed they had already countermeasured, putting safeguards and standards in place. Problem recurrence indicates either that the problem recurred because of a different root cause, the standards put in place were not followed, or they did not actually get to the true root cause the first time.
>
> Sorting out how many of what types of problems a management team is tackling says a great deal about the organization's problem-solving skills and/or discipline to adhere to standards.

There is no doubt that what we see at Toyota is the result of many years of experiments, both successful and failed. We must adopt best practices such as these, adapt them to our situation, and do *gemba kaizen* to build standards in our own way.

## *Kaizen* and International Quality Standards

Today, it has become almost a must for any company to apply for national or international certification of standards such as ISO 9000, QS 9000, and

AS 9000 or the environmental standards such as ISO 14000 if they are to stay in business and gain the confidence of their major global customers. These certification programs place much emphasis on standardization of the key processes and continual improvement.

In *kaizen* terms, the standards are the best way to do the job, and *gemba kaizen* such as *muda* elimination and housekeeping (5S in particular) should precede writing a standard. Writing down the working process of the *gemba* as it is now in great detail may be required for certifying the process but is useless if the current process contains much *muda* and variability.

Once standards have been established, improvement of those standards must follow. Thus it is imperative that *gemba kaizen* activities be carried out before applying for certification, as well as upgrading the standards after certification has been awarded.

Sometimes an executive preparing for certification of ISO 9000 or QS 9000 will say, "We are too busy preparing for the certification and have no time to do *kaizen*!" Nothing can be further from the truth. Unless *kaizen* is carried out concurrently, the ensuing standards with much variability owing to a lack of good housekeeping and *muda* elimination will be just a piece of paperwork far removed from the *gemba* and rarely practiced in daily work and will have no positive impact on improvement of the company's performance.

Thus *gemba kaizen* should become an integral part of getting international certification, and after having received it, *gemba kaizen* should be a means to upgrade such standards on a continual basis.

One of the *kaizen* consultants once shared his first encounter with the magic power of standardization as follows:

> In 1961, I was a manager for a large electronics company in Europe. I was responsible for transferring know-how and delivering machines from our factory to a Japanese electronics company with which we had a joint-venture agreement. Before we delivered the equipment, the Japanese company sent four operators into our factory to study our production process, where 20 fully automated lines were running on three shifts. Each line produced 2,000 semiconductor diodes per hour, with a yield of 98 percent.

About six months after the Japanese plant had begun operations, we received a letter from them thanking us for our cooperation and for the precision of our machinery. They also noted that their yield was 99.2 percent.

As a result, we went to Japan to study what had been done, asking our Japanese colleague, "What changes did you make to realize this higher yield?" His answer: "We made a study of your *gemba* and observed that you are following 60 different procedures (20 lines working on three shifts). We discussed this, and with mutual consent from the *gemba* observers who had gone to your country, we decided on the best way to standardize the process."

# CHAPTER FIVE

# The 5S: The Five Steps of Workplace Organization

It has become a popular saying among companies adopting a *kaizen* to say "the first step is 5S." This is true enough because at the most basic level, 5S requires that an organization ask the questions, "Do we have all that we need in the *gemba*?" and "Do we need all that we have in the *gemba*?" and then do *kaizen* whenever the answer is "No."

The five S (5S) are the five steps of workplace organization developed through intensive work in a manufacturing context. Service-oriented companies can readily see parallel circumstances in their own "production lines"—whether they come in the form of a request for proposal (RFP), the closing of a financial report, an application for a life insurance policy, or a potential client's request for legal services. Whatever triggers the process of work in the service company, conditions that exist in the work process *complicate* the work unnecessarily (Are there too many forms?), *impede* progress toward satisfying the customer (Does the size of the contract require signoff by three officers?), or actually *prohibit* the possibility of satisfying the customer (Does the company's overhead make it impossible to bid on the job?).

As Figure 2.3 shows, standardization, 5S, and *muda* ("waste") elimination are the three pillars of *gemba kaizen* in the commonsense, low-cost approach to improvement. *Kaizen* at any company—whether it is involved in a manufacturing or a service industry—should start with three activities: standardization, 5S, and *muda* elimination.

These activities involve no new management technologies and theories. In fact, such words as *housekeeping* and *muda* do not appear in management textbooks. They therefore do not excite the imagination of managers, who are accustomed to keeping abreast of the latest technologies. Those who attend my lectures sometimes wonder why these subjects have to be brought up. However, once they understand the implications of these three pillars, they become excited at the prospect of the tremendous benefits these activities can bring to the *gemba*. The case study "5S for the City" demonstrates how 5S activity can bring a group of people together toward a common goal and begin to build a *kaizen* culture.

## Good Housekeeping in Five Steps

The five steps of housekeeping, with their Japanese names, are as follows:

1. *Seiri:* Distinguish between necessary and unnecessary items in the *gemba*, and discard the latter.
2. *Seiton:* Arrange all items remaining after *seiri* in an orderly manner.
3. *Seiso:* Keep machines and working environments clean.
4. *Seiketsu:* Extend the concept of cleanliness to oneself, and continuously practice the preceding three steps.
5. *Shitsuke:* Build self-discipline and make a habit of engaging in 5S by establishing standards.

In introducing housekeeping, Western companies often prefer to use English equivalents of the five Japanese *S* words—as in a "5S campaign" or a "five Cs campaign."

### A 5S Campaign

1. *Sort:* Separate out all that is unnecessary, and eliminate it.
2. *Straighten:* Put essential things in order so that they can be accessed easily.
3. *Scrub:* Clean everything—tools and workplaces—removing stains, spots, and debris and eradicating sources of dirt.
4. *Systematize:* Make cleaning and checking routine.
5. *Standardize:* Standardize the preceding four steps to make the process one that never ends and can be improved on.

### A Five Cs Campaign

1. *Clear out:* Determine what is necessary and unnecessary, and dispose of the latter.

2. *Configure:* Provide a convenient, safe, and orderly place for everything, and keep it there.
3. *Clean and check:* Monitor and restore the condition of working areas during cleaning.
4. *Conform:* Set the standard, train and maintain.
5. *Custom and practice:* Develop the habit of routine maintenance, and strive for further improvement.

## 5S for the City: Civic Pride in Romania

Large parts of many cities, unfortunately, are not pleasant places to be. Garbage-strewn streets and parks, unclean and decaying public facilities, and vandalism and graffiti all contribute to a deteriorating visual environment, sending a message to visitors and residents alike that nobody cares. The situation is a vicious circle—the worse things look, the less people care—this is human nature, the same in the city as in a company.

In October 2011, three major Romanian cities decided to experiment with *kaizen* to reverse this trend. Their reasoning was summarized in a simple question: "If workers are able to use improvement methods to turn factories into clean, orderly environments, why can't civic workers and citizens improve their city using the same methods?" After only a few months, all three cities had shown that this certainly can be done and that *kaizen* and civic pride are a natural fit.

The driving force behind the 5S initiatives was to improve the public image of these cities. Organizers hoped to instill a sense of pride among all civic workers and citizens, as well as lend a sense of ownership that would encourage them to help keep their city facilities clean and tidy. They recognized that to really succeed with *kaizen*, it is necessary to get everybody involved. Under the guidance of the Kaizen Institute Romania, the cities held a series of training sessions, structured planning meetings, and a publicity campaign.

"We are using the 5S method to accomplish this public *kaizen* project because it is a simple method which can be easily applied to everyone, whether they are a school-age child or a pensioner," said Julien Bratu, country manager of the Kaizen Institute Romania in a countrywide television interview. "5S is a practical formula for order and beauty. In addition to basic cleaning principles, it also has a strong set of general rules and educational tools for continuous improvement and maintenance."

Companies with extensive experience in *kaizen* management were selected to help the three chosen cities with the project. Boosted by their own *kaizen* successes, these organizations were happy to provide material support, as well as volunteers to help implement 5S. Strong support also was obtained from the public authorities of the participating cities, including the municipalities of Brasov and Timisoara, as well as the county council and municipalities of Alba Iulia. Once their support was assured, specific areas were targeted for 5S, and 50 to 150 volunteers were placed in each city.

Renowned *kaizen* leader Yoshihito Tanaka, president of the Clean Up Japan Association and president of electronics company Tokai Shine Industrial Group, also was invited to join the project because of his extensive experience with 5S public cleaning events. Further support was provided for the city of Brasov by other Kaizen Institute representatives from Japan and Italy.

Each city was responsible for its own 5S project, and each city held three kinds of daily meetings inspired by a *gemba kaizen* work site.

- *Informational meetings*, held in the evenings, were focused on educating top management from all parties involved, including local public authorities, private-sector managers, and members of the media. Participants were encouraged to openly discuss various issues related to 5S.
- *Planning meetings* focused on taking specific action were held in the mornings. Volunteers would meet to receive training on explicit 5S methods and then would proceed to clean up areas designated by the local municipality.
- *Communication meetings* took place at lunchtime. Volunteers would provide feedback, press releases would be written, local media representatives were met with, and future initiatives were planned (see Figure 5.1).

The cleanup teams focused on high-profile areas in their cities. In Timisoara, the second largest Romanian city, 60 young volunteers, with the support of the mayor and deputy mayor, cleaned up the Bega River shore,

**Figure 5.1** 5S volunteers working to clean up Romania.

a beautiful green area that is a source of pride for the city. In Alba Iulia, a team that included 150 volunteers of all ages revitalized The Fortress, a national historical icon.

"I was very happy to contribute alongside my colleagues," said a high school student volunteer. "Our actions must continue so that what we have done will continue to be meaningful."

Over 250 volunteers were mobilized, including the mayors and deputy mayors of the three participating cities, and seven television stations have broadcasted the event. Most important, plans are in the works to build a long-term "5S for the city" strategy.

"This was one of the best projects of this kind that I have attended," said Yoshihito Tanaka. "I was amazed to see how involved local company managers were in their support of this event. Although economic conditions in Romania are tougher than those in Japan, I have rarely seen so many people as motivated to apply 5S in their city. I hope that this event encourages the expansion of *kaizen* culture worldwide."

## A Detailed Look at the Five Steps of 5S

Figure 5.2 shows the relationship of the five steps of 5S and how self-discipline and continual improvement is central to this approach.

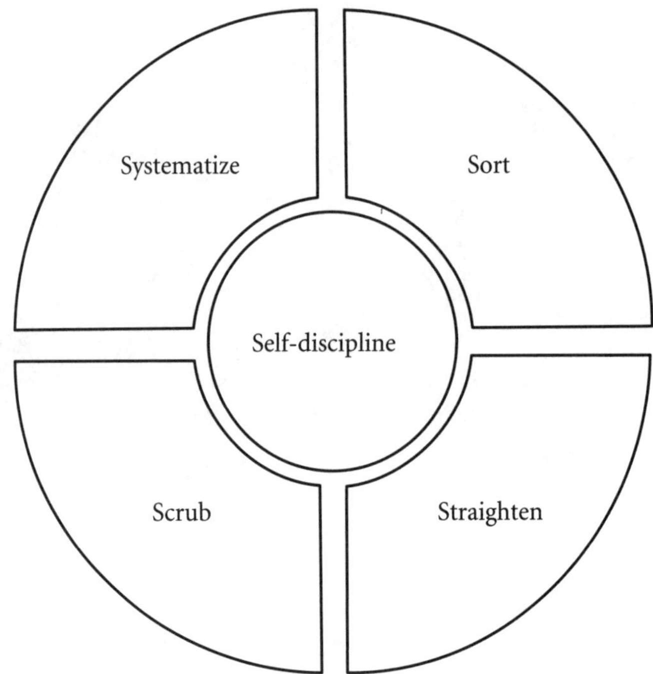


**Figure 5.2** The five steps of 5S.

## Seiri *(Sort)*

The first step of housekeeping, *seiri*, entails classifying items in the *gemba* into two categories—necessary and unnecessary—and discarding or removing the latter from the *gemba*. A ceiling on the number of necessary items should be established. All sorts of objects can be found in the *gemba*. A close look reveals that only a small number of them are needed in daily work; many others either will never be used or will be needed only in the distant future. The *gemba* is full of unused machines, jigs, dies and tools, rejects, work-in-process, raw materials, supplies and parts, shelves, containers, desks, workbenches, files of documents, carts, racks, pallets, and other items. An easy rule of thumb is to remove anything that will not be used within the next 30 days.

*Seiri* often begins with a red tag campaign. Select one area of the *gemba* as the site for *seiri*. Members of the designated 5S team go to the *gemba* with handfuls of red tags and place them on items they believe are unnecessary. The larger the red tags and the greater their number, the better. When it is unclear whether or not a particular item is needed, a red tag should be placed on it. By the end of the campaign, the area may be covered with hundreds of red tags, inviting comparison with a grove of maple trees in the fall.

Sometimes *gemba* employees may find red tags placed on items they actually need. In order to keep such items, employees must demonstrate the necessity of doing so. Otherwise, everything with a red tag on it is removed from the *gemba*. Things that have no reason to stay in the *gemba*, no apparent future usage and no intrinsic value, are thrown away. Things that will not be needed within the next 30 days but may be needed at some point in the future are moved to their rightful places (such as the warehouse in the case of supplies). Work-in-process that exceeds the needs of the *gemba* should be sent either to the warehouse or back to the process responsible for producing the surplus.

In the process of *seiri*, one can obtain valuable insights into how the company conducts its business. The red tag campaign leaves in its wake a mountain of unnecessary *gembutsu*, and employees are confronted with uncomfortable questions, such as, "How much money is tied up in prematurely manufactured products?" People ask themselves how they could have acted so foolishly. At one company, a red tag campaign unearthed enough supplies to last for 20 years!

Both managers and operators have to see such extravagance in the *gemba* to believe it. This is a practical way for managers to get a glimpse at how people work. On finding a heap of supplies, for example, the manager should be asking, "What kind of system do we have for placing orders to suppliers? What kind of information do our purchasing people use in placing orders? What kind of communication is maintained between production scheduling and production? Or do the staff responsible for purchasing just place orders when they think it is about time to do so?"

Managers should be equally rigorous when they find work-in-process made well in advance: "Why do our people keep producing work-in-process for which we have no immediate need? Based on what kind of information do they start production?" Such a situation indicates fundamental deficiencies in the system, such as having insufficient control between production and purchasing at the *gemba*. It also shows insufficient flexibility to cope with changes in production scheduling.

At the end of the red tag campaign, all managers—including the president and plant manager as well as *gemba* managers—should get together and have a good look at the heap of supplies, work-in-process, and other *gembutsu* and start making *kaizen* to correct the system that made this waste possible.

Eliminating unnecessary items via the red tag campaign also frees up space, enhancing flexibility in the use of the work area, because once unnecessary items have been discarded, only what is needed remains. At this stage, the maximum number of items to be left in the *gemba*—parts and supplies, work-in-process, and so on—must be determined.

*Seiri* can be applied to individuals working in offices as well. For example, a typical desk has two or more drawers. Items are often placed in these drawers indiscriminately; side by side in a single drawer one may find not only pencils, ballpoint pens, erasers, writing pads, rubber bands, business cards, and scissors but also toothbrushes, candy, perfume, aspirin, coins, matches, cigars, costume jewelry, Band-Aids, and other objects. These items first must be classified by use. In a desk with only two drawers, office supplies and personal items each should occupy one drawer.

Next, the maximum number of each item is determined. For instance, let's say we decide to place in the drawers only two pencils, one ballpoint pen, one eraser, one pad of paper, and so on. Any items beyond the maximum number are discarded—that is, removed from the drawer and taken to the office supply storage area in the corner of the room. Sometimes this storage area is called a *recycling bank*. When supplies in the drawers are exhausted, the employee goes to the recycling bank to replenish them. In turn, the employee in charge of the bank watches the inventory and, when it drops to the designated minimum, orders more supplies.

By paring to a minimum the supplies in our office drawers, we eliminate the need to shuffle through the collection of pencils, papers, and cosmetics to reach a desired item. This process develops self-discipline as well as improving recordkeeping and enhancing employees' ability to work effectively.

## Seiton *(Straighten)*

Once *seiri* has been carried out, all unnecessary items have been removed from *gemba*, leaving only the minimum number needed. But these needed items, such as tools, can be of no use if they are stored too far from the workstation or in a place where they cannot be found. This brings us to the next stage of 5S, *seiton*.

*Seiton* means classifying items by use and arranging them accordingly to minimize search time and effort. To do this, each item must have a designated name, address, and volume. Not only the location but also the maximum number of items allowed in the *gemba* must be specified. For example, work-in-process cannot be produced in unlimited quantities. Instead, the floor space for the boxes containing the work must be delineated clearly (by painting a rectangle to mark off the area, etc.), and a maximum allowable number of boxes—say, five—must be designated. A weight may be suspended from the ceiling above the boxes to make it impossible to stack more than five. When the maximum allowed level of inventory has been reached, the production in the previous process must stop; there is no need to produce more than what the following process can consume. In this manner, *seiton* ensures the flow of a minimum number of items in the *gemba* from station to station on a first-in, first-out basis.

Taiichi Ohno was once invited to visit the assembly line of another company. Asked to comment on the line, he said, "You have much too much work-in-process waiting by the line side. Leave a minimum number on the line side, and send back all the excessive items to the previous process." A mountain of pressed metal sheets had to be sent back to the press shop, and the workers there had to do their job surrounded by pressed metal sheets that created a prisonlike atmosphere. Ohno said, "This is the best way to show people that the harder they work, the more money the company will lose."

The items left in the *gemba* should be placed in the designated area. In other words, each item should have its own address, and conversely, each space in the *gemba* also should have its designated address. Each wall should be numbered, using designations such as wall A1 and wall B2. The location of such items as supplies, work-in-process, fire hydrants, tools, jigs, molds, and carts should be designated either by its address or by special markings. Markings on the floor or workstations indicate the proper locations of work-in-process, tools, and so on. Painting a rectangle on the floor to delineate the area for boxes containing work-in-process, for example, creates a space sufficient to store the maximum volume of items. At the same time, any deviation from the designated number of boxes shows up instantly. (Readers familiar with just-in-time will recognize that this is the first stage of introducing a pull production system.) Tools should be placed well within reach and be easy to pick up and put down. Their silhouettes might be painted on the surface where they are supposed to be stored. This makes it easy to tell when they are in use.

The hallway also should be marked clearly with paint. Just as other spaces are designated for supplies and work-in-process, the hallway is meant for transit: Nothing should be left there. The hallway should be completely clear so that any object left there will stand out, allowing supervisors to notice the abnormality instantly and take remedial action.

## Seiso *(Scrub)*

*Seiso* means cleaning the working environment, including machines and tools, as well as the floors, walls, and other areas of the workplace. There is also an axiom that goes, "*Seiso* is checking." An operator cleaning a machine can find many malfunctions. When the machine is covered with oil, soot, and dust, it is difficult to identify any problems that may be developing. While cleaning the machine, however, one can easily spot oil leakage, a crack developing on the cover, or loose nuts and bolts. Once these problems are recognized, they are easily fixed.

It is said that most machine breakdowns begin with vibration (owing to loose nuts and bolts), with the introduction of foreign particles such as dust (owing to a crack on the cover, for instance), or with inadequate oiling and greasing. For this reason, *seiso* is a great learning experience for operators because they can make many useful discoveries while cleaning machines.

I was once engaged in *seiso* activities at the plant of a wooden floor tile manufacturer that contained many woodworking machines such as power saws. All senior managers, including the president, joined in *seiso* with the operators. (This was said to have been the first time employees saw the president show up at the *gemba* wearing overalls and holding a broom.) While they were cleaning the exterior of the machines, as well as the walls and beams on the ceiling, they said over and over, "I can't believe it!" Thick layers of wood chips and dust clung to the walls. On removing the debris, the director of finance discovered naked electrical wires running along the walls. The vinyl cover had long since deteriorated. He marveled at the fact that a fire had never broken out in the plant.

## Seiketsu *(Systematize)*

*Seiketsu* means keeping one's person clean by such means as wearing proper working clothes, safety glasses, gloves, and shoes, as well as maintaining a clean, healthy working environment. Another interpretation of *seiketsu* is continuing to work on *seiri*, *seiton*, and *seiso* continually and every day.

For instance, it is easy to go through the process of *seiri* once and make some improvements, but without an effort to continue such activities, the situation soon will be back to where it started. To do *kaizen* just once in the *gemba* is easy. To keep doing *kaizen* continuously, day in, day out, is an entirely different matter. Management must deploy systems and procedures that ensure the continuity of *seiri*, *seiton*, and *seiso*. Management's commitment to, support of, and involvement in 5S becomes essential. Managers must determine, for example, how often *seiri*, *seiton*, and *seiso* should take place and who should be involved. This should become part of the annual planning schedule.

## Shitsuke *(Standardize)*

*Shitsuke* means self-discipline. People who practice *seiri*, *seiton*, *seiso*, and *seiketsu* continuously—people who have acquired the habit of making these activities part of their daily work—acquire self-discipline.

5S may be called a philosophy, a way of life in our daily work. The essence of 5S is to follow what has been agreed on. It begins with discarding what we don't need in the *gemba* (*seiri*) and then arranging all the necessary items in the *gemba* in an orderly manner (*seiton*). Then a clean environment must be sustained so that we can readily identify abnormalities (*seiso*), and these three steps must be maintained on a continuous basis (*shitsuke*). Employees must follow established and agreed-on rules at each step, and by the time they arrive at *shitsuke*, they will have the discipline to follow such rules in their daily work. This is why we call the last step of 5S self-discipline.

By this final stage, management should have established standards for each step of 5S and made certain that the *gemba* is following those standards. The standards should include ways to evaluate progress at each of the five steps.

A *gemba* manager of a chemical company once told me that when he asked his *gemba* operators to measure key parameters of the process and plot them on the control chart, the operators didn't take this task too seriously: The numbers always stayed in the center of the control chart. Once 5S was implemented successfully and everybody began to acquire self-discipline, however, the manager found that the operators' attitudes had changed: The data on the control chart began to show deviations.

There are five ways to appraise the level of 5S at each stage:

1. Self-evaluation
2. Evaluation by an expert consultant
3. Evaluation by a superior
4. A combination of the preceding
5. Competition among *gemba* groups

The plant manager should set up a contest among the workers; the manager then can review the state of 5S in each *gemba* and select the best and worst *gemba.* The best can receive some award or other recognition, whereas the worst receives a broom and a bucket. The latter group will have an incentive to do a better job so that another group will get those items the next time.

In order to review progress, evaluation must be conducted regularly by plant managers and *gemba* managers. Only after work on the first step has been approved can workers move on to the next step. This process lends a sense of accomplishment.

After *seiso* has been completed, management's attention should focus on a new horizon—namely, maintaining and ensuring momentum and enthusiasm. After working hard at *seiri, seiton*, and *seiso* and having seen improvements in the *gemba*, employees begin to think, "We've made it!" and relax and take it easy for a while (or even worse, cease their efforts altogether). Strong forces at work in the *gemba* try to push conditions back to their previous state, making it imperative for management to build a *system* to ensure the continuity of 5S activities.

## Introducing 5S

*Kaizen* values the process as much as the result. In order to get people involved in *continuing* their *kaizen* efforts, management must carefully plan, organize, and execute the project. Managers often wish to see the result too soon and skip a vital process. 5S is not a fad, a flavor of the month, but an ongoing part of daily life. Any *kaizen* project therefore needs to include follow-up steps.

Because *kaizen* addresses people's resistance to change, the first step is to prepare employees mentally to accept 5S before the campaign gets started. As a preliminary to the 5S effort, time should be allocated to discuss the philosophy behind and the benefits of 5S.

- ▲ Creating clean, sanitary, pleasant, and safe working environments
- ▲ Revitalizing the *gemba* and greatly improving employee morale and motivation
- ▲ Eliminating various kinds of *muda* by minimizing the need to search for tools, making operators' jobs easier, reducing physically strenuous work, and freeing up space

Management also should understand the many benefits of 5S in the *gemba* to the company overall:

- ▲ Helping employees to acquire self-discipline. Self-disciplined employees are always engaged in 5S, take positive interest in *kaizen*, and can be trusted to adhere to standards.
- ▲ Highlighting the many kinds of *muda* (waste) in the *gemba*. Recognizing problems is the first step in eliminating waste.
- ▲ Eliminating *muda* in the *gemba* enhances the 5S process.
- ▲ Pinpointing abnormalities, such as rejects and inventory surplus.
- ▲ Reducing wasteful motion, such as walking and needlessly strenuous work.
- ▲ Allowing problems associated with shortage of materials, line imbalances, machine breakdowns, and delivery delays to be identified visually and thence to be solved.
- ▲ Resolving outstanding logistical problems in the *gemba* in a simple manner.
- ▲ Making quality problems visible.
- ▲ Improving work efficiency and reducing costs of operation.
- ▲ Cutting down on industrial accidents by eliminating oily and slippery floors, dirty environments, rough clothing, and unsafe operations. *Seiso* in particular increases machine reliability, thus freeing maintenance engineers' time for working on machines that are prone to sudden

breakdown. As a result, engineers can concentrate on more upstream issues, such as preventive maintenance, predictive maintenance, and the creation of maintenance-free equipment in cooperation with design departments.

Having both understood these benefits and made certain that employees understand them, management then can move forward with the *kaizen* project.

# CHAPTER SIX

# *Muda*

One day, after attentively observing operators working in the *gemba*, Taiichi Ohno said to the workers, "May I ask you to do at least one hour's worth of work every day?" Believing themselves to have been working hard all day long, the workers resented this remark. What Ohno actually meant, however, was, "Will you do your value-adding work for at least one hour a day?" He knew that most of the time the operators were moving around the *gemba* without adding any value. Any non-value-adding activity is classified as *muda* in Japan. Ohno was the first person to recognize the enormous amount of *muda* that existed in the *gemba*.

The Japanese word *muda* means "waste," but the word carries a much deeper connotation. *Work* is a series of processes or steps starting with various inputs and raw materials and ending in a final product or service. At each process, value is added to the product (or, in the service sector, to the document or other piece of information), and then the product (or service) is sent on to the next process. The resources at each process—people and machines—either do add value or do not add value. *Muda* refers to any activity that does not add value. Ohno classified *muda* in the *gemba* according to the following seven categories:

1. *Muda* of overproduction
2. *Muda* of inventory
3. *Muda* of defects
4. *Muda* of motion
5. *Muda* of processing
6. *Muda* of waiting
7. *Muda* of transport

Since anything that does not add value is *muda*, the preceding list can be extended almost indefinitely. At Canon Company, *muda* is classified according to the categories listed in Table 6.1. The "Interpretation" column has been added to provide clarification.

## *Muda* of Overproduction

*Muda* of overproduction is a function of the mentality of the area supervisor, who is worried about such problems as machine failures, rejects, and absenteeism and who feels compelled to produce more than necessary just to be on the safe side. This type of *muda* results from getting ahead of the production schedule. When an expensive machine is involved, the requirement for the number of products is often disregarded in favor of efficient utilization of the machine.

In a just-in-time (JIT) system, however, being ahead of the production schedule is regarded as worse than being behind it. Producing more than necessary results in tremendous waste: consumption of raw materials before they are needed, wasteful input of personnel and utilities, additions of machinery, an increase in interest burdens, the need for additional space to store excess inventory, and added transportation and administrative costs. Of all *muda*, producing too much is the worst. It gives people a false sense of security, helps to cover up all sorts of problems, and obscures information that can provide clues for *kaizen* on the shop floor. It should be regarded as a crime to produce more than necessary. Overproduction stems from the following invalid assumptions or policies:

- Produce as many as we can in the process, disregarding the proper speed at which the next process or next line can operate.
- Give the operator enough elbow room to produce.
- Let each process or line have an interest in raising its own productivity.
- Speed up the line owing to a low first-pass yield ratio and few line stoppages. (*First-pass yield ratio* refers to the percentage of products completed without rework.)
- Allow machines to produce more than needed because they have excess capacity.
- Introduce expensive machines because they cannot be depreciated unless the operation ratio is improved.

**Table 6.1** Types of Waste and How to Eliminate Them

| Waste Category | Nature of Waste | How to Eliminate | Interpretation |
|---|---|---|---|
| Work-in-process | Stocking items not immediately needed | Streamline inventory | Removing false economies of scale, such as big batches caused by setup times |
| Rejection | Producing defective products | Reduce rejects | Remove defect causes at process |
| Facilities | Idle machinery, breakdowns, excessive setup time | Increase capacity utilization ratio | Reduce the losses to increase capacity |
| Expenses | Overinvesting for required output | Curtail expenses | Use additive equipment, the 3P process, and scalable process design |
| Indirect labor | Excess personnel owing to poor indirect labor system | Assign jobs efficiently | Thoroughly map the indirect processes to identify customer non-value-added steps |
| Design | Producing products with more functions or features than necessary | Reduce costs | Limit to features or functions desired by customers |
| Talent | Employing people for jobs that can be mechanized or assigned to less skilled people | Institute labor-saving or labor-maximizing measures | Use people where they are most needed—for work that requires human creativity and intelligence |
| Motion | Not working according to the work standard | Improve work standards | Make work standards reflect the safest, most effective known method, and train to it |
| New-product ramp-up | Slow start in stabilizing the production of a new product | Shift to full-line production more quickly | Simulate processes with cardboard mock-ups, use pilot runs, and reuse good design practices |

*Source: Adapted from Masaaki Imai,* Kaizen: The Key to Japan's Competitive Success *(New York: McGraw-Hill, 1986).*

## *Muda* of Inventory

Final products, semifinished products, or parts and supplies kept in inventory do not add any value. Rather, they add to the cost of operations by occupying space and requiring additional equipment and facilities such as warehouses, forklifts, and computerized conveyer systems. In addition, a warehouse requires additional personnel for operation and administration.

While excess items stay in inventory and gather dust, no value is added. Their quality deteriorates over time. Even worse, they could be destroyed by a fire or other disaster. If *muda* of inventory did not exist, much waste could be avoided. Inventory results from overproduction. If overproduction is a crime, inventory should be regarded as an enemy to be destroyed. Unfortunately, we all know managers who cannot sleep at night when they don't have "good inventory." Inventory is often likened to the water level that hides problems. When an inventory level is high, nobody gets serious enough to deal with problems such as quality, machine downtime, and absenteeism, and thus an opportunity for *kaizen* is lost.

Lower inventory levels help us to identify areas that need to be addressed and force us to deal with problems as they come up. This is exactly what the JIT production system is after: When the inventory level goes down and finally reaches the one-piece flow line, it makes *kaizen* a mandatory daily activity.

## *Muda* of Defects

Defects interrupt production and require expensive rework. Often the rejects must be discarded—a great waste of resources and effort. In today's mass-production environment, a malfunctioning high-speed automated machine can spew out a large number of defective products before the problem is arrested. The rejects also themselves may damage expensive jigs or machines. Attendants therefore must be assigned to high-speed machines, standing by to stop the machines as soon as a malfunction is identified. Dedicating an attendant to this task defeats the purpose of having a high-speed machine. Such machines at least should be equipped with mechanisms that shut them down as soon as a faulty product is produced (the concept of *jidoka*).

Suppliers often complain of too much paperwork and too many design changes when dealing with their customers. In a broader sense, both problems involve *muda*. The excess-paperwork *muda* could be eliminated by reducing red tape, streamlining operations, eliminating unnecessary processes, and speeding up processing decision-making time. The problem of excessive design changes results in *muda* of rework. If the designers did their work right the first time—if they had a better understanding of customer and supplier requirements as well as the requirement of their own *gemba*—they could eliminate the *muda* of design changes. *Kaizen* can be applied as effectively to engineering projects as to matters in the *gemba*.

## *Muda* of Motion

Any motion of a person's body not directly related to adding value is unproductive. When a person is walking, for instance, he or she is not adding any value. In particular, any action that requires great physical exertion on the part of an operator, such as lifting or carrying a heavy object, should be avoided not only because it is difficult but also because it represents *muda*. The need for an operator to carry a heavy object for a distance can be eliminated by rearranging the workplace. If you observe an operator at work, you will find that the actual value-adding moment takes only a few seconds; the remainder of his or her motions represent non-value-adding actions such as picking up or putting down a workpiece. Often, the same workpiece is first picked up with the right hand and then held with the left hand. A person working at a sewing machine, for example, first picks up a few pieces of fabric from the supply box, then puts them down on the machine, and finally picks up one piece of fabric to feed into the sewing machine. This is *muda* of motion. The supply box should be relocated so that the operator can pick up a piece of fabric and feed it directly into the sewing machine.

To identify *muda* of motion, we need to have a good look at the way operators use their hands and legs. We then need to rearrange the placement of parts and develop appropriate tools and jigs.

## *Muda* of Processing

Sometimes inadequate technology or design leads to *muda* in the processing work itself. An unduly long approach or overrun for machine processing, unproductive striking of the press, and deburring are all examples of processing *muda* that can be avoided. At every step in which a workpiece or piece of information is worked on, value is added and sent to the next process. *Processing* here refers to modifying such a workpiece or piece of information. Elimination of *muda* in processing frequently can be achieved with a commonsense, low-cost technique. Some wasteful processing can be avoided by combining operations. For instance, at a plant where telephones are produced, the receiver and the body are assembled on separate lines and later put together on the assembly line. To protect the surfaces of the receivers from scratches as they are being transported to the final assembly line, each receiver is wrapped in a plastic bag. By connecting the receiver assembly line and the final assembly line, however, the company can eliminate the plastic-wrapping operation.

Waste in processing also results, in many cases, from a failure to synchronize processes. Operators often try to engage in the processing work in a much finer degree than is necessary, which is another example of *muda* of processing.

## *Muda* of Waiting

*Muda* of waiting occurs when the hands of the operator are idle; when an operator's work is put on hold because of line imbalances, lack of parts, or machine downtime; or when the operator is simply monitoring a machine as the machine performs a value-adding job. This kind of *muda* is easy to detect. More difficult to detect is the *muda* of waiting during machine processing or assembly work. Even if an operator appears to be working hard, a great deal of *muda* may exist in the form of the seconds or minutes the operator spends waiting for the next workpiece to arrive. During this interval, the operator is simply watching the machine.

## *Muda* of Transport

In the *gemba*, one notices all sorts of transport by such means as trucks, forklifts, and conveyers. Transport is an essential part of operations, but moving materials or products adds no value. Even worse, damage often occurs during transport. Two separate processes require transport. In order to eliminate *muda* in this area, the so-called isolated island—any process that is physically distant from the main line—should be incorporated into the line, if possible.

Together with excess inventory and needless waiting, transport *muda* is a highly visible form of waste. One of the most conspicuous features of most Western manufacturing *gembas* is a heavy reliance on conveyer belts. Such layouts sometimes make me wonder whether the engineer who designed them is a model railway hobbyist. Whenever you notice a conveyer in the *gemba*, your first question should be, "Can we eliminate it?" The best thing a company can do with its conveyers is to sell them to its competitors. Better yet, the company should wrap them up in a gift package and send them to the competitor free of charge!

*Kaizen* consultant Greg Back recalls his experience when he was consulting for a well-known German automotive manufacturer. Back and his colleague were working at the press shop on a multidie press to reduce changeover and setup times.

At the start of the project, Back set a target of a 50 percent reduction in the setup time (which was then 10 hours) by the end of the week without any technological changes. Both the supervisor and the workers reacted with disbelief and anger ("We have not been sleeping all these years!").

By the end of the week, however, the time had been reduced to 5.5 hours, mostly through changes in the way of working, such as incorporating the five S (5S), shifting from internal to external work for changeover, and so on. By making additional minor technical changes and further standardizing procedures and practices in the following two months, the company, on its own, further reduced changeover time to 3.5 hours.

The press-line foreman later confessed to Back: "When you people showed up and told me what possibilities you saw, I was very angry. After all, I'm an expert at these things, and my people are very good. But then I said to myself, "Okay, let these *kaizen* consultants embarrass themselves! Now I have seen the results and how you did it, and I started thinking about why I hadn't seen all this *muda* before. And I thought about what I had been doing. When I was going through my line and saw that my people were busy and working hard, I was satisfied. I never really looked closely at *what* they were doing, *how* they were doing it, or *why* they were doing it that way! They were busy, complaining about the amount of work, the job was difficult and had always taken so long. I never really looked closely at the process in the *gemba*!"

Tomoo Sugiyama, director of Yamaha Engine Company, proposes *less* engineering in the *gemba* and has devised a "list of less-engineering items" as a means of highlighting things that should be eliminated. (Sugiyama came up with the term *less engineering* when he was searching for a catch phrase that would help *gemba* people identify problems more easily. Although the term does not sound like authentic English, it captured the imagination of his employees. The popularity of "endless tapes" and "tubeless tires" in Japan has made *less* a very familiar word.)

*Person (Worker)*
Look-less
Walk-less
Search-less
Block-less

*Machine*
Air-less
Conveyer-less
Air cut-less
Air press-less

*Material*
Bolt-less
Burr-less
Wait-less
Stop-less

*Methods*
Bottleneck-less
Stock-less

*Quality*
Reject-less
Careless mistake-less
Nonstandard-less

Sugiyama initiated *kaizen* activities in the name of "less engineering" and found that they were readily accepted. For example, the company developed the following three principles of air-less engineering:

1. Do not transport air.
2. Do not store air.

3. Eliminate space that does not create added value.

On determining that air constituted 93 percent of the company's packaging for motorcycle mufflers and exhaust pipes, Sugiyama targeted the waste for *kaizen* and achieved a great deal of savings. Later, air-less engineering also was applied to using warehouse space more efficiently. From this experience, Yamaha devised a formula for calculating space savings in monetary terms and embarked on a companywide air-less campaign.

In Sugiyama's view, anybody can add to the list of "less" engineering items simply by taking the trouble to identify *muda*.

## *Muda* of Time

Another type of *muda* observed daily is the waste of time, although it was not included in Ohno's seven categories of *muda*. Poor use of time results in stagnation. Materials, products, information, and documents sit in one place without adding value. On the production floor, temporal *muda* takes the form of inventory. In office work, it happens when a document or piece of information sits on a desk or inside a computer awaiting a decision or signature. Wherever there is stagnation, *muda* follows. In the same manner, the seven categories of *muda* invariably lead to the waste of time.

## Categorizing *Muda* in the Service Sector

The various *kaizen* transformations led by my colleagues at the Kaizen Institute have shown that *muda* exists no matter what type of process or business sector. Variations on the same seven types of *muda* have showed up in engineering, health care, software design, retail, farming, ports, local government, and so forth. Figure 6.1 shows a model for categorizing typical losses in service industries.

A series of studies conducted between 2004 and 2010 by the Kaizen Institute and the Fraunhover Institute revealed that white-collar workers were able to identify non-value-adding or *muda* activities taking up between 27 and 38 percent of their week. Figure 6.2 shows the survey questionnaire form used.

Often the *kaizen* ideas to save much of the lost time in office work are simple. "Meetings could be more efficient if you do standing meetings instead of sitting meetings," according to Sebastian Reimer, senior consult -

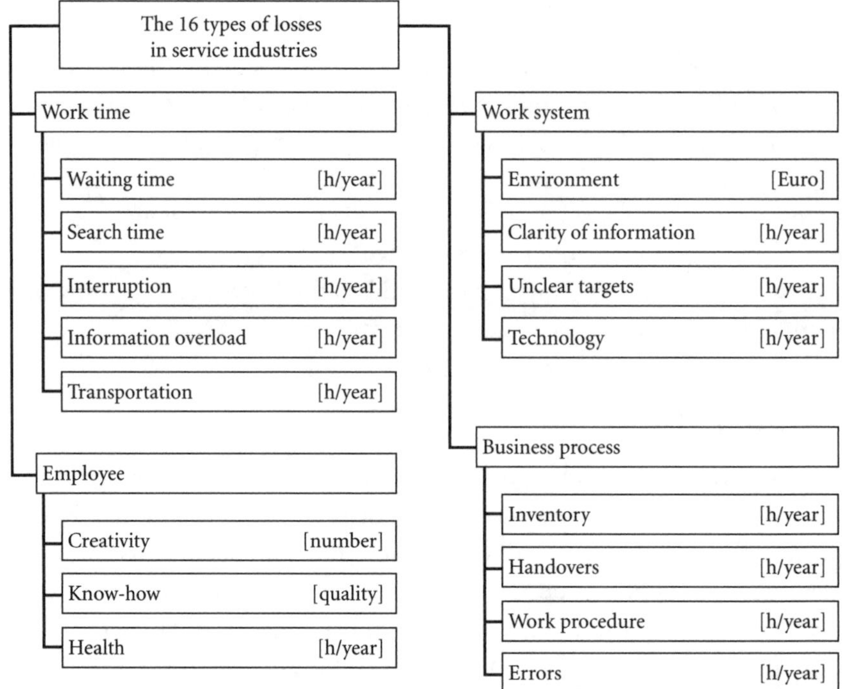


**Figure 6.1** The loss-tree model to quantify white-collar waste.

ant with the Kaizen Institute. "It is typical to see teams save nearly 50 percent of their time in meetings. Just calculate this by week and by year."

Even though information technology (IT) has made business more efficient, there is still much *muda* to be removed through *kaizen*. "You will be amazed if you measure and sum two numbers," says Reimer:

1. How long do you need per week to sort out all the information overload?
2. How long does it take to get all the necessary information for your processes?

Where the traditional approach to improvement of ever-better machines, software, and management programs too often adds cost, the elimination of *muda* focused on the *gemba* reduces cost while improving quality and delivery simultaneously.

This *muda* is far more prevalent in service sectors. By eliminating the aforementioned non-value-adding time bottlenecks, the service sector

Observed: weekly work time

| | h | % |
|---|---|---|
| Gross work time | 40.7 | 100% |

| Time losses due to | Description | h | % | Rank | Min | Max |
|---|---|---|---|---|---|---|
| Searching | For documentation, files, all types of information | 1.5 | 3.7% | 4 | 0.0 | 10.0 |
| Waiting | For IT programs, colleagues, signatures, etc. | 1.5 | 3.7% | 5 | 0.0 | 12.0 |
| Disruption | Of the actual work through interruptions | 2.5 | 6.2% | 1 | 0.0 | 11.0 |
| Pursuit | Of missing information, colleagues who cannot be reached | 1.8 | 4.5% | 3 | 0.0 | 15.0 |
| Participation | In inefficient, overly long meeting without results | 2 | 5.0% | 2 | 0.0 | 8.0 |
| Sorting out | Of excess information, advertising, email, spam, etc. | 1.2 | 2.9% | 8 | 0.0 | 13.0 |
| Clarification | Of badly delegated, unclear, or confusing tasks | 1.4 | 3.5% | 6 | 0.0 | 10.0 |
| Correcting | Of incorrect, incomplete entries/inputs | 1.4 | 3.4% | 7 | 0.0 | 12.0 |
| Tracking | Of complicated, redundant, or bureaucratic processes | 1.2 | 2.8% | 9 | 0.0 | 10.0 |
| Transporting | Of papers from and to the copier, post, etc. | 0.9 | 2.3% | 10 | 0.0 | 5.0 |

| | h | % |
|---|---|---|
| Total time losses | 15.4 | 38% |

For the actual productive work there is only:

| | h | % |
|---|---|---|
| Net work time | 25.3 | 62% |

**Figure 6.2** Study of time losses in the office: results after survey.

should be able to achieve substantial increases in both efficiency and customer satisfaction. Because it costs nothing, *muda* elimination is one of the easiest ways for a company to improve its operations. All we need to do is to go to the *gemba*, observe what is going on, recognize *muda*, and take steps to eliminate it.

## *Muda, Mura, Muri*

The words *muda*, *mura*, and *muri* are often used together and are referred to as the *three MUs* in Japan. Just as *muda* offers a handy checklist to start *kaizen*, the words *mura* and *muri* are used as a handy reminder to start *kaizen* in the *gemba*. *Mura* means "variation," and *muri* means "strain or overburden." Anything strenuous or irregular indicates a problem. Furthermore, both *mura* and *muri* also constitute *muda* that needs to be eliminated.

### Mura *(Variation)*

Whenever a smooth flow of work is interrupted in an operator's work, the flow of parts and machines, or the production schedule, there is *mura*. For example, assume that operators are working on the line, and each person is performing a given repetitive task before sending it to the next person. When one of them takes more time to do the job than the others, *mura* as well as *muda* results because everybody's work must be adjusted to meet the slowest person's work. The idea of *mura* also applies to the variation in quality of goods and services. Looking for such variation within the process or the results of processes becomes an easy way to start *gemba kaizen*.

### Muri *(Overburden)*

*Muri* means "strenuous conditions" for both workers and machines, as well as for the work processes. For instance, if a newly hired worker is assigned to do the job of a veteran worker without sufficient training, the job will be overburdened, and chances are that the worker will be slower in his or her work and even may make many mistakes, creating *muda*.

When we see an operator sweating profusely while doing a job, we must recognize that too much strain is required and remove it. When we hear a squeaking sound from a machine, we must recognize that it has been overburdened, meaning that an abnormality is occurring. Thus *muda*, *mura*, and *muri* combined are handy checks to identify abnormalities in the *gemba*.

Of all *kaizen* activities, *muda* elimination is the easiest to start because it is not too difficult to identify *muda* once one has acquired such a skill. *Muda elimination* usually refers to stopping something that we have been doing up until now and therefore costs little to implement. For these reasons, management should take the initiative in starting *kaizen* with *muda* elimination wherever it exists—in the *gemba*, in administration, and/or in the area of service provision.

## Removing *Muda* from Public-Sector Organizations

The African Union (AU) is an association of 54 nations that promotes peace, stability, and good governance in Africa through common social and economic policies. As with so many other public-service organizations, the resources it receives from its various donors are limited. To help make the most of what resources it had, the AU wanted to promote a culture of doing more with less by educating officials on the principles and methodologies of *kaizen*.

To help affect change, the Kaizen Institute customized the *gemba kaizen* principles for a public-sector environment. The 4P model for public-sector reforms is pictured in Figure 6.3. The "People Involvement" element involves sensitizing top management and training employees to become *kaizen* champions. Over 14 months, the Kaizen Institute certified more than 100 AU employees across four divisions, including administration and human resources, finance, conference services, and the medical center.

The second element, "Physical Workplace Improvement," uses methods such as 5S and visual management to transform the workplace by exposing and removing *muda*. This, in turn, helped to improve morale at the AU. The third element, "Process Improvement," and the fourth element, "Policy Review," are built on the solid foundation of the first two elements.

More than 100 *gemba kaizen* projects were implemented by AU staff. A major goal of those projects was to improve lead times, which are often very long in government and create an image of a slow, wasteful, and non-competitive public service. With the help of *kaizen* to reduce *muda* of transport, inventory, and waiting from various transactional processes,

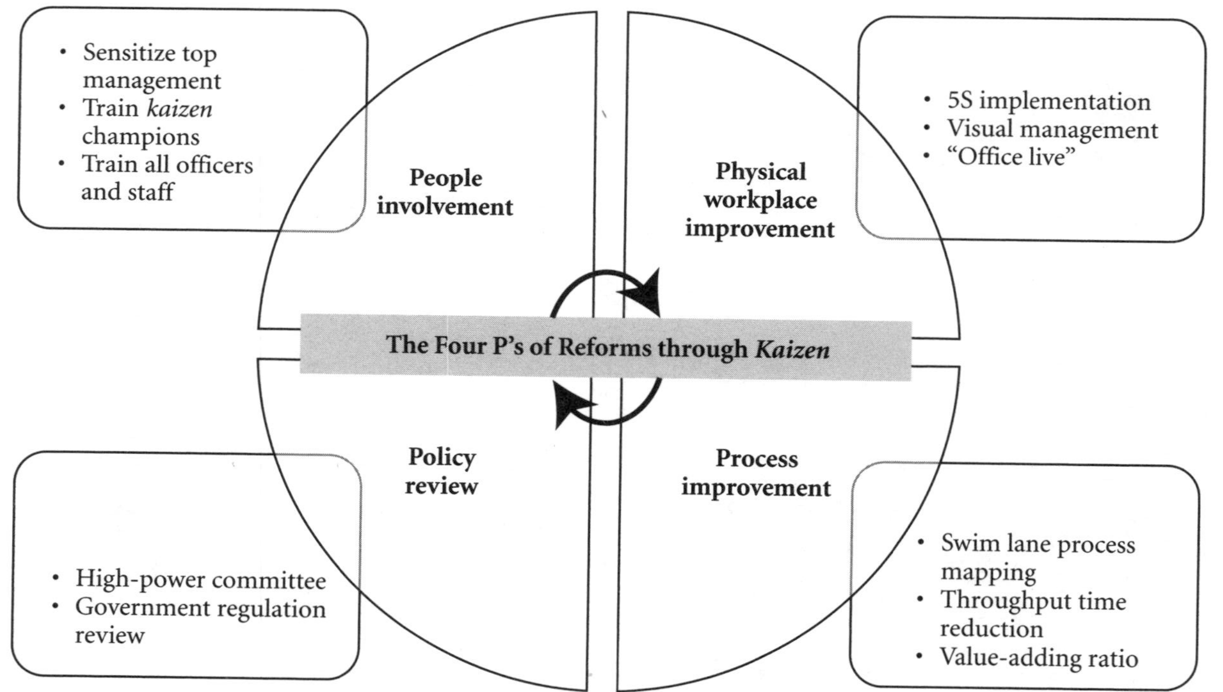


**Figure 6.3** The 4P model for public-sector reforms.

speed, efficiency, morale, and customer satisfaction were improved. The issuance of identity cards, for example, was shortened from 11 to 13 days to less than 1 day.

Lead times also were cut for staff payments, formerly a six-step process that was reduced to a one-step process. With the standardization of various activities and a load-leveled schedule, the lead time in this department was reduced from an average of 24 days to a maximum of 4 days.

Lead times in the procurement of medical supplies and other goods also dropped from 10 to 3 months. This was accomplished by 5S of information, categorizing goods based on usage and priority, and by modifying procurement policies to fit those demands.

Waste in government is often what people call "bureaucracy"—excessive paperwork, long lead times, and poorly coordinated subprocesses. All of this creates stereotypes of inefficient and uncaring government institutions. Removing this *muda* requires significant participation from the people who own the processes, which was enabled by the 4P model. At the AU, *kaizen* has put public servants on a path toward improving service by doing more with less.

## CHAPTER SEVEN

# The Foundation of the House of *Gemba*

According to Toyota, there are two pillars in their strategy to maintain its culture of sustained excellence, which it calls the *Toyota Way*. The two pillars are *kaizen* and *respect for people*. I have introduced, written in length, and spoken around the world about *kaizen*, and I also have tried to make it abundantly clear that *gemba kaizen* is a human system that works only when it is people-centered.

However, I have seen that many company initiatives such as lean manufacturing have emphasized the tools and techniques of *kaizen* and the Toyota Production System but have forgotten the vital importance of the people foundation. To all students and teachers of *kaizen*, I encourage a careful reading of this chapter as well as Chapters 9, 10, and 13 dealing specifically with the role of supervisors, managers, and the chief executive officer (CEO).

As shown in Chapter 2, the house of *gemba* rests on a solid foundation of employee involvement activities such as teamwork, morale enhancement, self-discipline, quality circles, making improvement suggestions, and related pursuits—communication, empowerment, and skill development, as well as visual management. Management must build a firm commitment to carrying out these activities continuously. Only when management demonstrates that it is highly motivated, self-disciplined, and *kaizen*-minded can *gemba* people do their job of maintaining and improving standards to satisfy customers by achieving the targets of quality, cost, and delivery (QCD).

Most companies that introduce *kaizen* unsuccessfully fail to build the necessary infrastructure first. Fortunately, we do not have to wait until the infrastructure is complete and everybody in the *gemba* has made the transformation to see improvement. People can begin to change their thinking and behavior as soon as they begin working on *kaizen*. For instance, by the time 5S is firmly established in the *gemba*, people will have the self-discipline necessary to follow through on what has been agreed on. *Gemba kaizen* yields such impressive results that *gemba* operators are the first to recognize its benefits.

Marina Calcagni, an operator at Giorgio Foods Company, offered the following comments on her personal experiences of *kaizen* in her company:

> When *kaizen* began, it was something that shocked everybody. It was something different. People came here just to work, to get their regular paycheck, and go home. And now instead *kaizen* opened our eyes—I think it really did. And it makes you think twice when you're doing something. I think we learned that if we're doing something better, it's for us; it's not for anybody from the outside. Our place looks cleaner and neater, [and] it's a safer place to work.
>
> Personally, I think that *kaizen* helped me not only here at work but even at home. I think it makes me think twice; I want to do better every day. We don't have to wait until there's a problem. We have to do something to make things better in whatever area. Not because it's a problem, just to make it better, and that's what *kaizen* taught us.

## A Learning Enterprise

Bill Ford, a visiting honorary professor at the Industrial Relations Research Center, University of New South Wales, Australia, advocates the concept of a *learning enterprise*. He quotes a saying by Dick Dusseldorp: "Training is for cats and dogs. People learn."

"A learning enterprise," says Ford, "is one where individuals, teams, and the enterprise itself are continually learning and sharing in the development, transfer, and use of knowledge and skills to produce continual improvement and the creation of a dynamic competitive advantage. Such enterprises are creating cooperative work environments in which the stakeholders in business—be they shareholders, managers, or the workforce—share in the development of common goals."

In building the foundation of *gemba kaizen*, we are pursuing the same goal—namely, building a learning enterprise involving both management and the workforce—to develop common goals and values. Here, improve -

ment is a way of life, and people take pride in their work, continually upgrade their skills, and they are empowered to solve problems in the *gemba.* The job is seen as a mission, a means to fulfillment and personal growth.

Thus the *gemba* should become a citadel of learning. In order to build a learning organization, management must empower *gemba* employees by providing learning experiences. As mentioned earlier, the tools for learning in the *gemba* rely heavily on common sense and simple checklists, such as asking "Why?" five times; the five steps of housekeeping; *muda, mura,* and *muri;* and following the axiom, "Don't accept poor quality. Don't make poor quality. Don't pass on poor quality."

Learning experiences in the *gemba* must be based on an appreciation of fundamental human values, such as respect for humanity, commitment, determination, economy (sensible use of resources), cleanliness, and order. *Learning* here should be synonymous with *doing.* Rather than being given too much teaching, *gemba* employees should be given opportunities to *learn by practicing and doing,* being physically involved, using their hands as well as their brains. After introduction of 5S and standardization at Tokai Shin-ei, President Yoshihito Tanaka said: "In hindsight, we have learned that our job is to do what we are supposed to do—namely, to do what we have agreed to do. In other words, a good company is the one where everybody is doing what he/she is supposed to do. We also learned that the best learning experience you can get is the one you gain through practicing, using your body, and learning by doing. Providing the concept alone is not enough."

This is why *gemba kaizen* activities in Japan have always stressed action. The following are the 10 basic rules for practicing *kaizen* in the *gemba:*

1. Discard conventional rigid thinking about production.
2. Think of *how to do it,* not why it cannot be done.
3. Do not make excuses. Start by questioning current practices.
4. Do not seek perfection. Do it right away even if for only 50 percent of target.
5. Correct mistakes at once.
6. Do not spend money for *kaizen.*
7. Wisdom is brought out when faced with hardship.
8. Ask "Why?" five times, and seek the root cause.
9. Seek the wisdom of 10 people rather than the knowledge of 1.
10. Remember that opportunities for *kaizen* are infinite.

People in the *gemba* are deeply ingrained in their old habits of working. When *gemba kaizen* is first introduced, strong psychological resistance must be overcome. The preceding 10 rules are employed by management as a guide to facilitate the introduction of *gemba kaizen.*

Just as Japanese companies faced obstacles in implementing *gemba kaizen,* Western management must be prepared for resistance and introduce *gemba kaizen* with firm determination.

Jim Crawford, vice president and managing director for value management and product research and development at Excel, offers the following observation on his personal transition through the *kaizen* process after several years of involvement in promoting *kaizen* at his company:

> The most profound personal change as a result of *kaizen* is understanding that our work processes are the delivery mechanisms for results. This understanding leads to the recognition that we can dramatically improve long-term results by improving our work processes.
>
> My past perspectives led me to believe that dramatic results could be achieved by working faster and harder. These efforts delivered disappointing results. Dramatic improvement was still elusive.
>
> Driven by the belief that quick, short-term dramatic improvement was still attainable, efforts were devised to approach the improvement of results by increasing manpower and capital resources. In retrospect, these efforts also fell short.
>
> The concept that we can only improve results by improving our work processes is a simple concept, yet it is often misunderstood. The key question is why do managers have such difficulty with this concept? Hopefully, by sharing my personal observations we can find an answer.

I have found through benchmarking that successful companies have managers who are committed to the *kaizen* process. These managers embrace the concept that results are not improved by management over the short term but by managers who support the long-term efforts of others in managing their work processes.

Transforming an organization from darkness to light also takes patience and courage. This change at Excel is painfully slow. Courage from within to support long-term efforts is difficult to come by given the pressure for dramatic improvements in results. The ability to lead and exhibit patience in order to achieve dramatic, long-term improvement only comes from understanding the concept that sustainable improvements in results come from long-term improvements in our work processes.

The case studies of both Leyland Trucks and Excel address the subject of how management went about building an internal structure to become a learning enterprise. According to one definition of the difference between *education* and *training*, education teaches what one does not yet know, whereas training teaches what one knows already—but teaches it in such a way that doing it right becomes almost second nature. In other words, in training, people learn by doing—by practicing repeatedly. Skills cannot be acquired simply by reading a book or listening to a lecture: They must be practiced!

## Suggestion System and Quality Circles

Important parts of the structure of the house of *gemba* are the suggestion system and quality circles—proof that employees are actively involved in *kaizen* and that management has been successful in building the *kaizen* infrastructure. There are marked differences between the suggestion systems practiced in Japan and those in the West.

Whereas the American-style suggestion system stresses the suggestion's economic benefits and provides financial incentives, the Japanese style stresses the morale-boosting benefits of positive employee participation. Over the years, the Japanese style has evolved into two segments: individual suggestions and group suggestions, including those generated by quality circles, *jishu kanri* (JK or "autonomous management") groups, zero defects (ZD) groups, and other group-based activities.

Suggestion systems are currently in operation at most large manufacturing companies and at about half the small and medium-size companies. In addition to making employees *kaizen*-conscious, suggestion systems provide an opportunity for the workers to speak out with their supervisors as well as among themselves. At the same time, they provide an opportunity for management to help the workers deal with problems. Thus suggestions are a valuable opportunity for two-way communication in the workshop as well as for worker self-development.

Generally speaking, Japanese managers have more leeway in implementing employee suggestions than their Western counterparts do. Japanese managers are willing to go along with a change if it contributes to any one of the following goals:

- Making the job easier
- Removing drudgery from the job
- Removing inconvenience from the job
- Making the job safer
- Making the job more productive
- Improving product quality
- Saving time and cost

The outlook of Japanese management stands out in sharp contrast to the Western manager's almost exclusive concern with the cost of the change and its economic payback.

The implications of standardization have been mentioned often in this book. When *gemba* employees participate in *gemba kaizen* and come up with new and upgraded standards, they naturally develop a sense of ownership of these standards and therefore will have the self-discipline to follow them.

If, on the other hand, the standards are imposed from above by management, *gemba* employees may show psychological resistance to following them. It becomes an "us versus them" issue. This is another reason why it is so crucial to involve *gemba* people in such *kaizen* activities as suggestion systems and quality circles.

## Building Self-Discipline

Needless to say, self-discipline is a cornerstone of the house of *gemba* management. Self-disciplined employees can be trusted to report on time to work; to maintain clean, orderly, and safe environments; and to follow the existing standards to achieve QCD targets. In my *kaizen* seminars, I often ask the participants to write down ways of helping employees acquire self-discipline. Here are some of the answers I have received:

1. Reward incremental steps.
2. Catch them doing it right.
3. Open yourself to questions.
4. Develop a culture that says let's do it!
5. Make the process known to improve standards.
6. Conduct assessment.
7. Encourage customer involvement.
8. Implement a suggestion system.
9. Establish quality circles.
10. Build in reward systems.
11. Communicate expectations clearly.
12. Conduct frequent reviews of the process.
13. Provide measurement feedback.
14. Foster a climate of cooperation.
15. Give specific instructions regarding criteria.
16. Be involved in setting standards.
17. Explain why.
18. Set a good example.
19. Teach how and why.
20. Make progress displays visible.
21. Remove barriers.
22. Encourage positive peer pressure.
23. Create a threat-free environment.

When employees in the *gemba* participate in such activities as housekeeping, *muda* elimination, and review of standards, they immediately begin to see the many benefits brought about by these *kaizen*, and they are the first to welcome such changes. Through such a process, their behaviors as well as attitudes begin to change.

For instance, as already mentioned in Chapter 5, the last step of 5S is *shitsuke* ("self-discipline"), and employees who have followed the five steps, up to the last step, are the ones who have acquired self-discipline. An employee who has participated in reviewing and upgrading the standard of his or her own work naturally develops ownership of such a standard as a result and willingly follows such a new standard.

In the same manner, employees eventually come to develop self-discipline as they engage in other *kaizen* projects and learn by doing such things as elimination of *muda* and visual management. Thus self-discipline translates into "everybody doing his or her own job according to the rules that have been agreed on." Self-discipline is a natural by-product of engaging in *gemba kaizen* activities.

Visual management is another key component of the foundation of the house of *gemba*, and it will be explained in detail in Chapter 8.

# CHAPTER EIGHT

# Visual Management

At the *gemba*, abnormalities of all sorts arise every day. Only two possible situations exist in the *gemba*: Either the process is under control, or it is out of control. The former situation means smooth operations; the latter spells trouble. The practice of visual management involves the clear display of *gembutsu*—the actual product, as well as charts, lists, and records of performance, so that both management and workers are continuously reminded of all the elements that make quality, cost, and delivery (QCD) successful—from a display of the overall strategy, to production figures, to a list of the latest employee suggestions. Thus visual management constitutes an integral part of the foundation of the house of *gemba*.

## Making Problems Visible

Problems should be made visible in the *gemba*. If an abnormality cannot be detected, nobody can manage the process. Thus the first principle of visual management is to spotlight problems.

If rejects are being produced by a broken die on a press and nobody sees the rejects, there will soon be a mountain of rejects. A machine equipped with *jidoka* devices, however, will stop the moment a reject is produced. The machine stoppage makes the problem visible.

When a hotel guest comes to the reception desk and asks for an aspirin or a list of good restaurants nearby, the hotel's inability to fulfill those needs constitutes an abnormality. By posting a list of the most frequent requests received from guests, the hotel's management can gain an awareness of service deficiencies that need to be addressed. This is visual management: making abnormalities visible to all employees—managers, supervisors, and workers—so that corrective action can begin at once.

Most information originating from the *gemba* goes through many managerial layers before reaching top management, and the information becomes increasingly abstract and remote from reality as it moves upward. Where visual management is practiced, however, a manager can see problems at a glance the moment he or she walks into the *gemba* and thus can give instructions on the spot in real time. Visual management techniques enable *gemba* employees to solve such problems.

The best thing that can happen in the *gemba* of a manufacturing company is for the line to stop when an abnormality is detected. Taiichi Ohno once said that an assembly line that never stops is either perfect (impossible, of course) or extremely bad. When a line is stopped, everyone recognizes that a problem has arisen and seeks to ensure that the line will not stop for the same reason again. Line stoppage is one of the best examples of visual management in the *gemba*.

## Staying in Touch with Reality

If the first reason for visual management's existence is to make problems visible, the second is to help both workers and supervisors stay in direct contact with the reality of the *gemba*. Visual management is a practical method for determining when everything is under control and for sending a warning the moment an abnormality arises. When visual management functions, everybody in the *gemba* can manage and improve processes to realize QCD.

When we take our customers on tours of Japanese factories, our hosts usually show us their display boards that allow everybody to see the production schedule and how the work is progressing. The formats are different for each plant. Some use whiteboards, whereas others use paper; some use magnets, but the display boards are always clear and easy to understand, serving the purpose of helping people by allowing them to stay in touch with reality on the *gemba*.

In an era of high-tech computer screens, these simple visual aids sometimes look arcane to an outsider, yet they are one of the most essential and powerful lean tools. Adopting this visual approach to management was an important step in the lean journey of Specialty Silicone Fabricators (SSF), a multisite high-technology contract manufacturer that has been working on a lean transformation for nearly 10 years.

In the mid-2000s, SSF was at a turning point. The company was facing a significant growth surge, but at the same time, the market was shifting to a lower-volume, higher-mix production. This created a *muri* situation where the existing scheduling methods were overburdened. Compounding the problem was an older facility with multiple rooms that made "seeing" the production process difficult and a customer care department that was located in another building.

According to SSF President Kevin Meyer, on-time delivery was poor, cycle time was approximately 10 times what it could be if non-value-added time was removed, and the customer care department spent most of its time explaining delays to customers. Meyer explains how SSF used *kaizen* to turn the situation around:

> We decided to try to capture all production work orders for one key value stream, and their status, on a manual magnetic whiteboard inside the production cleanroom. The initial format was chosen by the Value Stream Manager for that operation and was both time- and process-sequenced. Magnets were used to denote jobs and included basic customer, quantity, and due-date information. Soon afterwards, in an attempt to share this first pass at visual information, a webcam was installed that allowed for the whiteboard to be viewed across the company intranet. Customer care began using this to provide more accurate information to customers.
>
> By capturing all of the production work order information, combined with other lean tools such as value-stream mapping and 5S, we were able to rapidly reduce cycle time and improve on-time delivery. Customer care now focused on updating customers with good news, instead of bad. In fact, the webcam has been removed.
>
> The production whiteboards have been deployed in all value streams at all plants. Importantly, no single format was required—each value stream developed their format independently based on what was important to their processes, products, and systems. Also of critical importance was and is the concept that production operators own their boards and can continually evolve the format and types of information displayed. Ideas are shared between value streams, both informally and formally at quarterly Continuous

Improvement Team meetings. Operators are responsible for moving their magnets and can visually see and react to potential bottleneck situations. When combined with cross-training and TWI, this creates a very flexible operations organization. One value stream has even integrated leader standard work into the production control whiteboard.

This example shows that when employees "see" their contribution to the plant schedule, a process that they in fact own as a group, they collaborate better, and their work improves. This concept is so simple that it is easy to overlook, but it is part of the people foundation. It is important to never forget that visual tools have enormous power to drive improvement. This is why 5S, which sets standards and makes abnormality quickly visible, is always the first step and the last step in each *kaizen* journey.

## Visual Management in the Five Ms (5M)

In the *gemba*, management must manage the five Ms (5M): manpower, machines, materials, methods, and measurements. Any abnormality related to the 5M conditions must be displayed visually. What follows is a more detailed look at visual management in these five areas.

### *Manpower (Operators)*

- How is worker morale? This can be measured by the number of suggestions made, the extent of participation in quality circles, and figures on absenteeism. How do you know who is absent from the line today and who is taking their place? These items should be made visible at the *gemba.*
- How do you know people's skill level? A display board in the *gemba* can show who is trained to do what tasks and who needs additional training.
- How do you know that the operator is doing the job right? Standards that show the right way to do the job—for example, the one-point standard and the standard worksheet—must be displayed.

## Machines

- How do you know that the machine is producing good-quality products? If *jidoka* and *pokayoke* (mistake-proofing) devices are attached, the machine stops immediately after something goes wrong. When we see a machine that is stopped, we have to know why. Is it stopped because of scheduled downtime? Changeover and setup? Quality problems? Machine breakdown? Preventive repair?
- Lubrication levels, the frequency of exchange, and the type of lubricant must be indicated.
- Metal housings should be replaced by transparent covers so that operators can see when a malfunction occurs inside a machine.

## Materials

- How do you know the materials are flowing smoothly? How do you know whether you have more materials than you can handle and whether you are producing more products than you should? When a minimum inventory level is specified and *kanban*—attaching a card or tag to a batch of work-in-process as a means of communicating orders between processes—is used, such anomalies become visible.
- The address where materials are stored must be shown, together with the stock level and parts numbers. Different colors should be used to prevent mistakes. Use signal lamps and audio signs to highlight abnormalities such as supply shortages.

## Methods

- How does a supervisor know if people are doing their jobs right? This is made clear by standard worksheets posted at each workstation. The worksheets should show sequence of work, cycle time, safety items, quality checkpoints, and what to do when variability occurs.

## Measurements

- How do you check whether the process is running smoothly? Gauges must be clearly marked to show safe operating ranges. Temperature-sensing tapes must be attached to motors to show whether they are generating excess heat.

- How do you know whether an improvement has been made and whether you are on the way to reaching the target?
- How do you find out whether precision equipment is properly gauged?
- Trend charts should be displayed in the *gemba* to show the number of suggestions, production schedules, targets for quality improvement, productivity improvement, setup-time reduction, and reduction in industrial accidents.

## Visual Management to Manage Complexity

During a recent trip to New Zealand, I visited Stainless Design Limited in Hamilton. I saw how workers used a sequence of *heijunka* ("leveling") boxes to process customer orders, helping them to plan their work, level their flow, and manage their resources accordingly. This was an excellent example of using visual management to manage complexity in a simple way.

Being at the heart of the metal fabrication industry, the company had an extra challenge because it did not produce a range of the same product; over 80 percent of what it manufactured for its customers was one-off orders.

Team leaders reviewed each order and then used the *heijunka* box to determine the date by which the order needed to be ready for dispatch. They explained to me that any additional outwork or internal production could be sequenced visually using the box.

After the work had been scheduled, the respective team leader took the work order information—printed on a simple card—reviewed the daily workload with the team, and sequenced the work for manufacture.

Both the *heijunka* box and the sequencing board were reviewed by the team to ensure that the work was done on time, and any problems were identified and resolved. A role called the "water spider" (*mizusumashi*) was created to enable the frequent pickup and delivery of materials within the plant. The "fabrication water spider" delivered the component's pieces and manufactured parts to the final assembly/fabrication operator without any complications, using simple visual management and effective communication.

The "logistics water spider" made rounds four times per day and could use the *heijunka* box to easily see what was ready to be collected and ready for final dispatch.

The message here is simple: A *gemba* team can use simple cardboard and paper labels to manage a very complicated production schedule and ensure good flow.

## Visual Management with 5S

You probably have realized that visual management also has a lot to do with the five steps of housekeeping. When we engage in 5S, we find that its outcome is better visual management. Better housekeeping helps to make abnormalities visible so that they can be corrected.

The 5S methods can be organized from the perspective of visual management:

▲ *Seiri* ("discarding unnecessary things"). Everything in the *gemba* should be there if, and only if, it is needed now or will be used in the immediate future. When you walk through the *gemba*, do you find unused work-in-process, supplies, machines, tools, dies, shelves, carts, containers, documents, or personal belongings that are not in use? Throw them away so that only what is needed remains.

▲ *Seiton* ("putting in order the things that remain"). Everything in the *gemba* must be in the right place, ready for use when you need it. Everything should have a specific address and be placed there. Are the lines on the floor marked properly? Are the hallways free of obstacles? Once *seiton* is being practiced, it is easy to identify anything out of order.

▲ *Seiso* ("thorough cleaning of equipment and the area"). Are equipment, floors, and walls clean? Can you detect abnormalities (e.g., vibrations, oil leakage, etc.) in the equipment? Where *seiso* is practiced, any such abnormality should soon become apparent.

▲ *Seiketsu* ("keeping oneself clean and working on the three preceding items daily"). Do employees wear proper working clothes? Do they use safety glasses and gloves? Do they continue their work on *seiri*, *seiton*, and *seiso* as a part of their daily routine?

▲ *Shitsuke* ("self-discipline"). Each individual's 5S duties must be specified. Are they visible? Have you established standards for them? Do workers follow such standards? The workers must record data on graphs and check sheets on an hourly, daily, or weekly basis as requested. As a means of fostering self-discipline, management may request that workers fill in data each day before going home.

Good 5S in the *gemba* means that as long as the machines are in operation, they are producing good-quality products.

## Posting Standards

When we go to the *gemba*, visual management provides performance measures. We see an abnormality when we find excessive boxes of supplies on the line side, when a cart carrying supplies is left outside its designed area, and when a hallway is filled with boxes, ropes, rejects, and rugs. (A hallway is meant to serve only as a passage, not a storage area.)

Displaying work standards in front of the workstation is visual management. These work standards not only remind the worker of the right way to do the job but, more important, enable the manager to determine whether the work is being done according to standards. When operators leave their stations, we know there is an abnormality because the standards displayed in front of the workstation specify that the operators are supposed to stay there during working hours. When the operators do not finish their work within cycle time, we cannot expect to achieve the day's production target.

While standards delineate how workers should do their jobs, they often do not specify what action should be taken in the event of an abnormality. Standards should first define abnormalities and then outline the steps to follow in response.

Daily production targets also should be visible. Hourly and daily targets should be displayed on a board alongside the actual figures. This information alerts the supervisor to take the measures necessary to achieve the target, such as shifting workers to the line that is behind schedule.

All the walls in the *gemba* can be turned into tools for visual management. The following information should be displayed on the walls and at the workstations to let everybody know the current status of QCD:

- *Quality information*—daily, weekly, and monthly reject figures and trend charts, as well as targets for improvement
- *Gembutsu* ("actual pieces") of rejects—for all operators to see (These *gembutsu* are sometimes referred to as *sarashikubi*—a word from medieval times meaning the "severed head of a criminal on display in the village square." These rejects are often used for training purposes.)
- *Cost information*—productivity figures, trends, and targets

- *Worker hours*
- *Delivery information*—daily production charts
- *Machine downtime figures, trends, and targets*
- *Overall equipment efficiency (OEE)*
- *Number of suggestions submitted*
- *Quality-circle activities*

For each particular process, any number of additional items may be required.

## Setting Targets

The third purpose of visual management is to clarify targets for improvement. Suppose that external requirements have prompted a plant to reduce the setup time of a particular press within six months. In such a case, a display board is set up next to the machine. First, the current setup time (e.g., six hours as of January) is plotted on a graph. Next, the target value (one-half hour by June) is plotted. Then a straight line is drawn between the points showing the target to aim for each month. Every time the press is set up, the time is measured and plotted on the board. Special training must be provided to help workers reach the target.

Over time, something incredible takes place. The actual setup time on the graph starts to follow the target line! This happens because the operators become conscious of the target and realize that management expects them to reach it. Whenever the number jumps above the target, they know that an abnormality (e.g., missing tools, etc.) has arisen and take action to avoid such a mishap in the future. This is one of the most powerful effects of visual management. Numbers alone are not enough to motivate people. Without targets, numbers are dead.

Yuzuru Itoh, former director of the quality-control center at Matsushita Electric, made the following comments (quoted in my book, *Kaizen: The Key to Japan's Competitive Success*, pp. 64–65) about the power of targets to motivate people:

> One of the more interesting experiences I had involved the soldering workers at a television plant. On average, each of our workers soldered 10 points per work-piece, 400 work-pieces per day, for a daily total of 4,000 soldering connections. Assuming a

soldering worker works 20 days a month, that's 80,000 soldering connections per month. One color TV set requires about 1,000 soldering connections. Of course, nowadays most soldering is done automatically, and soldering workers are required to maintain a very low defect rate of no more than one mistake per 500,000 to 1 million connections.

Visitors to our TV factory are usually quite surprised to find workers doing such a monotonous job without any serious mistakes. But let's consider some of the other monotonous things humans do, like walking, for example. We've walked practically all our lives, repeating the same motion over and over again. It's an extremely monotonous movement, but there are people such as the Olympic athletes who are intensely devoted to walking faster than anyone has ever walked before. This is analogous to how we approach quality control in the factory.

Some jobs can be very monotonous, but if we can give workers a sense of mission or a goal to aim at, interest can be maintained even in a monotonous job.

The ultimate target of improvement is top management's policies. One of top management's roles is to establish long- and medium-term policies as well as annual policies and to visibly show them to employees. Often such policies are shown at the entrance to the plant and in the dining room as well as in the *gemba.* As these policies are broken down into subsequent levels of management that finally reach the shop floor, everybody will understand why it is necessary to engage in *kaizen* activities.

*Kaizen* activities become meaningful in the minds of *gemba* people as they realize that their activities relate to corporate strategies, and a sense of mission is instilled. Visual management helps to identify problems and highlight discrepancies between targets and current realities. In other words, it is a means to stabilize the process (maintenance function) as well as improve the process (improvement function). Visual management is a powerful tool for motivating *gemba* people to achieve managerial targets. It provides many opportunities for workers to reinforce their own performance through displays of targets reached and progress made toward goals.

# CHAPTER NINE

# The Supervisors' Roles in the *Gemba*

One of the most important emerging developments in management since the year 2000 has been the rediscovery and spread of *Training Within Industries* (TWI). The introduction made to TWI in this chapter in the first edition helped to spark the interest of a few people who worked to make this valuable program available again, nearly a half-century after it was created and forgotten. The TWI program was the foundation of the structuring and development of the roles, responsibilities, and skills of supervisors at Toyota, and without TWI, it is safe to say that their *gemba* would not be as excellent as it is today.

Frequently, supervisors in the *gemba* do not exactly know their responsibilities. They engage in such activities as firefighting, managing personnel issues, and achieving production quotas without regard to quality. Sometimes they don't even have daily production quotas in mind; they just try to produce as many pieces as possible while the process is under control—between the many interruptions caused by machine downtime, absenteeism, and quality problems. This situation arises when management does not clearly explain how to manage the *gemba* and has not given a precise description of supervisors' roles and accountability.

## Training Within Industries

Supervisors' roles have evolved in Japan over the past five decades. Japan owes much to the Management Training Program (MTP) and the Training Within Industries (TWI) program. These programs came to Japan from the United States and were designed to help the Japanese develop their own managerial and supervisory training programs. The MTP trained primarily middle managers, whereas the TWI programs trained supervisors.

The following is a summary of these programs' origins and development from Alan G. Robinson and Dean M. Schroaeder, "Training, Continuous Improvement, and Human Relations: The US TWI Programs and the Japanese Management Style" [*California Management Review* 35(2), 1993]*:

> W. Edwards Deming, Joseph Juran, and other American experts have rightfully earned their place in the history books for their significant contributions to the industrial development of Japan. However, the U.S. Training Within Industries (TWI) programs, installed in Japan by the occupation authorities after World War II, may well have been even more influential. At least ten million Japanese managers, supervisors, and workers are graduates of the TWI programs or one of their many derivative courses, all of which remain in wide use in Japan in 1992. TWI has indeed had a strong influence on Japanese management thought and practice: a number of management practices thought of as "Japanese" trace their roots to TWI.
>
> The TWI programs were developed in the United States more than half a century ago. They were designed to play a major role in boosting industrial production to the levels required to win the Second World War. Even though TWI did this very successfully, after the war the programs' usage dropped off until, in 1992, they are hardly used or even known in the United States.
>
> The story is quite different in Japan. After the war, Japanese industry was running at less than 10 percent of its 1935 to 1937 levels. Faced with the threat of widespread unrest, starvation, and social disorder, it was only natural for the Occupation authorities to think of TWI, a set of programs specifically designed to boost productivity and quality on a national scale. While TWI had an impact on many countries around the world, it undoubtedly had its greatest effect on Japan. In 1992, even though the programs

*Copyright © 1993 by The Regents of the University of California. Reprinted from the *California Management Review* 35(2) by permission of The Regents.

> have changed little since their arrival in Japan, they are well-respected in Japanese management circles and are viewed as important enough to the national interest to be overseen by the Ministry of Labor, which licenses instructors and upholds training standards throughout the country.
>
> TWI provided three standardized training programs for supervisors and foremen. The first, Job Instructional Training (JIT), taught supervisors the importance of proper training of their workforce and how to provide this training. The second, Job Methods Training (JMT), focuses on how to generate and implement ideas for methods improvement. The third, Job Relations Training (JRT), was a course in supervisor-worker relations and in leadership.

The Japan Industrial Training Association and various professional organizations conducted these training programs. At the same time, many leading Japanese companies internalized the programs to meet their own requirements to train supervisors.

The United States Air Force (USAF) initiated, developed, and introduced MTP to Japan during the occupation period following World War II. Japan's Ministry of International Trade and Industry (MITI) and Nikkeiren, the Japan Federation of Employers' Associations, have jointly overseen the course for close to fifty years, strongly influencing Japanese management thought and practice. And yet, even though many of the management practices commonly thought of as Japanese trace their roots back to MTP, the course is hardly known in the West. As Alan G. Robinson and Sam Stern point out in "Strategic National HRD Initiations: Lessons from the Management Training Program of Japan" [*Human Resource Development Quarterly* 6(2), 1995]:

> By the end of 1994, more than three million Japanese managers will have graduated from the Management Training Program or one of its derivative courses. In many Japanese companies, successful completion of MTP has become mandatory for promotion into middle management.
>
> MTP taught a significant percentage of several generations of Japanese managers three things:

1. The importance of human relations and employee involvement
2. The methodology and value of *continuously improving* processes and products
3. The usefulness of a scientific and rational "plan-do-see" approach to managing people and operations

The first point—the importance of human relations and employee involvement—bore fruit in Japan in the formation of quality circles; the development of internal facilitators such as big brothers, big sisters, and junior leaders and the like; and the organization of employee involvement programs such as sports clubs and book clubs to promote mutual self-enlightenment among employees.

The second point—the methodology and value of *continuously improving* processes and products—matched perfectly with the *kaizen* way of doing business that was emerging in Japan at the time and helped managers and supervisors to review and improve their work processes.

The third point—the usefulness of a scientific and rational plan-do-see approach to managing people and operations—has come to be well known in Japan, together with the plan-do-check-act (PDCA) cycles of Deming's teachings, and has helped to deeply instill the mind-set of the PDCA of never-ending improvement. Even to this day, many Japanese executives prefer to use the term *plan-do-see* as a model.

The curriculums of MTP and TWI produced another forerunner of something that has come to be known and practiced by every Japanese manager even to this day: the so-called five W's and one H—why, what, where, when, who, and how—otherwise known as 5W1H. The 5W1H approach provides a widely used checklist when quality-circle members go about solving problems, as well as when managers are engaged in a *kaizen* project.

While the original formats of MTP and TWI have remained almost the same during the last 50 years, new subjects have been added or incorporated into the curriculum, particularly for companies that have developed their own training programs. These subjects include the concepts of quality, cost, and delivery (QCD), standardization, visual management, *muda* elimination, the five S (5S), and *takt* time (the theoretical time it takes to produce a piece of product ordered by the customer), reflecting the transformation of Japanese management over the years as a result of various *kaizen* practices and the introduction of such new practices as total quality control (TQC), total productive maintenance (TPM), and just-in-time (JIT).

The transformation of the TWI program has firmly established the roles of a typical Japanese supervisor in the *gemba.*

## Managing Input (Manpower, Materials, and Machines)

A supervisor is a person who has a line responsibility for directly supervising 20 or so operators in the *gemba* and has accountability for the outcome. The span of control of supervision may differ from industry to industry and from company to company. Also, the title of such a person's job may vary; the person may be called *group leader, foreman, hancho,* or (in Germany) *meister.* (By the way, *hancho,* originally a Japanese term meaning "chief or boss," means "supervisor" when used in the *gemba.*)

In the *gemba,* the supervisor manages inputs to produce outputs. The inputs are the so-called three Ms—namely, manpower, materials, and machines. (Sometimes methods and measurements are added to these three, and the list is collectively referred to as the five Ms.) The output is quality, cost, and delivery, or QCD. (Sometimes morale and safety are added to these three, and the list is referred to as QCDMS.)

A company's supervisors are held accountable for achieving the outputs of QCD, but they must manage the basic three Ms—manpower, materials, and machines—in order to do so.

First and foremost, supervisors must manage their people. Yet supervisors often say, "Yes, I know that I am supposed to make good products on schedule, but you see, my people are not motivated to do a good job. They are poorly trained, and they don't even follow established standards. That is my problem!"

No supervisor has any business making such a statement. If his or her people are not motivated, the supervisor must introduce various programs to motivate them. If people don't follow standards, countermeasures must be developed. Perhaps the current standards are outdated and impractical, or the operators lack the training to follow them. Or there may be too much *muda, mura,* and *muri* in the work environment, making standards too difficult to follow. Supervisors who blame their people are abdicating their role.

In a plant producing electronic devices that employs housewives part-time in the afternoon, management found that the part-timers made far more rejects than regular employees. Data revealed that most of their mistakes occurred around 3 p.m. When management asked the housewives what sorts of things were on their minds around that time of day, the typical answers were as follows:

"At about that time, I am suddenly reminded that it is time for our children to come home from school, and I start wondering if they can find the cookies I left in the refrigerator."

"I start thinking about the dinner and wonder which nearby store I should go to in order to buy fish. I want to know which store offers the best loss leader. Ms. A on the next line is knowledgeable on such matters, and maybe I should go meet her after work."

The insights gleaned from interviewing these members of its part-time workforce prompted management to set aside a large meeting room for the employees' use during their 3 p.m. coffee break. Management told the employees that they could talk about cookies, fish, loss leaders, or other subjects to their hearts' content, but after the break, they should concentrate on their work. Eventually, the company saw a dramatic reduction in its reject rate.

Figure 9.1 shows a cause-and-effect diagram of a supervisor's work. This type of diagram is called an *Ishikawa diagram*, after its developer, Professor Kaoru Ishikawa. Because of its shape, it is also called a *fishbone diagram*. The effect (result) is quality, cost, and delivery (QCD). The causes (processes) are materials, machines, manpower, measurements, and methods (the five Ms). Depending on the circumstances, more causes (in this case, the environment) can be added to the diagram.

By managing the causes, supervisors can realize the goal of their work: QCD. The diagram shows that just as supervisors must manage materials and machines in the *gemba*, they also must manage manpower (personnel). To do this, they must manage several smaller "bones" of the fishbone: training, communication, quality circles, suggestions, rewards and awards, absenteeism, and morale. Whenever they find a human-related problem, they are supposed to find a solution.

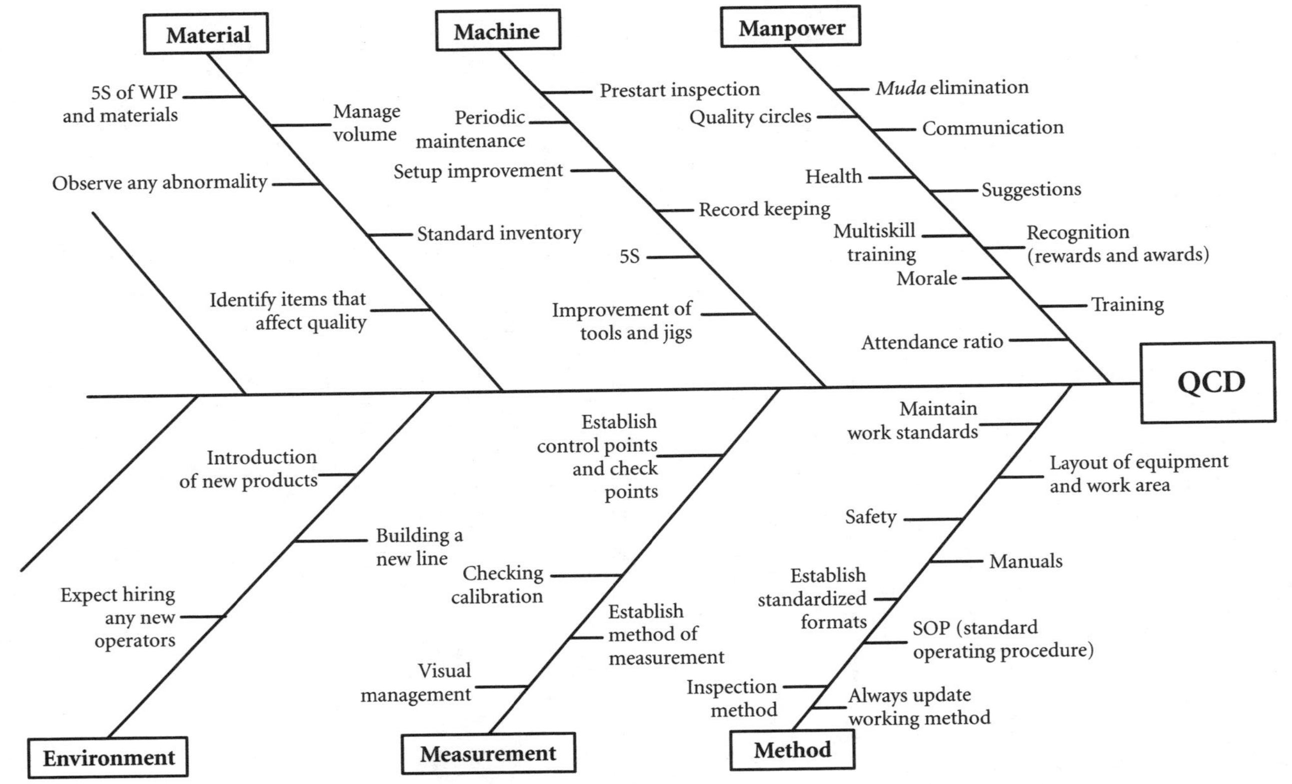


**Figure 9.1** A cause-and-effect diagram of a supervisor's work.

## A Day in the Life of a Supervisor at Toyota Motor Manufacturing Kentucky

Toyota Motor Corporation defines its priorities of floor management as safety, quality, and delivery in that order. It is a must to achieve safety in order to succeed to quality; the success with safety and quality will feed delivery. In this instance, it is assumed that the cost targets are achieved together with the volume. The second major task for supervisors is to develop the team members. To do this, the supervisor must be able to transfer his or her Toyota Production System (TPS) knowledge and management skills to team leaders, which will allow them to step into the supervisor's role when needed.

Steve Burkhalter was a group leader at Toyota Motor Manufacturing Kentucky. Since leaving Toyota, he has been a consultant actively helping companies implement Toyota-style frontline supervision systems. Steve shares his firsthand experience of the frontline supervision system, required training, and supervisor's work routine.

The TPS is culture-driven. It requires mutual respect and trust among team members and the management team. The system succeeds through what could be called *mini-company operations* managed by the group leader. The metrics are aligned with overall company goals. The mini-company could have up to five teams in a group. For example, a group may have one group leader, five team leaders, and based on the department, up to seven members on each team.

At Toyota, team members who wish to be promoted must take a core set of TPS classes to qualify for team leader, group leader, or manager positions. The job requirements include practical knowledge of visual management, standard work, problem solving, SMED, TPM, *kanban*, *heijunka*, and *jidoka*, just to name a few.

Standard work is one of the foundation tools of the TPS, creating key points to safety and quality standards. These quality and safety key points guide team members through the process. Figure 9.2 provides an example of quality and safety key points documented on a standardized worksheet.

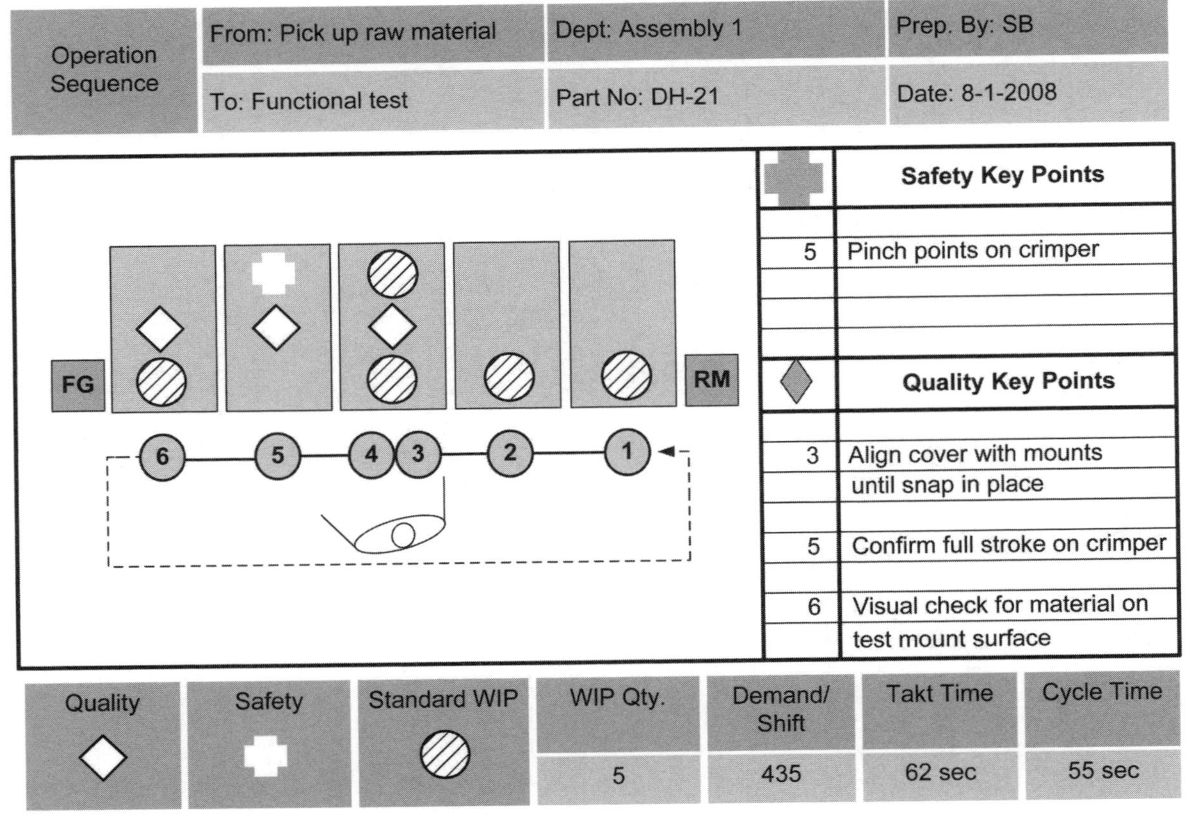


**Figure 9.2** Quality and safety key points.

Once the team leaders were qualified in "job instruction training," they graduated to "train-the-trainer" by the group leader transferring this knowledge by training the team members. This approach to teaching was instrumental in the development of "quality built into standard work." When team members were taught this standard, they develop "three ways" of key points: safety, quality, and a knack for the one work element they delivered. This served to make checks via their senses as they completed each work element.

The group leader starts his or her day with a stand-up meeting to deliver company information, today's production targets, quality alerts (supplier and last-shift quality information), any needed changes to staffing, ongoing projects, and recognition of team member efforts.

The group leader already has planned his or her daily schedule and is actively auditing safety conditions, completing quality audits, ensuring support systems are in compliance, following up on projects with which the team leaders are involved, and completing reports requested by the manager before the daily staff meeting. The group leader carries a two-way radio (a pager for outside calls) to keep in constant contact with the team leaders and group leaders to respond to problems as they happen.

Group leaders can be found on the *gemba* at all times. The standard is to "go see" to gather facts and solve problems. The term *managing by walking around* describes the group leader's approach to visual management at the highest level. Lamps or other visual signals called *andon* are used by team members and team leaders to call for help. The *andon* call happens dozens of time each day.

A team member is instructed to make an *andon* call any time abnormalities such as these are found:

- Standard work cannot be performed correctly.
- There is a deviation for raw material from specification.
- Machine cycle time is over *takt* time.
- A supplier part shortage is causing the customer to wait.
- An error is found in quality of the previous processes during an incoming parts check.

- Machines stop owing to defect detection by *pokayoke*.
- Broken tools cause scrap and downtime.

All problems are recorded with root-cause analysis and the countermeasure that sustains it as well. The support teams (maintenance and engineering) understand their role to respond to production needs.

During the managing by walking around cycles, the group leader engages with the members not just on the business of the day but also on their personal interests outside work. The group leader also arranges social events with team members and their spouses, engaging the group on a quarterly basis. This helps connect the individual to the team on a social level.

Mentoring is another role as the group leader coaches the team members toward the next step in the promotion path. The group leader will conduct a performance review yearly that is metric loaded to gauge the development of the member. It is a valued tool of recognition that places a lot of emphasis on the Toyota Production System and personal skills of communication.

The major daily task for the group leader is to track group performance and to develop projects to get to or exceed yearly goals. The key areas the group leader tracks are the following:

## Safety

- Safety near misses and root-cause analysis and countermeasures to them.
- Safety lost time (accident) and root-cause analysis to eliminate them. Group leaders complete a full report of such safety issues resolved and present it to senior management.
- These reports are forwarded throughout the organization to investigate similar conditions elsewhere in a process called *yokoten*.
- Put tracking mechanisms in place for safety awareness and suggestions to targeted safety.
- Establish personal protective equipment (PPE) standards, and instruct members on the proper use of the equipment.

- Train team leaders specifically to audit the safety in their work areas daily and to immediately countermeasure all discrepancies.
- Schedule team members to do daily safety talks discussing near misses on the plant floor or areas "outside the walls" in their personal life.
- Schedule monthly waste walks with the team leaders not only to investigate the seven wastes but also to do audits regarding 4S (what Toyota calls 5S) principles of the TPS—everything in its place and a place for everything.
- Audit members for safety key points from the standard work, and correct the nonstandard.
- Update the safety calendar to track daily performance, and audit the countermeasure activity to ensure that it has been sustained.

Creating a culture of safety awareness in the group is the highest priority for the group leader at Toyota.

## Quality

- Put a tracking mechanism in place for quality defects found inside the group leader's area of responsibility, and confirm root-cause analysis and countermeasures.
- All scrap is categorized in a Pareto chart, and priority is used to correct and eliminate the problem.
- A tracking mechanism for supplier defects is created, visually showing a Pareto chart of performance.
- Maintain various boundary samples on display.
- Suppliers will be aware of this system and will demonstrate the ability to countermeasure problems.
- Put tracking mechanisms in place for machine scrap and root-cause analysis with countermeasure activities to eliminate.
- Perform quality checks daily with samplings from all production lines for which you are responsible; track the performance trends and stability of the process.

The group leader is responsible for all quality conditions throughout the lines of his or her responsibility; auditing on the *gemba* to track real-time performance is a key to sustaining the quality the customer (the next process) expects.

## Delivery

- Machine cycle time analysis is constantly monitored to ensure delivery.
- Monitor by-hour or by-minute delivery status boards for problems.
- Audit standard work-in-process (SWIP) before and after, and shift maintaining levels needed.

The group leader communicates with support groups and directs the activity to reduce downtime generated by root causes related to manpower, materials, machines, or methods.

## Cost

- Understand the cost of tooling in the processes, using tool cost per unit as a measure tracking trends to the data.
- Overall budget costs tracking data for consumables and outside vendor costs to continuously investigate performance opportunities to cost.
- Understand the opportunities to improve person-hours per unit.
- Implement a team member cross-training schedule.
- Measure operation rate and investigate opportunities to improve.
- Train team leaders on the problem-solving process using the 4M method of analysis.
- Schedule weekly problem-solving meetings for each team, coaching and mentoring the teams to apply successful countermeasures.

The group leader must motivate, train, and coach team members on the *kaizen* program system so that each person is able to submit at least 12 ideas per year to improve safety, quality, productivity, and as a result, cost.

The group leader also will audit specific systems (e.g., TPM, *kanban* system, adherence to standard work, setup times and procedures, variation in machine capability, and buffer levels), tracking performance in those areas of the TPS. In this way, not only does the group leader check the results of his or her team's work, but he or she also checks the process and standards that ensure those results. The audits are performed using simple cards that identify where to go and what standard to check. These checks are performed not only by the group leader but also by all levels of management. This system of linked checking by layered audits is called *kamishibai*.

The performance of all metrics is what drives the TPS. At Toyota, performance management is driven at the group leader level, with the team members given authority to improve the process.

At first glance, the typical group leader or supervisor's day at Toyota may seem very full or even overwhelming. In fact, the day is highly structured, freeing the person from the need to make detailed decisions and choices because standard work has been established to guide the supervisor through the critical steps of each day.

It is clear that the supervisor must wear many hats and support all processes in his or her area of responsibility and beyond. Through a *yokoten* process supported by management, the supervisor also supports the *kaizen* efforts of other plants to improve performance of specific metrics by sharing best practices and learning. The expectations from a *gemba* supervisor are high, and the supervisor expects the same from the group he or she is supervising.

## Morning Market (*Asaichi*)

The *morning market (asaichi)* is an activity employed in Japanese companies as a part of daily activities to reduce rejects in the *gemba* by supervisors and operators on the line. The morning market derives its name from the markets where farmers bring their daily fresh produce to sell. The Japanese word *asaichi* means "the first thing in the morning." The *gemba*'s morning market displays rejects on the table the first thing in the morning of the day after they are made so that countermeasures may be adopted on the spot and as soon as practicable based on the *gemba-gembutsu* principles. All participants in this activity stand up. Morning market differs distinctly from other types of quality-related problem-solving activities involving staff in that the supervisor and operators must play a leading role, with a commitment never to carry forward the same problem to the next day.

Nobody in the *gemba* produces rejects out of a desire to do so. And yet they continue to occur. The causes are many, including the following:

- Abrupt breakdown of equipment
- Forced equipment deterioration that goes beyond the specified allowances
- Failure to follow standards
- Failure of materials and parts to meet specifications
- Failure to maintain 5S
- Careless and absentminded mistakes

Unless management determines the root causes of these problems one by one, the *gemba* soon will be filled with a mountain of rejects. The morning market in *gemba* consists of the following steps:

- An operator tags and places in a red box all rejects in a particular process and lists the rejects in the quality morning market report.
- The next morning, the supervisor in charge brings both the reports and *gembutsu* to the morning market corner and displays the rejects on the desk.
- The supervisor reviews the rejects with the operators and discusses countermeasures.
- The rejects are classified according to three categories (types A, B, and C), and countermeasures are adopted as soon as practicable (Figure 9.3).

It is important that both the supervisor and the workforce touch and hold the *gembutsu* (in this case, the rejects) themselves. They should see them, smell them, taste them (if necessary), and discuss how they were made at the specific work site (*gemba*) and what equipment (also *gembutsu*) was used.

After type A problems have been solved, countermeasures to prevent recurrence must be adopted. As for problems of types B and C, the supervisor must report them to the section manager, who will hold a meeting later to devise solutions and will present the results to the plant manager.

When a company holds its first morning market, participants may find that there are too many rejects to fit on one table. However, if the morning market continues for three months, the rejects, as well as the time for the

| Type | Nature | Percentage | Examples |
|---|---|---|---|
| Type A | Causes are clear. Countermeasures can be taken immediately. | 70–80% | Standard was not followed. Out of spec materials and supplies. |
| Type B | Causes are known but countermeasures cannot be adopted. | 15–20% | Occurs at the time of setup adjustment. Occurs during frequent stoppages of equipment. |
| Type C | Unidentifed causes | 10–15% | Situation suddenly went out of control. |

**Figure 9.3** How rejects can be sorted into three very distinct categories.

meeting, will be greatly reduced. In the meantime, plant productivity and profitability also will improve.

The plant manager should attend morning markets at different sites within the plant each day in order to become familiar with the problems encountered in each place. Figure 9.4 shows an example of a morning market report.

At Toyoda Automatic Loom Works, a booklet containing all relevant standards is handed to each operator shortly before the operator begins tasks of mass production. The booklet—used for training at first and as a reference after production has started—contains the following standards:

- ▲ Organization chart and layout
- ▲ Operational safety rules (which also show what happens if the rules are not followed)
- ▲ Information on how to build quality into the process
- ▲ Work-sequence table
- ▲ Standard operation procedures (SOPs)
- ▲ Abnormality-handling procedures (which include a definition of abnormality and instructions on how to detect abnormality and whom to report abnormality to)

| Plant | Dept. | Manager | Supervisor | Operator |
|---|---|---|---|---|

1. *Date of occurrence:* October 9, 1995 14:00 PM
2. *Part number:* 123456-G1002
3. *Process and machine:* Key groove process (F-3214)
4. *Number of occurrence:* 4 pieces
   — Number of processed pieces for the day: 920
   — Reject rate: 0.43%
5. *Description of the reject:* [Draw sketches whenever possible.]
6. *Causes (confirmed/assured/not identified):*
   One out of four bolts in the machine was loosened, causing vibration.
7. *Countermeasures:*
   Tightened the four bolts with the right torque. Have not seen occurrences of the same problem since.
8. *Prevention of recurrences:*
   Requested the preset group to add a new standard.
   "Confirm the torque of tightening bolts of _____ equipment."

**Figure 9.4** An example of a morning market report.

- Definition of rejects (quality-related problems)
- Rules on the use of *kanban*

At one time, *gemba kaizen* activities were promoted uniformly at all factories of Toyoda Automatic Loom Works. Later, management realized that it should live up to the philosophy behind the Pareto diagram (graphical tool used for solving problems) and establish priorities in selecting *kaizen* projects. Thus, instead of promoting *gemba kaizen* indiscriminately throughout all areas in the plant, management decided to select one line as a model and provide the line with all the help and resources it needed from the corporate office as well as from plant management. Once visible progress had been made, improvements were extended to other lines. Top management visited the model line once a month to review daily management and *kaizen* activities. The review covered the following major points:

- What kinds of standards are installed?
- How are standards adhered to?
- Who manages the standards?

- Who is engaged in *kaizen*?
- What roles do the line managers play?

Since the managerial hierarchy included foremen, supervisors, and group leaders in the factory, top management also monitored the roles of these managers and the items for which they were responsible.

## Best-Line Quality-Assurance Certification

For more than 30 years now, the so-called acceptable-quality-level (AQL) approach has not been followed in Japan. AQL is a system of quality assurance (QA) that allows suppliers to deliver a certain percentage of rejects to customers. For example, a customer might allow a supplier to deliver rejects at a rate of up to 1 percent, provided that the supplier compensated the customer according to terms previously agreed on.

However, major Japanese companies have long since discarded this approach to QA. Except for the first lot, companies accept supplies without inspection. During that initial inspection, if even one reject is found, the whole lot is returned to the supplier.

As highly automated high-speed production lines have come into wide use, the emergence of even a single reject in the process has meant serious consequences for companies. Even small errors are quickly multiplied, resulting in large economic losses. The Japanese automobile manufacturers therefore have stepped up quality requirements—from the previous 0.1 percent to between 30 and 50 parts per million (ppm). In order to achieve this level of quality, it was essential to eliminate defects in the process itself.

Suppliers have been obliged to review the QA practices of their production lines. The line supervisors took on the challenge of improving the process capability of their lines to the levels requested by their customers. The real test of the success of these efforts was the in-process rejects rate and customer returns. This system is known among Nissan's suppliers as *best-line quality assurance*, or *QA best line.*

After a line has reached a certain level of quality, the supervisor decides to apply for best-line certification. Toward this end, the supervisor and production manager together conduct a diagnosis of the line. The supervisor also asks the corporate director in charge of the corporate QA department to visit and diagnose the line based on certain pre-established criteria, including various statistics on rejects and customer returns.

Figure 9.5 shows a certification awarded for QA best lines in one supplier to Nissan Motor Company. The idea behind QA best-line certification is that first one line achieves the necessary improvements, and then the process is extended to other lines, until every line in the plant has attained the same level of QA and received certification to that effect.

Chart 1 Requirements for Approval

| QA Performance | | Grade | | |
|---|---|---|---|---|
| | | C | B | A |
| Customer returns | | No. of Returns/ 3 Months | No. of Returns/ 6 Months | No. of Returns/ 12 Months |
| Final inspection rejects ppm | | < 500 | < 50 | < 10 |
| In-process Rejects % | Repair | < 0.5 | < 0.1 | < 0.01 |
| | Scraps | < 0.1 | < 0.05 | 0 |

Chart 2 Evaluation Items for Certification of QA System

| | |
|---|---|
| Standardization | 55 checkpoints |
| Work to standards | 34 checkpoints |
| Quality confirmation of designing/process changes | 16 checkpoints |
| 5S | 31 checkpoints |
| Education and training | 6 checkpoints |
| Problem solved | 7 checkpoints |

**Figure 9.5** Certification system for QA best lines.

## Defining Challenges

In today's dynamic and competitive environment, management faces increasingly stringent requirements from customers who want better quality, a lower price, and prompt delivery. Only a clear management plan for improving on QCD all the time will keep up with this demand. Management therefore must keep setting higher QCD targets and challenging subordinates to attain them. As soon as a new target has been achieved, management must establish the next one, thus continuously urging subordinates along the never-ending road of improvement. Successful companies continue their success because managers lead subordinates in this manner and build a corporate culture of challenge. Such companies also know that once they lose this spirit, particularly at the *gemba* level, there will be no future for them. In today's companies, whether or not management possesses a spirit of challenge makes the difference between success and failure. Such a spirit of challenge should be the backbone of the *gemba.*

However, most managers today have lost the enthusiasm to challenge. In particular, many *gemba* supervisors settle for trying to maintain the status quo and working hard and loud and running around throughout the day without having any clear idea where they are going.

Setting challenges is the key element of a successful supervisor's job. The supervisor must possess sufficient understanding of the current process to establish appropriately challenging targets.

## Pseudomanagerial Functions of Supervisors in the *Gemba*

As stated earlier, managers' jobs boil down to two major functions in the *gemba*: maintenance and improvement. *Maintenance* refers to preserving the status quo—that is, making certain that subordinates follow current standards to achieve expected results. The objective of maintenance—to make certain that things do not go out of control—takes a lot of effort. Without maintenance, everything in the *gemba* will deteriorate over time.

*Improvement*, meanwhile, refers to enhancing and upgrading current standards by continually establishing new and higher targets. Improvement can be further broken down into *kaizen* and innovation. Simply stated, *kaizen* means making better use of the existing internal resources of the five

Ms of manpower, machines, materials, methods, and measurements. *Kaizen* is accomplished by changing the way people do their jobs rather than by spending large amounts of money. It takes a challenging spirit to bring about *kaizen* because people are always more comfortable with the way they have been doing their job in the past.

I believe that a surfeit of resources has unforeseen drawbacks: There is no impetus for *kaizen*, no incentive to rack our brains and look within for ways to improve—and before we know it, the competition has passed us.

In the context of *kaizen* philosophy, supervisors' jobs also should be broken down into two functions: (1) maintenance, the task of stabilizing and preserving the current process and, whenever an abnormality is detected, bringing the process back under control, and (2) improvement, which is as important as maintenance. In the improvement function, management must check to determine whether supervisors have attained management-imposed targets. Maintenance is sometimes referred to as *daily activities* and improvement, as *kaizen activities.*

Supervisors must carry out all these activities in order to realize QCD. Chapter 3 points out that the real challenge for management is to manage quality, cost, and delivery simultaneously. Supervisors should not confine their concern toward meeting production volume nor sacrifice quality and/or cost to meet production targets. The supervisor in the *gemba* should always strive to realize QCD by attaining targets set by management and demanded by customers.

The properly trained supervisor participates in policy deployment by always keeping in mind two or three annual targets for *kaizen*, such as halving rejects and reducing inventory. In the process of assuming responsibility of this kind, supervisors come to regard themselves as members of the management team—in spirit, if not in fact.

CHAPTER TEN

# *Gemba* Managers' Roles and Accountability

## *Kaizen* at Toyota Astra Motor Company

Chapter 9 outlined the *roles* of a supervisor. Another crucial subject is the *accountability* of a supervisor. Every large industrial complex has several management layers in the *gemba*, and defining their respective roles and accountabilities is often an issue. The following case study of *kaizen* at Toyota Astra Motor Company vividly illustrates the value of clarifying *gemba* managers' areas of accountability.

Toyota Astra Motor Company, a joint venture of Toyota Motor Company and P.T. Astra International, produces passenger and commercial cars in Indonesia. It began operations in 1971 and today has 5,000 employees.

Although it had been operating in Indonesia for many years, Toyota Astra Motor Company recognized an acute need to clarify the roles of its *gemba* managers around 1991. The company had such *gemba* managers as supervisors, foremen, and group leaders, but confusion often arose as to their respective roles. When a given problem arose, the question often asked was who among these managers should address the particular issue. Who should devise a temporary countermeasure and standardize the new method to prevent the problem from recurring?

In addition, many other issues needed attention—there were problems with systems and procedures and problems concerning human resources—and many different areas required managing—among them quality, safety, cost reduction, 5S, and productivity.

Toyota Astra Motor Company (TAM) had sent many *gemba* trainee managers of different ranks to Toyota Motor Company (TMC) in Japan.

However, when the managers came back to Indonesia and tried to implement what they had learned, the ambiguity of *gemba* managers' roles there remained unresolved. Finally, in 1992, the company began serious efforts to redefine the roles and accountabilities of each level of *gemba* managers. As a first step, Eddie Paino, manager of TAM's *kaizen* implementation office, went to TMC to learn in depth how TMC defined the roles of each level of manager.

TAM had the following managerial layers in the *gemba*: group leaders, foremen, supervisors, and section managers. The ratio of subordinates to managers in each category was as follows:

| | |
|---|---|
| Group leader | One for every 8 operators |
| Foreman | One for every 2 group leaders |
| Supervisor | One for every 2 to 3 foremen |
| Section manager | One for every 2 to 4 supervisors |

One of the first tasks the company tackled was to clarify its various managers' roles and prioritize them in order to avoid conflicts and ambiguities. At that point, many of these managers took part in both pre- and postpromotion training courses developed jointly by TAM's *kaizen* office and human resources division, as well as in problem-solving sessions and group discussions. All of this contributed greatly to clarifying the managers' roles. As a result, the prioritized roles of *gemba* managers at each level came to be defined as explained in Table 10.1.

## Role Manuals at TAM

Once these roles were defined in order of their *priority*, management began to develop a system to evaluate the *performance* of managers at each level. To do this, the personnel division created and published pocketsize role manuals detailing the responsibilities of *gemba* managers of each rank and distributed the manuals to every manager.

As a general rule, the manuals divide managers' tasks into two parts: (1) the roles managers are expected to play (the activities they are expected to carry out) throughout the day and (2) the items for which managers are held accountable. For both group leaders and foremen, the manuals provide a list of daily activities to be carried out during working hours. (The manuals contain no such list for supervisors and section managers because their daily

**Table 10-1** Roles Defined in Order of Priority at TAM

| Manager | Roles in Order of Priority | Qualifications |
|---|---|---|
| Group Leader | • Quality and defects oversight<br>• Line stops responsibility | • Must be able to help operators to follow standard operating procedure (SOP) and standard worksheet (SWS) in work area and assist foreman in developing and implementing work standard and quality standard.<br>• Must be responsible for preparing standard worksheet. |
| Foreman | • Productivity improvement<br>• Cost reduction | • Must be able to improve working conditions (productivity, cost, and quality) and increase subordinates' skills and capabilities.<br>• Must prepare activity plans for the above and discuss them with the supervisor. |
| Supervisor | • Human resources management<br>• People-related problem solving | • Must be able to assist section manager in improving a system of production control, standard operating procedure, quality control, safety, training, and development of multiskilled and thinking employees. |
| Section Manager | • Policy deployment<br>• Deal with specific problems brought up by subordinates<br>• People-related problem solving<br>• New product development coordination | • Must establish challenging enough target for quality, cost, delivery, safety, and morale (QCDSM).<br>• Must oversee line stop of more than 20 minutes, safety violation, accidents, and chronic defects. |

activities cannot be defined in the same manner and detail as those of group leaders and foremen. This means that the supervisors and section managers have more freedom for carrying out their daily responsibilities.)

The managerial roles/activities and accountabilities are also directly related to managers' performance appraisals and salaries. For instance, group leaders are requested to monitor defects and abnormalities, to keep records, to enter the data on certain checklists or graphs, and to post those checklists or graphs on large display boards in the *gemba*. Each group leader and foreman has his or her own display board, which is shared with the second and third shifts. (Often operators' skill inventory tables and other tables and graphs are displayed on the same boards.)

The manuals clearly define the items to be monitored and the types of data to be collected, as well as the kinds of checklists to use. The items to be monitored for appraising purposes are not always the same for every process, but they always refer to such vital functions as quality, safety, productivity, cost reduction, training, and total productive maintenance (TPM).

By looking up and filling in data on such display boards daily, both group leaders and foremen can focus on the items requiring immediate attention. Their supervisors, in turn, can look at the display boards and instantly evaluate the initiatives, maintenance, and improvement of these items by their group leaders and foremen. The supervisors then can post a summary of the data obtained from their group leaders and foremen on their own display boards. Section managers have similar display boards. In effect, the series of checklists and graphs on the display boards serves as a *visual reporting system* among managers, allowing quick identification of certain prioritized activities that need immediate attention. The display boards also serve as a visual monitoring system of the meeting of the minds of managers.

Before every shift starts, everybody gathers around the display boards for a five-minute talk in which group leaders can explain specific problems, drawing data from the display boards. (Similarly, the boards offer a means of telling visitors what is going on in the *gemba*. When guests or senior managers conduct a "*gemba* walk," they can glance at the boards to find out what is going on or to update their knowledge of the various lines' progress.)

At the end of every month, the section managers and supervisors get together and evaluate the work of their subordinates (group leaders and foremen). Their evaluations then are posted on the display boards. The items to be evaluated are drawn from the role manual notebook and are categorized as *activities*, *initiatives*, *contributions*, or *efforts* related to maintaining and improving the items for which their subordinates are responsible. The following are the major items to be evaluated in the case of a group leader, for example:

- Line stop for a cause
- Checking and identification of safety items

- Quality defects
- *Hiyari* ("near miss") reports
- Idea suggestions
- Quality circles
- 5S

For TAM foremen, the items to be evaluated are as follows:

- Safety awareness
- Absenteeism
- Line stop for an external cause
- Worker-hours per unit
- Quality system
- Idea suggestions
- Quality circles
- *Hiyari* ("near miss") reports
- 5S
- Cost-down activities

## TAM Group Leaders' Responsibilities

Group leaders are promoted from among the operators and receive additional allowances for their work. If they have done a good job as group leaders, they have a good chance of being promoted to a higher managerial level. The major responsibilities of group leaders are maintaining quality and managing line stops. Group leaders are requested to have at least one quality circle in their group, and each quality circle must complete two themes per year. The circle members meet twice a month. Another of the group leaders' duties is taking care of absenteeism. If someone in the group is absent, the group leader either finds a replacement or takes the person's place himself or herself.

Group leaders must fill out checklists daily and, on critical items such as line stoppages, every hour. Group leaders hold a five-minute talk with their groups before each shift starts. The following subjects are discussed:

- Accidents that took place the day before
- Problems encountered during the night shift
- Targets that were not achieved
- Any electrical or mechanical failures that took place

Group leaders also must manage line stops. Whenever a problem is found, operators are allowed to stop the line by pushing a nearby button. The moment the line is stopped, a clock showing aggregate stoppage time begins to tick. The group leader checks the clock every hour. He or she also checks for any problem or abnormality if line stoppages are higher than normal. Based on such data, the target for line stops for each month is determined. Needless to say, the total stoppage time affects the work-hour productivity for the group.

## TAM Foremen's Responsibilities

A foreman's main job is improving productivity and reducing cost. To do this, a foreman is expected to reduce worker-hours (called *kosu* in Japanese), as well as eliminating all sorts of *muda*. *Kosu* is defined as total worker-hours in a particular process multiplied by actual working time and divided by the units produced. For instance, if 10 people worked in one process for 9 hours, including overtime, and produced 200 units, the *kosu* would be calculated as follows:

$$\frac{10 \times 9}{200} = 0.45$$

Each working group must calculate its *kosu* per unit produced. Each group leader, foreman, and supervisor must set monthly targets for reducing *kosu*.

At TAM, a long time passed before *kosu* began to be used as a criterion of productivity improvement and cost reduction. Today, however, *kosu* is a very realistic indication of productivity improvement and cost reduction for each manager there, right down to the lowest-level group leader, and the relationship between *kosu* improvement and available data is very clear to everyone at the company. At all levels, TAM employees can see how their actions contribute to *kosu* reduction.

## TAM Supervisors' Responsibilities

Supervisors' main tasks are people-related:

- Developing multi-skilled workers

- Quality circles
- Safety, etc.

Foremen and supervisors hold weekly meetings and discuss the following subjects:

- Safety
- Productivity
- Cost
- Quality
- Absenteeism
- Suggestions
- Quality circles

Every foreman and supervisor must submit weekly reports to his or her boss.

## Items That Need to Be Managed in the *Gemba*

Generally speaking, items that need to be managed in the *gemba* include the following:

- Productivity
- Cost reduction, including *kosu* reduction
- Safety
- Personnel training
- *Kaizen* activities
- 5S
- Improving employees' skills
- Quality
- Line stops

As already mentioned, TAM has developed a manual describing the roles and accountabilities of *gemba* managers. All the managers' jobs are broken down into two parts: (1) their daily activities and (2) the specific actions for which they are held accountable. The daily activities column contains a detailed description of what managers are expected to do throughout the day, item by item. The items for which managers are accountable fall into the following categories:

- ▲ Production
- ▲ Cost
- ▲ Housekeeping
- ▲ Quality
- ▲ Personnel and training
- ▲ Safety

For each category, the manual provides a list of activities for the manager to perform. All these subjects are shared by the group leader, foreman, and supervisor, although the activities for each subject and the level of involvement differ among the three levels of management.

## *Group Leaders' Daily Schedule of Activities: Examples from the TAM Manual*

A. Before start of work:
   1. Enter factory and go to the *gemba.*
   2. Review the report from the previous shift.
   3. Preparations before work:
      a. Prepare the work team, and check the readiness of all equipment, jigs, tools, and other auxiliary material.
      b. If someone is absent, fill in the report and find a replacement through the foreman.
   4. Morning exercise and five-minute talk.

B. Morning working hours:
   1. Start of work: Confirm that everybody has started working on time.
   2. Change of work procedures: Help foreman to teach newly developed work procedures.
   3. Check the production process: Lead the line operators in observing standard worksheets.
   4. During morning break:
      a. Perform sampling checks on several predetermined quality items.
      b. Lead and guide operators to counter any abnormalities during their work.
      c. Assist in or perform foreman's job when he or she is at a meeting.
   5. Conduct on-the-job training to develop multi-skilled workers.

C. Afternoon working hours:
   1. Check results of inspections:
      a. Check result of morning inspection by quality-control people, and ask for guidance on improvement from the foreman.
      b. As instructed by foreman, perform temporary countermeasures against problems, and ask for further guidance for permanent solution.
   2. Help operators engage in repairs and rework, and check and evaluate results.
   3. Investigate the cause of line stoppage: Propose temporary countermeasures and preventive measures to foreman.
   4. Give instructions for overtime work, if necessary.
   5. Lead operators in performing 5S activities in the *gemba*.
D. After working hours:
   1. Write the shift report and leave any pertinent information for the next shift.
   2. Lead quality circle meetings: Actively promote quality circle activities and boost morale of workers.

## *Group Leaders' Activities: Production, Cost, and Quality Examples from the TAM Manual*

The TAM manual also describes in detail the activities for which group leaders are accountable within the aforementioned categories—namely, production, cost, total productive maintenance (TPM), quality, personnel, training, and safety. As an example, the following are the group leaders' activities related to production, cost, and quality:

A. Production:
   1. Implement monthly production plan:
      a. Assign workers for smooth production flow.
      b. Train and assist new workers in their jobs.
   2. Prepare for daily production:
      a. Check equipment, tools, parts, and materials.
      b. Perform the task as instructed by the foreman.
      c. Switch on the machines, and confirm that they are functioning properly.

3. Follow-up:
    a. Investigate the causes of abnormalities.
    b. Report to the foreman.
    c. Take temporary actions.
    d. Devise permanent countermeasures.
    e. Report any actions taken to the foreman.
    f. Help foreman as instructed.
4. After the operation:
    a. Prepare for the next shift; inform the next shift if any abnormalities have been found.
    b. Confirm that every switch is in the "off" position.
    c. Assist superior in preparing daily reports.
5. Handle line stoppages:
    a. Investigate external line stoppages.
    b. Investigate internal line stoppages.
    c. Determine the causes, and take countermeasures.
6. Prepare for the introduction of new models on the line:
    a. Help the foreman.
    b. Learn the new model, and guide the operators.

B. Cost:
1. Plan cost improvements: Voice opinions and suggestions on improvement plan to the foreman.
2. Reduce labor costs: Propose ideas and help superior in implementing labor cost reduction.
3. Reduce direct costs:
    a. Record usage of materials.
    b. Study the cause of increase in material usage and propose countermeasures.
4. Save energy:
    a. Identify any leakage, such as air and/or water.
    b. On identification, decide whether to act alone or seek help.
5. Improve on a daily basis:
    a. Prepare for improvement.
    b. Assist foreman in guiding subordinates' improvement efforts.
6. Other:
    a. Meet with subordinates to explain the results of cost reduction.
    b. Take every opportunity to enhance operators' cost awareness.

C. Quality:
   1. Maintain and improve quality level:
      a. Clarify current quality levels versus targets for team members.
      b. Monitor and control the process of inputting quality data.
      c. Analyze cause and take countermeasures.
   2. Pursue daily "built-in quality":
      a. Inspect the first and last product of every working day.
      b. Perform scheduled inspection to prevent defects.
      c. Monitor workers to see that they perform their jobs according to work standards.
   3. Take countermeasures when defects are found:
      a. For internally produced defects: Repair the defects, and report to foreman while proposing countermeasures.
      b. For externally produced defects: Report to foreman, and ask for instructions on repair.
   4. Other: Meet daily with the team; inform members about quality problems, and discuss with them; elevate the quality awareness of all members.

### *Foremen's Activities: Cost-Reduction Examples from the TAM Manual*

The foreman's activities in the area of cost reduction are as follows:

A. Plan *kaizen*:
   1. Prepare the schedule of cost-reduction programs after discussions with group leaders.
   2. Coordinate activities within sections, and request specific *kaizen* items (e.g., new tools, etc.) from other sections.
   3. Monitor and follow up on progress of cost-reduction schedule.
B. Reduce labor cost (*kosu*):
   1. Monitor monthly *kosu* reduction activities and follow up on progress.
   2. In case target was not met, study the cause and take action.
C. Reduce direct costs:
   1. Monitor actual usage of material, consumable tools, supplies, oil, etc. against planned usage.

   2. If usage is greater than planned, study the cause of the increase, and take countermeasures.

D. Save energy:
   1. Identify leakage of such items as pressurized air and water, and institute action programs to stop it.
   2. Train and motivate workers to always switch off any equipment after use.

E. Daily *kaizen*:
   1. Prepare charts and monitor *kosu* for *kaizen* activities.
   2. Give instructions for *kaizen* activities based on problems identified.

F. Other:
   1. Lead group meetings and explain progress of cost-reduction activities.
   2. Encourage everyone to increase cost consciousness.

## *Supervisors' Activities: Personnel and Training Examples from the TAM Manual*

Supervisors' activities in the personnel and training areas are as follows:

A. Train and develop subordinates:
   1. Inform all subsection members of the current status of the company, its environment, and its management policies. Subordinates also should be informed of such important matters as new market development and new products.
   2. Prepare long-term training programs for individual members.
   3. Maintain and update records of staffing capabilities and improvement status.

B. Develop multi-skilled workers:
   1. Monitor training schedules and programs to train multi-skilled workers.
   2. Monitor the way the multi-skilled training program is carried out, and follow up on it.

C. Teach skills:
   1. Provide skills training through on-the-job training (OJT).
   2. Lead and guide the process of skill standardization required for each workstation based on past experience and practices.

D. Enhance knowledge of equipment:
   1. Gain better understanding of equipment structures, functions, and manual of operations.
   2. Guide foreman and group leader to better understanding of equipment.
   3. Check and revise operation manuals as needed.

E. Guide new workers and transferees:
   1. Explain organization of the subsection to the newly recruited or transferred workers.
   2. Give guidance on job items in the subsection.
   3. Evaluate, prepare, and revise "new worker guide" to be used by foreman.
   4. Monitor and follow up on orientations of new workers based on the manual guide.

F. Pursue human-relations activities:
   1. Follow up and advise on informal activities such as personal touch activities (PTA) [each group is entitled to hold a meeting during working hours every month to enhance human relations, recreation, and free talk].

G. Pursue quality circle activities:
   1. Act as senior facilitator and assist and give advice on quality circle activities.
   2. Assist and give guidance in quality-control meetings, seminars, and training sessions.
   3. Give advice and follow up on smooth advance of quality circle activities within subsection.
   4. Conduct activities to further enhance understanding of quality circle activities.

H. Encourage suggestions:
   1. Promote and guide idea suggestion programs to meet the target of number of suggestions in each group.
   2. Monitor development and give guidance.
   3. Provide individual counseling for less active members.
   4. Review suggestions.

I. Build work discipline:
   1. Organize meetings and provide counseling to build a more positive working atmosphere.

2. Confirm if all rules and regulations are observed. Give feedback if nonconformance is observed.
3. Provide individual counseling to members who routinely violate rules and regulations.
4. Check implementation status of rules and regulations within subsection.

J. Other:
1. Give approval and instructions to engage in overtime work.
2. Monitor and follow up on annual leave status.
3. Provide individual counseling to operators with special problems.

### *Section Managers' Roles and Accountabilities: Examples from the TAM Manual*

While the roles of group leader, foreman, and supervisor can be spelled out in specific action programs, the role of the section manager—building better internal systems and procedures—is less concretely defined. For instance, the section manager on quality at TAM is expected to shoulder the following responsibilities:

A. Policy and target setting:
1. Define and relay to each foreman targets for quality improvement on each item.
2. Devise strategies to achieve the targets.

B. Follow up on the progress toward targets:
1. Conduct periodical review of section targets.
2. Take problem-solving countermeasures.
3. Follow up on the results of countermeasures.
4. Support subordinates if results are unsatisfactory.
5. Take up serious problems directly under section manager's authority.

C. Improve quality-assurance system: Build quality into the process and achieve 100 percent assurance.

## The Conditions Necessary for Successfully Defining Roles and Accountability at TAM

The TAM manuals clearly outline the roles and accountabilities of the various *gemba* managers at Toyota Astra Motor Company in Indonesia.

However, in order for the manuals to be effective, two fundamental conditions must be satisfied:

1. There must be training programs to help managers to acquire the necessary skills to perform their respective roles.
2. There must be systems and procedures to manage such items as quality, cost, and delivery so that every manager knows exactly what he or she is supposed to do.

A group leader accountable for reducing quality problems, and a foreman who has to reduce *kosu*, must know exactly what to measure, using what kind of checklists, and must know how to calculate and report the data. Group leaders also must be equipped with problem-solving capabilities.

TAM's 25 years of effort in building its internal systems and procedures have enabled the company to define the respective roles of managers successfully. The managerial training programs that achieved this success have been closely related to the elimination of *muda* (waste), *mura* (irregularity), and *muri* (strain) that are often associated with *kaizen* activities on the shop floor. When these three elements are applied to staff development, *muri* may be understood to mean "human strain on the job." Workers not equipped with enough skills to perform their jobs will feel strained. When such workers do not have sufficient and timely information about their jobs, they probably will make mistakes. When the workers do not understand the value they are adding for their customers, they will create more waste and cost. To eliminate *muri*, workers as well as managers should be trained to perform their jobs. In particular, the ability to adapt to changes in the business environment is regarded as one of the most important traits. Toyota believes that staff development is so important that it should be carried out continuously.

## Staff Development

### *On-the-Job Training*

The mainstay of Toyota's training is *on-the-job training* (OJT), which builds the *skills* of the worker. For its OJT program, Toyota developed a program called *Toyota Job Instruction* (TJI).

The training materials were derived initially from the Training Within Industries (TWI) program, which includes job relations, job improvement (*kaizen*), and job instructions.

### *Formal Classroom Training*

Various subjects are taught by certified trainers within Toyota. The TJI program, for example, is taught by a certified trainer in the classroom. Other examples of formal classroom training at the company are the Toyota Production System (TPS), problem solving, pre- and postpromotion training, safety training, and technical training.

### *Voluntary Activities*

These training activities are less structured than the other two, and participation is not compulsory. Activities falling into this category are quality circles, suggestion systems, *hiyari* ("near miss") reports, and quality *hiyari* reports. Management feels that these activities stimulate employees' minds and teach them something of great value. Following is a description of one area of voluntary activities, *hiyari* reports.

## The Identification of Potential Problems

### Hiyari *Reports*

TAM has two special programs directed at anticipating problems in advance. One is called the *hiyari* ("near miss") *report*, and the other is called the *quality hiyari*, or *kiken yochi training* (KYT)–*anticipating danger in advance, report*. The *hiyari* report points out unsafe conditions or actions that eventually could lead to accidents in the workplace, whereas the quality *hiyari* report anticipates conditions that could lead to such quality problems as defects.

The two *hiyari* report forms are used regularly in conjunction with the submission of suggestions identifying potential problems. Such suggestions are more likely to receive a positive evaluation if a *hiyari* report form or a quality *hiyari* report form is attached to them. In other words, management rewards the efforts of employees in anticipating or detecting problems in advance and solving them before they become a reality. Management regards such an approach to problem solving as more valuable than addressing the problem *after* it has become a reality. In one such case, an operator in the paint shop worried that the hoist might strike his head. He suggested removing the visual blockage so that he could see the incoming hoist chain more clearly. Another case concerned detection of defects during trial production of a new model (land cruisers). A metal finish worker had identified the possibility of dents when the operator opened the back door of the vehicle. The proposal was to install a door stopper for both side doors.

### *Training in the Anticipation of Problems*

TAM has a special training program devoted to anticipating dangers in which such subjects as safety, identification of potential problems, and *hiyari* reports are dealt with. The program elevates workers' awareness of unsafe conditions and behaviors, enhances their sensitivity about safety matters, and helps to increase the number of *hiyari* reports.

## The Benefits of *Kaizen* at Toyota Astra Motor Company

After 25 years in operation, it appears that the *kaizen* culture has been firmly established at Toyota Astra Motor Company in Indonesia. In 1995, the average number of suggestions was seven per person per month, which was better than that at most Japanese companies. Management estimates that savings made by the suggestions were $5 million for the year. Since 1990, the target for *kosu* reduction or productivity improvement has been 10 percent every year, and this has been achieved every year. In the early 1980s, Astra had a "car hospital" on site housing as many as 400 current "car patients." Today, *all finished cars are delivered directly to the customer and an average inventory of finished cars at the factory is six hours. Kaizen* consultant Kristianto Jahja, who used to work at TAM, remembers that in the early days he used to carry a plastic bag in his hand when he went to the *gemba* and pick up the nuts and bolts that littered the floors. Nuts and bolts—and even machine part labels and soft-drink bottles—were sometimes found inside fully assembled cars as well! Today, such conditions are a thing of the past. Obviously, it takes many years of firm determination of management to bring about such a change, but it has been done at TAM—and with workers who typically earn as little as $150 per month.

Today, manufacturing companies are seeking new horizons outside their own countries. After Singapore, Indonesia, Malaysia, and Thailand, these companies are looking to Vietnam, Myanmar (Burma), China, and India, where workplace cultural transformations such as the one achieved at Toyota's Astra Motor Company plant in Indonesia are not unlikely. This is bound to present a real challenge to companies in North America, Europe, and other industrial regions, where workers earn 10 times more and are deeply imbued with traditional Western approaches to supervision and management.

## CHAPTER ELEVEN

# From Just-In-Time to Total Flow Management

Broadly speaking, there have been three major streams of manufacturing process excellence, sometimes competing but ideally collaborating. They are *quality management*, *flow management*, and *asset management*. The names of these programs have evolved, such as the shift from total quality control to total quality management (TQM) to six sigma or the evolution of total productive maintenance (TPM) to total productivity management. Likewise, our work has seen the development of just-in-time (JIT) to a more comprehensive system of total flow management.

In order to achieve successful quality, cost, and delivery (QCD) and satisfy the customer as well as itself, a manufacturing company must have all three major systems in place: (1) total quality control (TQC) or total quality management (TQM), (2) total productive maintenance (TPM), and (3) just-in-time (JIT) production. Under Taiichi Ohno, the Toyota Motor Company originated JIT. Along with the *jidoka*, or built-in quality pillar, JIT is the second pillar of the Toyota Production System. Many companies prefer to use *lean production* or *[Company Name] Production System*. This chapter will use the term *JIT production system* and also introduce total flow management to provide a more detailed view of this system.

Each of the three major systems necessary for achieving QCD has different targets: TQC has overall quality as its major target, whereas TPM addresses the reliability and quality of equipment. JIT, meanwhile, deals with the other top priorities of management—namely, cost and delivery. Top management must firmly establish both TQC and TPM before JIT production is introduced. Many people have misinterpreted JIT. In one of the most common misunderstandings, a company expects its suppliers to deliver just-in-time. In order to benefit from a supplier's just-in-time delivery, a company must first establish the best possible efficiencies in its own internal processes. JIT is a revolutionary way to reduce cost while at the same time meeting the customer's delivery needs.

## Just-In-Time at Aisin Seiki's Anjo Plant

A visit to Aisin Seiki's Anjo plant in Japan will help the reader understand JIT. This plant produces such products as bed mattresses, industrial sewing machines, gas heating pumps, and air conditioners. On entering the mattress production area, one would expect to find a huge space where many employees—surrounded by stacks of frames, springs, and fabrics—assemble mattresses. However, what the visitor sees instead is a compact scale of operations. In a space no larger than a high school basketball court, seven dedicated lines produce mattresses of 750 different colors, styles, and sizes per day.

The machines on each line, except for quilting machines, are laid out in the order of processing. The major processes include spring-coil forming, spring-coil assembly, multi-needle quilting, cutting, flange sewing, padding, border sewing, tape-edge sewing, and packaging. Each process connects to the next, allowing no room to place extra work-in-processes. Only one work-piece at a time flows between the processes. The quilting process makes only one piece of cloth for one mattress at a time. Each work-piece moves through the workstations while being processed. Twenty minutes after the weaving machine starts weaving the mattress cover, the mattress is completed and ready to be shipped to the customer, one of about 2,000 furniture stores scattered throughout Japan that serve the company's dealers.

For the most popular models, a small storeroom at the end of the line holds a standard inventory of between 3 and 40 mattresses (the number depends on daily sales), each placed in a given location and with a *kanban* tag (production order slip) attached. Every time an order comes in and a mattress is shipped, the *kanban* that had been attached to that mattress is sent back to the starting point of the line and serves as an order to start production. This system ensures that the minimum required number of the popular models is always in stock. For nonstandard types of mattresses, no storeroom exists because the mattresses are shipped directly from the production line to the furniture store that placed the order.

Aisin Seiki begins producing a mattress the day after receiving an order from a dealer; this is made possible by the very short lead time of production (two hours). Sometimes the company receives large orders from hotels and vacation resorts; when this happens, the company spreads the production, manufacturing a given number of mattresses each day. It fits this production evenly in between the production of other models so that the normal production schedule is not disturbed. This is called *heijunka* or *leveling.* Large orders of this kind require the company to secure outside storage space until the shipping date. Although JIT is sometimes referred to as a *non-stock production system,* it is not always either possible or practical to keep a zero inventory.

Such a production system yields many insights. First, one can sense an invisible line connecting the customer and the production process. The short lead time allows production to begin *after* an order has been received, and *gemba* employees can keep the customer in mind while making the product. It is almost as if the customer is waiting in the next room to receive the finished mattress.

Second, this system allows great flexibility to meet customer needs. With the use of *kanban,* popular models are replenished as soon as they are sold, thus minimizing inventory.

Third, this kind of production system can respond quickly to abnormalities on the line. If a reject is produced, the whole line must be stopped because there will be no replacement. In other words, management has to make a concerted effort to address problems on the line so that the line never stops. Every quality problem, every equipment malfunction, every problem related to human error must be dealt with and settled so that the line is not stopped. JIT necessitates ongoing *kaizen* activities in the *gemba* and calls for rigid self-discipline on the part of both management and workers. The fact that Aisin Seiki has received both the Deming Prize and the Japan Quality Control Award attests to the company's commitment to quality.

Fourth, JIT permits flexible production scheduling. Aisin Seiki produces only as many mattresses as are ordered by customers. Even for the most popular models, the company does not start production in anticipation of future demand and before the daily minimum allowable inventory is determined. On the other hand, once production begins, stagnation in the form of work-in-process is not allowed, and the product must be finished within the shortest possible time and shipped directly to the customer right away. For most products, a warehouse is not needed, and the truck running on the street serves as a warehouse.

Fifth, this kind of production system helps companies to forecast the market more accurately. In an ideal world, production would not begin until all the orders had been received. However, this is not possible in reality. Because Aisin Seiki has learned from experience that the daily demand for its most popular model is about 40, that is the number of mattresses kept in inventory for that particular type. Depending on the popularity of each model, the daily inventory ranges between 3 and 40. The *kanban* system is used to make only as many as have been sold every day. For other types of mattresses, the company starts production only after the order has been received. Bear in mind that one of the definitions of JIT is "making only as many products, and in the same sequence, as ordered."

These are some of the visible features that we can readily identify by observing Aisin Seiki or any other JIT-based operation. Some additional features that may not be as visible but are present nonetheless include:

- *Takt* time versus cycle time (theoretical time versus actual time for completing one work-piece)
- Pull production versus push production (producing only as many items as the next process needs versus producing as many as can be produced)
- Establishing production flow (rearranging equipment layout according to work sequence)

## *Takt* Time versus Cycle Time

*Takt time* is the total production time divided by the number of units required by the customer. The figure is expressed in seconds for mass-produced items. For slower-moving items, the *takt* time may be expressed in minutes or even hours, as is the case in shipbuilding, for example. If line A produces 80 mattresses in one day and the workers work for eight hours, the *takt* time is calculated as follows:

(8 hours/day x 60 minutes/hour) /
80 mattresses/day = 6 minutes/mattress

This means that if each process within line A completes its work every six minutes, the finished mattresses go out the door every six minutes, and 80 mattresses will have been produced by the end of the day.

The word *takt* comes from the German word for the baton used by an orchestra conductor. The *takt* time is a magic number because it is the pulse of the market. This is the number everybody in the company must live by. Just as a conductor's baton sways between andante and crescendo, the *takt* of the market keeps changing, and the *gemba* must respond accordingly. If each process exceeds the *takt* time, a shortage of products will result; if each process is faster, a surplus will occur. When *takt* time is observed properly, the *gemba* is moving ahead with the same pulse as the market. Once management has achieved sufficient flexibility, the *gemba* can respond instantaneously to changes in the pulse of the market, producing only as many pieces as are ordered.

*Takt* time is a *theoretical figure* that tells us how much time is needed to make one product at each process. *Cycle time*, on the other hand, is the *actual time* required for each operator to complete the operation. In the *gemba*, abnormalities are a fact of life, and each time they arise, the cycle time is prolonged. The idea behind JIT is to bring the cycle time as close as possible to the *takt* time.

To achieve this ideal, abnormalities of all types must be addressed. When the cycle time is compared with the *takt* time in a company that has not adopted JIT, the cycle time is much shorter—in many cases, half the *takt* time—producing a buildup of work-in-process and finished products that become surplus inventory.

The lines also should be reviewed for uniformity of cycle times. No matter how quickly a particular line may produce, total efficiency will not improve if the other lines operate at slower cycle times.

## Push Production versus Pull Production

Most manufacturing companies today are engaged in *push production*. Every process produces just as many units as it can and sends them to the next process, whether the next process needs them or not. This stems in part from the following line of thinking: "As long as the processes are in order, let's make as many units as we can because we never know when things might go wrong again."

In a mattress company, this way of thinking translates into weaving as many mattress covers as possible at the weaving machine or making as many springs as possible at the spring-making machine. In a conventional company, such processes usually are located separately from the assembly line. Chances are that there are several weaving machines or spring-making machines located in one corner of the plant—far removed from the final assembly line—and that they continuously produce works-in-process that are sent first to the warehouse and later to the final assembly line.

Operators in such an environment do not know, and do not need to know, the volume and time requirements of their customer. This is a typical example of a push production system. At the final assembly, too, chances are that operators are assembling as many products as the line can churn out; the finished products find their way to the warehouse and wait for the order to arrive. A push system necessitates batch production, creating *muda* of transport and inventory.

Production at Aisin Seiki's Anjo plant, in contrast, is based on the *pull* of the market; the entire plant springs into action with the receipt of an order from a customer. Rather than build up inventory in anticipation of orders, the company makes every effort to anticipate customer demand for the immediate future and to build flexibility within the plant to cope with fluctuations.

Toshihiko Mitsuya, project general manager at Aisin Seiki's Anjo plant, says that mattress production is quite different from automotive production in that there are no fixed daily production volumes. In other words, there is no production planning; the only planning for mattresses is the orders received from the customers. For automobiles, daily production volumes are at least uniform.

A customer who comes to one of the 2,000 furniture stores wants a mattress and wants it right away. Unlike a person shopping for an automobile, a person shopping for a mattress is not willing to wait long. Although the furniture stores carry competitive models, Aisin Seiki's production method gives the company the flexibility to offer its full range of products in the shortest possible lead time. Today, a customer at any one of the 2,000 furniture stores can select any of the 750 different models and have it delivered the next day, as long as the customer lives within 100 kilometers of the plant. If Aisin had not developed such a production system, the alternative would have been to build a large inventory.

## Establishing Production Flow

In *pull production*, all processes should be rearranged so that the work-piece flows through the workstations in the order in which the processes take place. Because some equipment is too large or too heavy or is used for multiple purposes, it is not always possible or practical to rearrange the equipment into exact workflow order. However, isolated machines should be moved and incorporated into the line as much as is practicable.

Once the line is formed, the next step is to start a one-piece flow, allowing only one piece at a time to flow from process to process. This shortens lead time and makes it difficult for the line to build up inventory between processes.

An aircraft component plant conducted a simulation of such a one-piece flow, assuming that all processes were connected and that only one piece would flow between the processes according to *takt* time. At the time the plant conducted this simulation, its total lead time was eight weeks from start of production until the finished product went out the door. The simulation revealed that the lead time should take no more than four hours. The layout has since been changed to accommodate one-piece flow, and *kaizen* activities have solved many bottleneck problems.

Before starting a production line with one-piece flow, however, such problems as quality, machine downtime, and absenteeism must be addressed. One-piece flow production cannot begin until these problems are resolved because each time a problem arises, the line must be stopped, and the problems that up to now have been regarded lightly become visible. The company loses money when the line is stopped. Precisely for this reason, management has to address the problem, and thus a line with one-piece flow makes it mandatory to identify and solve problems.

Besides shortening lead times and cutting excess inventory, one-piece flow also helps workers to identify quality problems right away because any problem in the previous process can be detected in the next process. One-piece flow also allows 100 percent inspection because every piece goes through the hand of every operator.

Yet another positive merit of the one-piece-flow line is that it does not require large equipment. A machine need only be large enough to process one piece at a time within the *takt* time. Conventional production processes based on the batch concept, meanwhile, require large machines to process large batches of work-pieces at a time. Furnaces and painting units are a good example. In one plant, I saw a heating furnace as large as an indoor swimming pool.

At one of the plants of Matsushita Electric Works, a large oven was used for treating microswitches on the main line. When a one-piece-flow line was introduced, however, the company found that a toaster purchased at a nearby supermarket was sufficient for this purpose. As I mentioned earlier, machines are usually working too fast. A machine in a one-piece-flow line can be much smaller than a machine used in conventional batch production. It also operates more slowly, making it suitable for a slower *takt* time. Such a machine may be purchased at a much lower price; even better, the company itself may be able to design and produce the machine. If a company with an expensive high-speed line producing a large quantity of products wishes to increase flexibility, it is often possible for the company to create, at little cost, an additional line to accommodate small or urgent orders. Because such additional lines do not require much investment, management can afford to dedicate them to small or urgent orders while using the existing main line for large production runs. This minimizes the need for frequent setup changes on the main line. Often it is possible to arrange these small new lines in a U shape; the operator working inside the U can readily move from one process to the next. This arrangement makes it possible to manufacture products according to cycle time when needed, giving the manufacturer more flexibility to cope with diversified customer requirements.

## The Introduction of JIT at Aisin Seiki

Until the mid-1980s, each of the eight sales offices of Aisin Seiki kept its own inventory of mattresses and delivered them to furniture stores. In those days, the company offered 220 different types of mattresses and required 30 days' inventory. The factory produced 160 mattresses per day with 20 operators, and the *kosu* (work hours) of production per mattress was 75 minutes.

"Salespeople," said Toshihiko Mitsuya, "gave us monthly sales projections, but they never turned out to be accurate. It resembled looking into a crystal ball. The plan changed all the time, giving our suppliers a difficult time keeping up with our changing orders as well. We had shortages of some supplies on the line all the time, and yet we had a mountain of inventory."

In those days, the plant had a warehouse with a capacity of 2,200 square meters to meet fluctuations in sales as well as shortages of special types of mattresses. The company's production system, based on market projections, had the following shortcomings:

- ▲ It was difficult to estimate demands accurately. Given the long lead time, it was necessary to venture a long-term forecast, and the plan was not very reliable.

- ▲ Production schedules had to be changed frequently. Responding to the changing information was difficult because it involved changes in production planning in many processes.
- ▲ Much *muda* was created in the *gemba*. Since the *gemba* people did not want to be accused of being short of inventory, they tended to plan monthly production in a large lot.
- ▲ A warehouse was necessary to avoid shortages of work-in-process; managing the warehouse entailed additional costs.

## *The First Step of* Kaizen *at Aisin Seiki*

In 1988, Aisin Seiki decided to produce mattresses only in response to orders rather than in anticipation of orders. The first step eliminated the warehouse. The question at that time was which type of inventory to address first: finished mattresses or work-in-process. The company chose to start with finished products, which included the accumulation of all costs incurred, such as labor, materials, processing, and utilities.

*Kanban* was introduced to maintain only the number of mattresses typically ordered every day. This meant carrying inventory for the popular models only—in proportion to their daily sales. For storage of the most popular mattresses (those with sales of over three units a day), a "store" was created immediately adjacent to the end of the production line. When popular models left the store, the *kanban* (production order slip) attached to each mattress went back to the start of the line in preparation for production of the units just sold to begin the next day.

Until Aisin implemented *kanban*, it had produced different types of mattresses, such as single, double, and semi-double models, on a weekly schedule. Under the new system, though, what had once been a weekly cycle of production was reduced to a daily cycle. Today, the cycle has been further reduced to two hours.

The key point at this first stage of *kaizen* was to start producing the different models of mattresses in the same sequence as the orders were received. To do this, shortening the setup time became a critical task. By shortening the setup time, the company increased by sixtyfold the number of setup changes needed for the quilting machines.

In 1986—two years before the first phase of its *kaizen* effort—Aisin Seiki made 220 different types of mattresses. After the first phase of *kaizen*, the number jumped to 335. Yet finished-product inventory was reduced to 2.5 days, compared with 30 days before. Although the company has grown by only one employee during this time, it now produces 70 more mattresses per day than before. And the *kosu* per mattress dropped to 54 minutes, from the previous 75 minutes.

## *The Second Step of* Kaizen *at Aisin Seiki*

The use of *kanban* eliminated inventory at both the factory and the furniture stores in 1988. In 1992, Aisin Seiki was ready to tackle the second step of *kaizen*: eliminating excess inventory within the plant.

In an effort to reduce work-in-process, the company developed a tool it called the *assembly-initiation sequence table* that specified the sequence in which to initiate production of 750 different types of mattresses to meet delivery dates. Customer orders are sent to the plant online from Aisin Seiki's 2,000 dealers and eight sales offices throughout Japan. The production line receives urethane, cotton, felt, and textiles just-in-time from the suppliers and assembles them into mattresses. Each of the seven lines has been arranged in such a way as to produce any type and size of mattress in a one-piece flow.

The plant has five quilting machines serving seven lines. The orders received and delivery dates specified determine the quilting sequence, and the quilting machine follows the sequence, producing only one quilt unit for each mattress. Since 750 different types of quilts must be produced, the system does not function without the sequence table, which allows the quilting process to keep two hours' worth of inventory. In other words, the quilting process is allowed to do only two hours' advance production; it does not know what other types of quilts will be required by the assembly line after that.

The daily production schedule, including the number and types of units to be produced, is not provided to the *gemba* in order to keep the *gemba* people from producing the mattresses at their own convenience rather than to accommodate customer orders. The system provides the *gemba* with two hours' advance notice—a sufficient amount of time to cope with any urgent orders. One part-time worker using a PC prepares the sequence tables based on the day's orders.

Aisin's Anjo plant has introduced many other features of JIT production, such as leveling, in its mattress manufacturing. The plant has achieved its great flexibility to meet customer needs while reducing cost to a minimum.

## Spreading the Benefits of JIT to Other Industries

Aisin Seiki has been successful in introducing JIT in mattress manufacturing, a business that is highly seasonal and characterized by diverse demand. The company has taken on the challenge of delivering product just-in-time, immediately after receiving orders, and starting many *kaizen* activities. Today, Aisin Seiki has two approaches to production: (1) producing only items that replace inventory and (2) producing only in response to orders. The latter approach can be further broken down into two components: (1) producing for the day and (2) producing in response to advance orders. With zero inventory, daily production takes first priority in each day's schedule. Large advance orders from hotels and the like, meanwhile, take second priority. Because orders tend to be concentrated on weekends, large fluctuations in demand sometimes occur. By spreading out the additional production over a given period, the company maintains a steady production level and thus avoids disrupting the production line.

Since 1986, when it introduced JIT production, Aisin Seiki has increased its productivity by a factor of 4.5 and its gross sales by a factor of 1.8. The number of different types of products it makes has grown from 220 to 750, whereas inventory turnover has plunged to 1.8 days, one-seventeenth the original time. And the *kosu* per unit has dropped from 75 to 42 minutes. To realize production in a small lot, the number of setup changes had to be increased 40 times while the total setup time was reduced. This was made possible because *kaizen* was launched based on real need.

JIT production has yielded other benefits as well. It has substantially reduced not only setup times but also *kosu* and lead time. It has eliminated the warehouse. Mattresses now can be delivered immediately, and the sales staff can offer customers a full range of choices. Furthermore, eliminating the need to store finished products for long periods of time also has eliminated such quality problems as stains, soiling, and color fading.

Aisin Seiki is one of the major suppliers of automotive parts and components to Toyota Motor Company and has been engaged in JIT production for its automotive products for many years. The fact that the company succeeded in realizing JIT in an unrelated field such as mattress manufacturing points out that JIT production techniques and know-how can be applied with equal success to many different types of production lines. Aisin Seiki has implemented JIT technology in its production of industrial sewing machines as well. The company also provides a consulting service, called Toyota Sewing Products Management System, that assists the apparel industry in solving plant design, operation, and management problems.

Those who ignore such new trends soon will find themselves left out of the competition as their competitors start to take advantage of this wonderful production system. They would do well to heed the following remarks made by Chie Takagi, supervisor at Matsushita Electric Works, after the company had introduced JIT production: "As I look back on those old days, I wonder how we could have conducted our business that way. The way we produced our products then was—almost a crime!"

## Total Flow Management*

Based on the 25 years of experience implementing *kaizen* and lean logistics principles, the Kaizen Institute has developed what is called *Total Flow Management* (TFM). This detailed model allows a smooth implementation of the Toyota Production System not only inside manufacturing plants but also covering the integrated supply chain.

TFM is a *kaizen* strategy based on the creation of *pull flow*, a new operations system paradigm that is by far the best way of designing and managing the operations and supply chains of any company. *Creating a flow* means creating a movement both of materials and information all across any supply chain. This movement of materials and information should be driven by real customer orders or real customer consumption. Movement of materials and information should be understood in a supply-chain environment starting with final customers buying (pulling) products (materials) from the retail stores, the retails stores pulling from the product distribution centers, the distribution centers pulling from the manufacturing companies, and the manufacturing companies pulling from their network of suppliers. This should be the flow in a simplified supply chain (a

*Adapted from Euclides A. Coimbra, *Total Flow Management: Achieving Excellence with Kaizen and Lean Supply Chains* (Switzerland: Kaizen Institute, 2009).

real supply chain may have many elements in the chain both before and after the final manufacturing facility).

This is what Toyota developed and implemented in all its supply chains starting in the car dealers and going back to all its suppliers. TFM is a system where pull flow (one-piece flow pulled by consumption) and a strong engagement in *kaizen* every day, everywhere, and by everybody are the main principles applied.

The starting point for the design is the point where you are located in the supply chain. This may be you as a manufacturing facility or may be you as a product distribution facility. By applying the model, you will be looking at creating your internal pull flow system and also looking at how you can expand this model downstream on your supply chain. This is what I call the *delivery side* of the supply chain. You also will be looking to expand the model to the upstream side of your supply chain, or what I call the *source side* of the supply chain. The general principles of the model are the ones depicted in Figure 11.1.

The scope is the supply chain with you in the middle. The main target of TFM is the reduction in total lead time in the supply chain. The measure of lead time is the inventory coverage all across the supply chain, which can be measured in days. Reducing lead time also results in eliminating the *muda* of waiting and really means creating a material flow. The systems, processes, and standards necessary to create and maintain this flow require a high level of rigor and bring about very important results in terms of

- Cost reduction
- Working capital reduction
- Increased productivity
- Increased quality
- Increased customer service and satisfaction

This is achieved by creating a flow all across the supply chain and starting this flow with customer consumption. We will see that we can start with real orders or inventory-replenishment orders. Physically, it will be necessary to create one-piece flow, one-container flow, and one-pallet flow and to accelerate this flow by using the concept of "milk-run loops" in transportation (another polemic and hard-to-believe solution by many managers). Forecasts will not be used for creating production or distribution orders, but they will serve the purpose of capacity management.

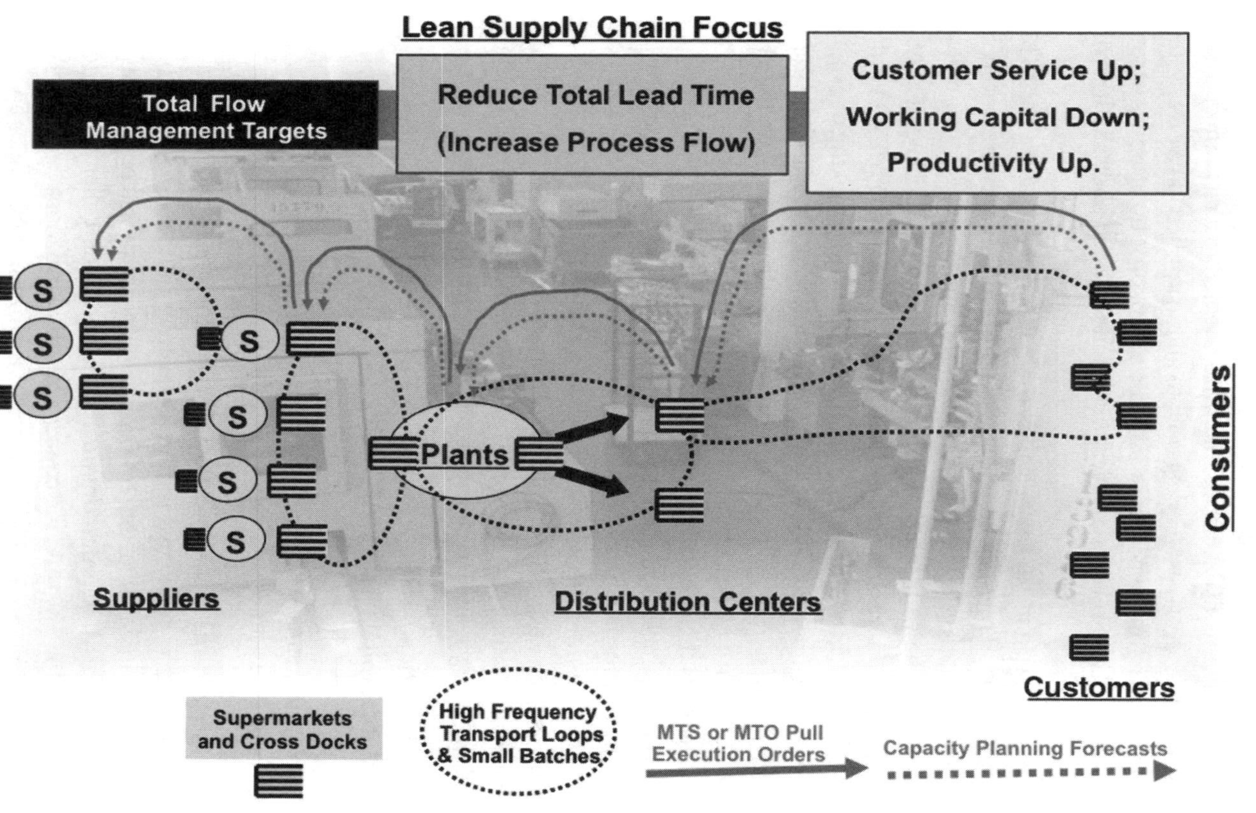


**Figure 11.1** Total flow management (TFM) model.

In fact, we are talking about a TFM model based on the process of creating a flow of materials and information and aiming at getting breakthrough results in terms of quality, cost and delivery (QCD) and at the same time changing the company culture to a culture based on *kaizen* every day, everywhere, and everybody spirit. Figure 11.2 shows the relationship of the production, internal logistics, and external logistics systems within TFM.

Figure 11.3 shows the three types of pull logistic loops, and you can see that the external logistics pillar is divided into two symmetrical sides: the *source flows* and the *delivery flows.* In each of these flows we can find the need to create flow in storage design, inbound, outbound, milk-run, and logistics pull planning.

## TFM Transformation in Company A

Company A is a member of a well-respected global manufacturing corporation. The company produces water-heating devices, such as water heaters and boilers, for the household market. The company was founded in 1977 in a small European country and started operating under a license of the corporation. At that time, the company belonged to the founding family. In 1988, the company was bought by the corporate parent.

The founding family had strong hopes for the development of the company and worked hard for many years. From 1977 until 1988, the company became a market leader in the country and established a sound business and a sound trademark. The owners were very concerned with quality, so this was one of the main areas for improvement, and the company did develop very well in this area. The local university was engaged in building a database for quality defects, and diligent work in attempting to find and eliminate internal and external failures gave very good results for many years.

This is a company that had always been very profitable since its foundation. By joining the corporate group, new horizons were born for Company A. It quickly became a product-development center for the corporation and started to export to all the European markets, soon becoming the European market leader.

We can say that since the beginning, this company was a model company with a very good social climate where everybody worked hard for the success of the company. The CEO, a member of the founding family,

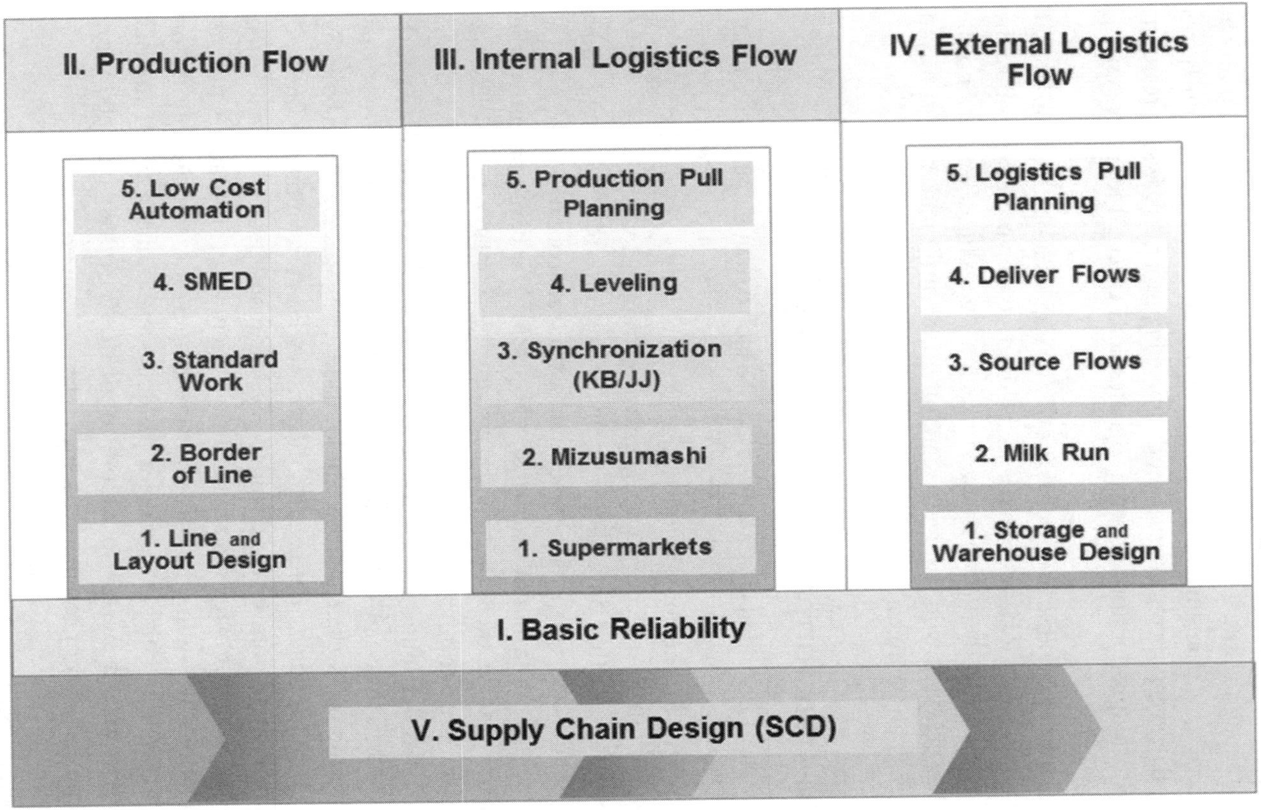

II. Production Flow

5. Low Cost Automation
4. SMED
3. Standard Work
2. Border of Line
1. Line and Layout Design

III. Internal Logistics Flow

5. Production Pull Planning
4. Leveling
3. Synchronization (KB/JJ)
2. Mizusumashi
1. Supermarkets

IV. External Logistics Flow

5. Logistics Pull Planning
4. Deliver Flows
3. Source Flows
2. Milk Run
1. Storage and Warehouse Design

I. Basic Reliability

V. Supply Chain Design (SCD)

**Figure 11.2** Integration of TFM elements.

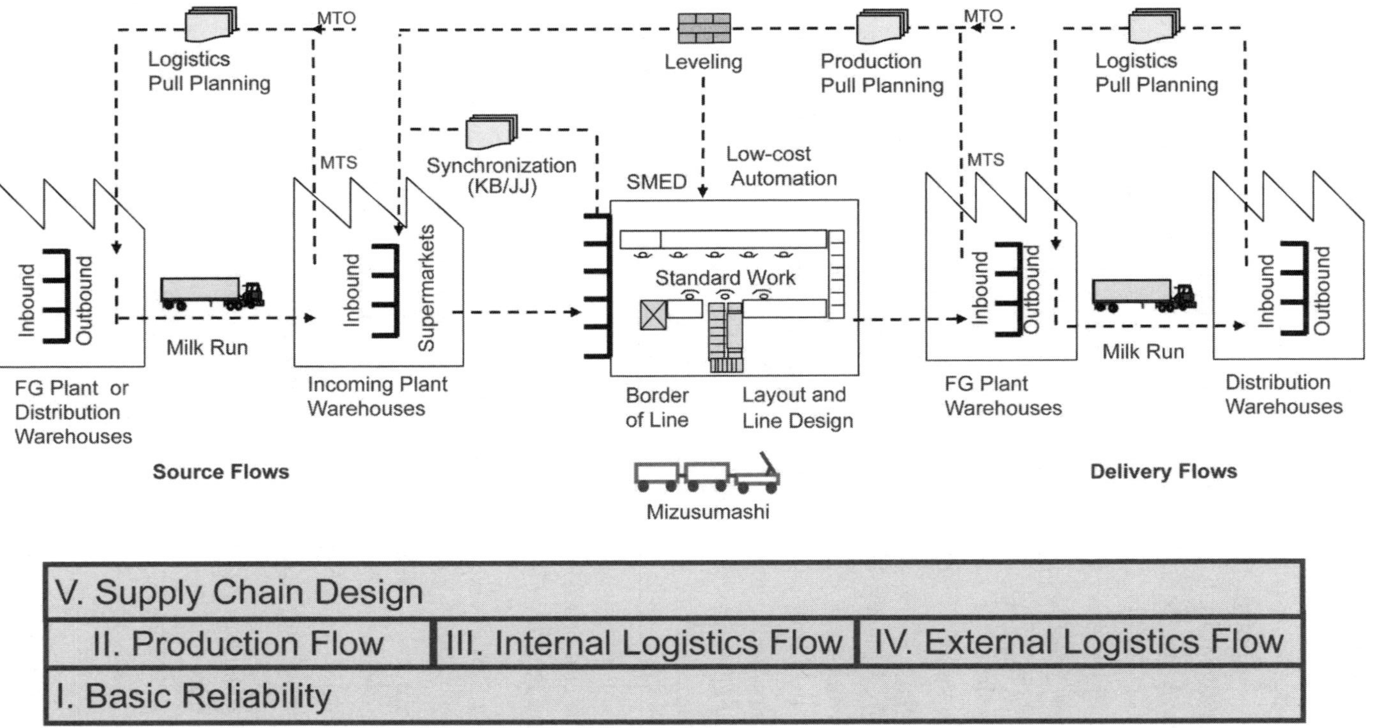


**Figure 11.3** Pull logistics loops.

became the plant manager of the biggest and best plant of the group after sale of the company to the multinational group, and he continued his policy of excellence, now reinforced by the values, mission, financial power, and technical and organizational know-how coming from the corporate group.

After the initial years of investing in quality-improvement efforts, Company A started looking at other ways to increase its performance. In the beginning of the 1990s, a two-bin system was established to reduce the lack of components supplied to the assembly lines, and some small productivity improvements in the same assembly lines were started.

At the same time, an important changeover time-reduction project was started in the press section. At that time, it was possible to cut the changeover time by half, from about 2 hours to 1 hour, in the stamping presses. Some projects also were initiated to implement one-piece-flow cells, and a good level of integration of operations was achieved, namely, some bending cells and subassembly cells were created with very good results.

Thus the improvement strategy went on steadily from the beginning of the company to the end of the nineties. Every year, the company saw overall productivity improvement of around 10 percent, always improving quality and customer service. Meanwhile, a big drive on product development with the launch of many new products served to establish the plant as the biggest and most profitable of the corporation.

By the end of 1999, the main key performance indicators (KPIs) of the plant were the following:

- Total inventory (raw material, work-in-process, finished goods): 50 days
- Internal defects rate: 12.000 parts per million (ppm)
- Customer-service level: 91 percent
- Achievement of the assembly production schedule: 50 percent
- Productivity: 70 parts/operator
- Final assembly-line efficiency: 75 percent

All KPIs had a good trend until the end of 1999, but since 2000 it had become more and more difficult to improve, and all the KPIs had become stable. It appeared that the improvement streak had come to an end.

Until then, a lot of *kaizen* tools were used, the main ones being the following:

- Quality problem solving
- Single-minute exchange of die (SMED)

- Integration of operations into one-piece-flow cells of lines
- Two-bin system (full box/empty box *kanban*)
- Maintenance improvement
- Scheduling and synchronization system (I will speak about this system in more detail later)

At the beginning of 2000, the headquarters of the corporation decided to launch a corporate continuous-improvement (CI) initiative. A corporate team was set up, and the CI model was developed. The reason behind this initiative was that all the plants applied a collection of tools, but there was not a common strategy or a common language. Also, there was not a measuring system to quantify the degree of development in terms of improvement-process development.

The first audit applied to Company A showed a score of about 28 percent. This was a surprise to many people inside the plant because they thought that they had tried all sorts of improvement tools, and they couldn't see where else to improve.

One of the paradigms existing was the very small involvement of *gemba* people, the operators and team leaders. All the improvement activities done until then were executed by project teams that seldom included operators. The *gemba kaizen* event approach was not being used and the improvements were being managed by project teams and mainly the engineering department.

Another existing paradigm was that everybody was convinced that they already had a pull system. In particular, especially the production manager argued that they were working according to hourly batches in the final assembly (this was a reality because they were defining assembly batches that took one hour to assemble in most of the cases). The problem was that the information used to calculate the hourly batches was coming from the forecasts, and this is not exactly a pull system according to customer needs or consumption. Another argument was that the company was using in most cases the customers' orders to plan assembly.

This paradigm was reinforced by an internally developed synchronization system between the final assembly and the preassemblies (as well as the manufacturing sections). It was argued that this system, working accordingly to a central materials requirements planning (MRP) algorithm, was pulling the assembly supplies on an hourly basis.

The problem was that the quality of the synchronization was far from good, and the reason why was not only the MRP system but also the ineffective synchronization system in terms of material movements on the shop floor. The very complex logistics of supplying hundreds of components was not effective at all.

A time came when the current information and materials movement system reached its limits, and a new paradigm had to be installed if the plant wanted to go over the current paradigm fence. An improved pull system had to be tried (in fact, a system change from a push system to a pull system with improved flow on the shop floor was a real need).

If looked at with *kaizen* eyes that spot all *muda* (waste), there were plenty of opportunities to improve. Basically, all operations could be subject to *muda* elimination activities. The problem is that it is relatively easy to say that we have much *muda*, but it is harder to say it with the belief that we can eliminate it. We can only achieve *muda* elimination if we are backed up by a strong conceptual model of TFM improvement and some experience in implementation.

Time went by, with more of the same improvement projects and a lot of training (this was an advantage of the corporate CI initiative). We were now at the end of 2004. In the last four years, the plant continued to train people and deploy improvement projects in many areas. But it never changed the push system, and it never redesigned the flows in a significant way.

As a result, all the main KPIs were evolving very slowly, and the plant was no longer capable of showing the improvement pace and vitality it had in the past. The time had come to try a system change. The targets had to be the following:

- Reduction of the finished goods inventory
- Achievement of over 98 percent OTIF in final customer service
- Achievement of over 98 percent OTIF in assembly-plan fulfillment
- Reduction of parts and raw materials inventory
- Achievement of over 98 percent OTIF in suppliers' deliveries
- Overall productivity increase of 10 percent minimum every year
- Continuation of quality defects reduction
- Improvement in the corporate CI audit score of 10 to 20 percent points every year

This was the challenge that finally was accepted by Company A. Nowadays, the competition is so strong in any market that only the best have a chance. Continuous improvement, better defined as every day, everybody, everywhere improvement, is a key competitive advantage. Accustomed to being the best in its field, Company A couldn't imagine stopping the improvement pace. A new paradigm had to be implemented. A system change had to be tried. The old system had reached its limits.

Then a new officer came in, responsible for finance and logistics, and together with the production and engineering officer, he decided to do something different and innovative. The first step was to convince the CI corporate team to let the company contact outside experts. This turned out to be a hard job, and it took about one year to succeed.

In the beginning of 2005, the planning phase of a pull flow project was started. This job consisted of analyzing the current state using *value-stream mapping*, defining a *future-state vision*, and organizing a project to implement *pull flow* based on the TFM model.

The design project team consisted of the heads of the production, logistics, engineering, maintenance, and continuous-improvement departments and some key deputies in those departments. The outside experts were two *kaizen* coaches from the Kaizen Institute, and the team leader was the production manager. The team mapped the flows of one of the most important product families and agreed on a common understanding of the current way of doing things and all the *muda* and improvement opportunities available. Figure 11.4 shows the current-state mapping done by the team to analyze the value stream.

A summary of the main issues is as follows:

- ▲ Too much finished goods inventory (15 days)
- ▲ Dysfunctional finished goods inventory (final customer-service level of 93 percent)
- ▲ Order planning based on sales forecasts
- ▲ Planning department overloaded with planning tasks, especially at the end of the week (preparing next week's plan)
- ▲ Low fulfillment of the assembly schedule (50 percent)
- ▲ Poor assembly-line efficiency: Operators isolated from each other, back supply, supply of big pallet-size containers, bad operator standard work, and line balancing not very good
- ▲ Many line stops and schedule changes owing to lack of parts; many difficulties in line supply and synchronization

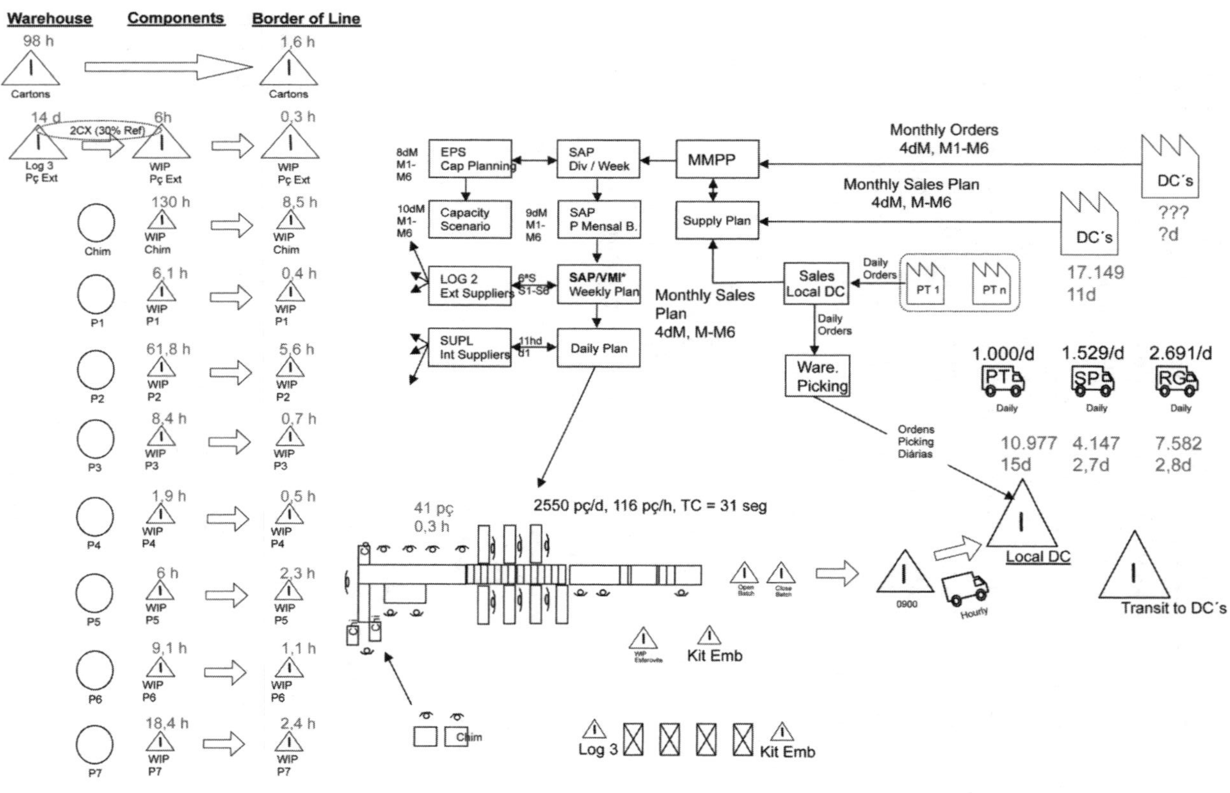


**Figure 11.4** Current-state mapping of Company A operations.

- Big inventory of bought materials and parts
- A lot of management time dedicated to daily fine-tuning and firefighting; lots of stress and no time for *kaizen*

After the current-state mapping, the team received training using the TFM simulation games and applied the score cards in order to have a full awareness of the TFM concepts to change the current system and design a practical pull flow system. Recognition by the team that the situation could be improved drastically was a surprise to everybody, and a sense of hope and challenge began to emerge.

The *future-state vision map* began to be discussed by looking at the final assembly line. This was the line that had been diagnosed some time ago as not having a big potential for improvement (remember that an expert from the corporation had said the improvement potential was only 3 percent), so everybody was anxious to understand how the evident *mudas* (using *kaizen* eyes) could be eliminated.

The design team spent four days in current-state mapping and training activities and another three days designing the future-state vision. Another two days were dedicated to planning the implementation. The whole project planning phase was done spending nine days spread over a period of one month. The first implementation phase took 10 months. Let's now examine other features of the project and how the implementation proceeded.

The team started applying the lean production flow concepts of line design, border of line, standard work, SMED, and low-cost automation. It quickly became evident that a real one-piece flow was not working, although the line had a conveyor. Small batches were resulting from accumulation owing to different worker speeds and other line issues. It also was evident that the workers were separated from each other and that the supply of parts was done from behind.

The lines could be described as "fast-cycle-time lines with isolated worker islands." Cycle time was 30 seconds, and setup time was about 5 minutes owing to a small press die changeover at the beginning of the line as well as the difficulty of changing the parts.

It was evident that a leaner line could be designed and classified as follows: "lower cycle time and high-efficient standard work." Thus the line should have the following features:

- Less variety of product references
- Small containers and hand reach and in a fixed location

- Better balancing
- Zero changeovers and low-cost automation

A new line was designed to have a cycle time of 60 seconds (this implies having two lines instead of one) and zero changeover time (by moving the small press upstream in the process). The first workshop after the planning phase was dedicated to designing a line in detail and building a mockup of the new line. The test of the mockup showed a productivity increase of 25 percent. Figure 11.5 shows a comparison between the old and new lines.

It is also of the utmost importance to define the characteristics of the border-of-line "supermarkets." In this case, about half the parts could be supplied using *kanban*. The border of line was designed with flow racks to enable pull flow replenishment. For the other half, there was no space owing to an extremely large number of variants. The high-variety parts had to be supplied using a sequential replenishment (*junjo*) system.

Having *junjo* parts means that the synchronization system must be totally foolproof; otherwise, the wrong sequence will be supplied to the line, and you have the old problems of stoppages and assembly schedule changes.

The TFM system also looks closely at the planning method. The current-state map showed that two types of customers were being planned: a distribution center (DC) in the country and DCs abroad. Both types of customers were sending monthly forecasts. The DC in the country was a warehouse located on the plant premises that also was sending daily orders to be picked from the stores and delivered the next day. The planners were managing the stock in this DC but had no information about the stock in the DCs abroad. In both cases, the planners were relying on the monthly forecasts sent by the sales department to plan production.

The process was explained earlier and consists of these main steps:

- Maintaining a monthly master production schedule (MPS) based on forecasts
- Using this monthly plan to decide monthly capacity
- Using this monthly plan to decide the weekly assembly schedule (one week frozen)
- Using the weekly assembly schedule to decide the daily assembly schedule
- Using the weekly assembly schedule to synchronize subassemblies and internal suppliers
- Using the monthly plan to order from external suppliers

- From isolated islands and fast cycle time...
- ...to lower cycle time and high efficiency standard work:
  - Reduction of variability;
  - Small containers at hand reach and fixed location;
  - Better balancing;
  - Manpower flexibility;
  - Zero changeover;
  - Low-cost automation.

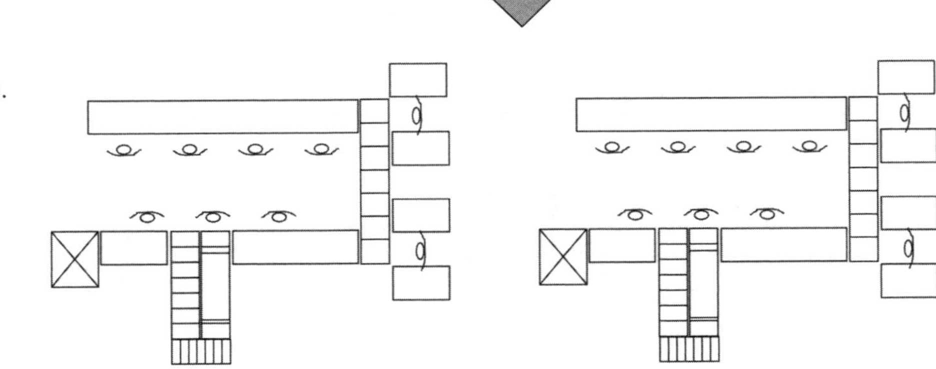


**Figure 11.5** Layout change based on TFM principles.

This was a typical "plan from plan from plan" MRP-based process. The name *master production schedule* (MPS) tells you everything, starting from the monthly demand forecasts. The first solution was deciding to use the forecast only to do the monthly capacity planning exercise.

The second was applying a pull planning algorithm on a daily basis to compare a replenishment level with the current stock of finished goods. If the actual stock is below the replenishment level, a replenishment order is generated. This was applied only to the DC in the country. Later, the same solution could be adapted to the foreign DCs.

Thus the planning system was divided in two blocks: business and capacity planning and order planning system. The order planning system is a daily vendor-managed inventory (VMI) system in the sense that Company A checks the inventory and customer orders every day and decides what to replenish. This is done for the country DC. Plans were made to later extend the VMI to the foreign DCs.

The new order planning process can be summarized as follows:

- Calculate replenishment needs every day.
- Maintain a production order list with the replenishment needs and the DC orders.
- Transform the production order list into *kanbans.*
- Assign *kanbans* to the production day using a logistic box.
- With this system, the source of the data is no longer forecasts but real pull orders. The daily assembly schedule then is decided by freezing one day of production on the logistic box. Every day, the *kanbans* are transferred to a leveling box. The rules for leveling are as follows:
    - Fill the day with the available *kanban* orders.
    - If the quantity is not enough to fill the contracted capacity (production-logistic contract), anticipate some MTO orders from the DCs abroad.
    - If the quantity is still not enough, make some MTS high runners up to a maximum stock level defined.
    - If the quantity is still not enough, stop the process (implies working less time).
    - If the quantity is too much for the day, postpone some MTO orders (if possible, depending on the final delivery date).
    - If the quantity is still too much, delay some MTS high runners down to a minimum stock level defined.

- If the quantity is still too much, increase capacity by working during weekends for the MTO orders.

This leveling represents a stable daily schedule for production, and the extreme solutions of having to stop the line or working overtime on the weekend rarely appeared.

Three types of internal logistic shuttle lines were established: one for the purchased parts, another for the subassemblies, and another for the finished goods and packaging. All of them handled both downstream replenishment (*kanban*) and sequential replenishment (*junjo*) parts. Regarding the *kanban* parts, the process is simpler and consists of just exchanging empty containers with full containers in the supermarkets.

Each material runner, called a "water spider" or *mizusumashi*, has supermarkets available in each supplier with an area for *kanban* parts (the high runners) and a lane for *junjo* parts (with four sequenced trolleys or containers). According to the information received, every cycle, the supervisor picks one sequenced trolley (or containers) and delivers to the line in sequence.

Figure 11.6 illustrates the future-state vision implemented during the first year of the project.

The productivity of the line started growing steadily and, after the initial training month, reached the target of a 27 percent increase. The company already had in place a daily *kaizen* meeting between workers and supervisors done at an information corner close to the line, and the workers could see the results of their efforts on a daily basis regarding output, productivity, quality, and compliance to schedule.

The logistics pull planning subproject consisted of changing the method of ordering from suppliers. The previous system issued weekly or monthly call-off orders together with a six-week forecast. The forecast information was increased to 8 weeks but stayed very similar to what was being done, so the suppliers were able to do their own capacity planning. The weekly and monthly call-off connected to the MPS was eliminated and transformed into a daily call-off based on the results of a pull planning algorithm. The parts inventory was checked on a daily basis, and if it was below the replenishment level, an order was generated, typically equal to the daily consumption. At the same time, a pilot local milk run was established with suppliers located fewer than six hours' travel distance from Company A. Most of the suppliers were already delivering something on a daily basis, so it was not difficult to ship daily orders on a daily basis.

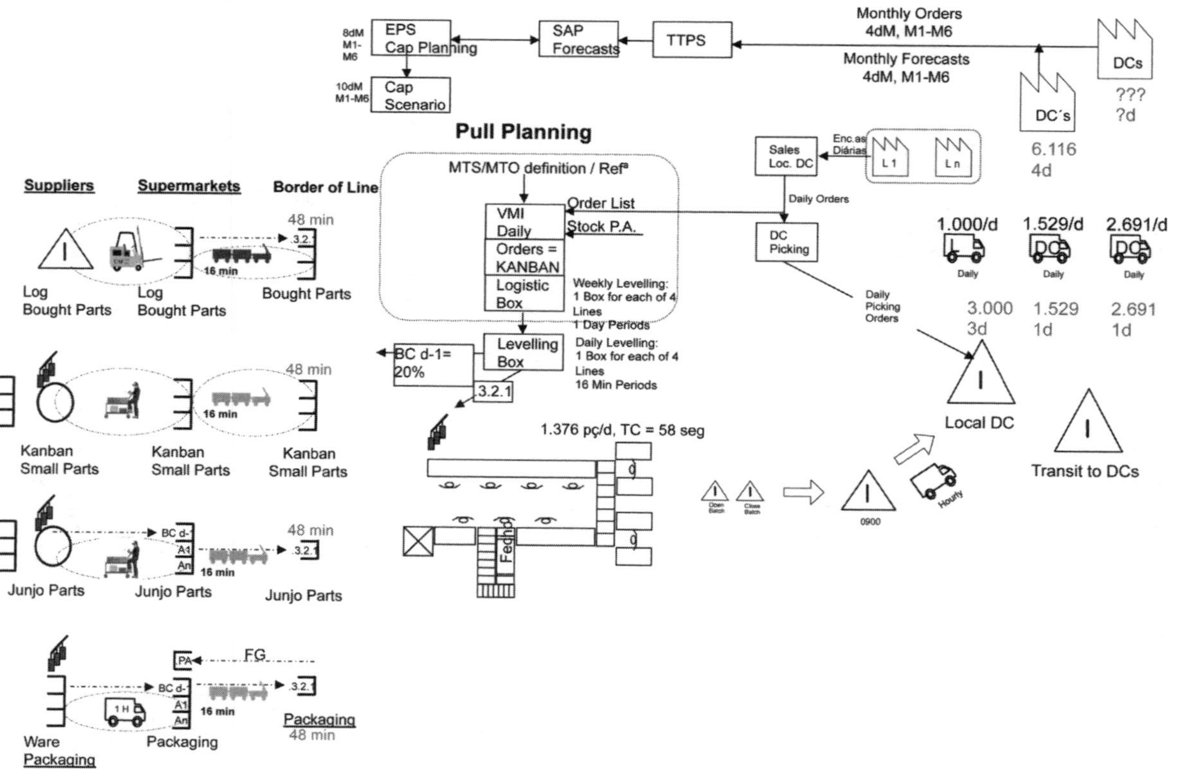


**Figure 11.6** TFM internal logistics replenishment.

In some suppliers located far away, the shipping frequency could go up to one week.

The supply-chain design work was done in January 2005, and this gave origin to an initial plan covering the year 2005. At the end of the year, the project was implemented on behalf of the assembly lines. In 2006, the TFM transformation was expanded to all the final assembly lines. The evolution of the main KPIs from the end of 2004 to the end of 2006 was as follows:

- Total inventory (i.e., raw materials, work-in-process, and finished goods): From 50 to 30 days
- Internal defects rate: From 12.000 to 5.750 ppm
- Customer-service level: From 93 to 98.5 percent
- Achievement of the assembly production schedule: From 50 to 92 percent
- Productivity: From 70 to 94.5 parts/operator
- Final assembly line efficiency: From 75 to 101 percent

By the end of 2005 (at the end of the first year of implementation), it became clear that new horizons could be seen for the supply chain based on the pull flow possibilities discovered. Company A finally understood a new operations system paradigm, and the preceding results were a big breakthrough from the stagnation observed from 2000 to 2004.

The project gave new breath to Company A's improvement strategy, and by the end of 2005, a new strategy was defined covering the next seven years until the end of 2012. This strategy was divided in the following components:

- *Pull make strategy.* Including all the production and internal logistics flow domain tools. The first two years of implementation focused a lot in the final assembly, and the system has to be extended to all the internal suppliers. The goal is to have all the logistic loops type 2 and type 3 (presented in Chapter 2) perfected with the TFM tools.
- *Pull deliver strategy.* Including all the external logistics flow domain tools applied to the delivery side of the supply chain. The first two years focused on the local DC. The goal is to have all foreign DCs managed by VMI. The plan also includes extending the model to the DCs and perfecting the logistic loop type 1 (from final customer order to final customer delivery and satisfaction). This means that all transport and the DCs operation will be improved.

- *Pull source strategy.* Including all the external logistics flow domain tools applied to the source side of the supply chain. The first two years focused on the bought parts for the final assembly lines and the suppliers involved. The plan includes extending the model to all suppliers and creating logistic cells in the warehouse for all parts.

Company A is now sure again that every year will be a better year in terms of *kaizen* results. The company is confident about the future because a clear model started to be deployed, and all the employees were fully aligned with this model. The *kaizen* pull flow system is no longer a theoretical model or something to be wished but is now made of practical *gemba* activities that make the company more competitive on a daily basis.

CHAPTER TWELVE

# Just-In-Time at Wiremold

This chapter is a case study showing how one company, Wiremold, tackled the problem of building in flexibility by introducing just-in-time (JIT). Wiremold is an approximately 100-year-old company in West Hartford, Connecticut. Its two major product lines are wire-management devices and power-conditioning products. Under the leadership of Art Byrne, who arrived as the new president in September 1991, the company started an all-out effort to introduce *gemba kaizen* with particular emphasis on JIT.

I had an opportunity to visit the company five months after the introduction of *gemba kaizen* and had lengthy discussions with Art Byrne and Frank Giannattasio, vice president of operations. During the first five months, the company had freed up approximately 40 percent of the floor space in one plant. The inventory level, including raw materials, work-in-process, and finished goods, was down more than 20 percent. The working capital turns had been increased by 30 percent, and the company was expecting a 25 to 30 percent productivity improvement by the end of the year. After five months, a lead time of approximately six weeks to make a product now took one week. Clearly, introducing *gemba kaizen* had brought about enormous improvements in a very short period of time. My conversations with Byrne and Giannattasio provided a vivid picture of JIT in action:

**Imai:** What kind of changes have taken place in the way you do your business since JIT was introduced?

**Giannattasio:** Wiremold previously operated on the premise of a forecast. We would generate a presumed consumption level of our various products, at which point we would schedule batches and produce several months' consumable volume of various products, and store them in a warehouse. Then they would be consumed over a period of time—some in line with our forecast, some at an amount in excess of our forecast, and some at a volume less than our forecast. The result was slow-moving, expensive inventories. To begin JIT manufacturing, we looked more at near-term consumption levels. Looking at the past six months and, more specifically, at the last 10 days, we used the information to continuously adjust our daily product-mix schedule and were able to run most of our products at least every week, many every day. We used *kaizen* to improve setup times and eliminate waste, which gave us the flexibility to achieve those very rigorous daily schedules. As we developed our daily product-mix schedules to meet customer demand as opposed to a forecast, we made several changes. First, after evaluating customer demand, we determined our *takt* time, or the demand frequency of each of our products. Based on that *takt* time, we then managed our production system cycles to meet *takt* time, whereas previously we would set up in batch production and build as much as we could of whatever product we had in production. We really did not have a measurement of how efficiently we were producing. Now, with daily demand known and established cycle time, we evaluate our production against the expectation of building a predetermined or expected quantity of each product. We gauge ourselves against our ability to produce the volume in the appropriate time based on our cycle time, and when we are unable to do that, we are very quick to identify what problems exist, either with our supplier's product quality or with downtime within our own production system. We are much quicker to identify problems that are affecting our output and, ultimately, our customer service.

**Imai:** What kind of role did you play as president in making such a change?

**Byrne:** I think the role of top management in making such a change is really to be the key driver for that change. I've often told companies that have asked me, "Gee, what should we do to start just-in-time?" First, they have to have total commitment from top management. If they don't, and they try to implement it from the bottom up, they're basically bound to fail. In our particular case, I was not only the driver and leader for change, but because I didn't have anybody else here when I came who really understood just-in-time, I also was the initial trainer. I wrote the manual and conducted the first training classes. I trained perhaps 200 people in two-day sessions. But that certainly gave people the clear impression that this is what we're going to do and that I was really very much behind it. It wasn't something that I tried to delegate to somebody else or hand off to a committee. I think the key is that senior management has to be leading it and has to be very involved, be willing to go out on the shop floor, be willing to work on *kaizen* teams themselves, and be willing to pick up a machine and move it. It really needs total commitment from the top or it'll never really happen.

**Imai:** What kind of organizational changes were needed to introduce JIT *gemba kaizen*?

**Byrne:** Well, before we started *kaizen*, we did do a number of organizational changes. First of all, we tried to flatten out the organization. We had a very traditional vertical organization that we flattened out by putting everybody on teams. Our old organization typically was based on a type of process, for example, stamping, painting, etc. We organized the new teams based around a product so that each team leader had all the resources needed to make the complete product, plan that product, and get that product to market. That was our first step. To support that activity, we've created a fairly extensive just-in-time promotion office. We have about six of our brighter, high-potential people in the just-in-time promotion office, and their job basically is to teach just-in-time, to train it, to follow up on the *kaizen* that we do, and to make sure that all the things that we said we were going to do get done. They also work with the team leaders on the shop floor and help to implement *kaizen*.

**Imai:** How do you relate the *kaizen* JIT approach to the conventional strategic planning?

**Byrne:** One thing that struck me in my previous role as a strategic planner was that many companies can put together really wonderful strategic plans, but they have an awful lot of difficulty in implementing those plans. The reason for this is that their fundamental delivery systems (the way they make a product, the way they design and introduce a product) don't function very well. Thus, even though the strategic plan may have been good, the time it took to develop the product, the time it took to make the product, and the quality of the product that came out was such that the plan's implementation never really worked. With just-in-time, it's really akin to building a house. If you're going to build a house, you want a really solid foundation. Just-in-time is the solid foundation that we want to put under Wiremold, where we can reduce our lead time from six weeks to six hours, where we have the flexibility to make every product every day, and do it with very little inventory. That's quite a task when you have over 1,400 products. If we can build that foundation, however, and then marry it with drastic reductions in the time it takes to design and introduce new products, then we're going to be able to service our customers much better than any of our competitors. We are going to be able to introduce products much faster, and I think the combination will give us strategic alternatives that we wouldn't have had if we'd started simply with strategy but without a solid foundation. Thus I see *kaizen* as being part of strategy. It's not just some manufacturing thing but is integral to your strategy.

**Imai:** Why is it that it is taking such a long time before American businesspeople begin to realize the challenge as well as the benefits of JIT/*kaizen*?

**Byrne:** U.S. management has for the most part neglected this and tried to solve problems with short-term things or strategic things, not really looking at the fundamentals of what makes the business go. As a result, we find ourselves way behind in a lot of areas. Despite this, it's still very difficult to get a lot of U.S. companies to change. Most feel that, well, we're already good enough, we're making money, we're holding our market share. All this may be true until somebody comes along who plays by a different set of rules.

The U.S. auto companies thought that it was doing a great job until all of a sudden the Japanese auto companies came along and played by a different set of rules, and the U.S. auto makers found themselves over a period of time way, way behind. Now they've caught up a lot in quality but still trail in other areas, such as productivity. The area that they're really way behind in is in the speed that it takes to develop new cars. If your competitor can develop a product in two years and it takes you six, you have to guess pretty accurately to even have a chance of being competitive by the time your new car comes out.

The auto industry tends to get a lot of publicity because of its size, but the same thing is true in most other industries in the United States. I think that Japan bashing is unfortunate because it tends to allow people to hide behind their current way of doing things and doesn't force them to change.

I think that about 90 percent of all U.S. businesspeople don't understand the depth of the disparity in the Japanese versus American manufacturing system. Almost all U.S. manufacturing managers that you talk with today will tell you that they are doing just-in-time, but most of them are way off track. Half of them think of just-in-time as some sort of inventory system that, in turn, means that what they should do is beat up their suppliers. Others have read enough books to understand that they should be creating a *flow* system. Very few understand how fundamental and detailed you have to get, and even fewer understand that it's really a people thing. You have to change people's attitudes, and that takes time, a lot of commitment, and a lot of education. We tend to be a country that is oriented toward making this quarter or making this month. This, in turn, makes it difficult for long-term change to occur. We talk about it, we publicize it, but when it comes right down to actually doing it, most companies balk.

**Imai:** How much money did you spend to achieve these results?

**Byrne:** We really didn't spend any money in terms of capital investment. Perhaps some small things here and there, but for the most part, the money that was spent was people-time—people-time invested in training and *kaizen* activities to change the factory around, change the flow, and improve the way we work.

**Imai:** Do you believe that JIT is a Japanese method?

**Byrne:** It's a fallacy to say that just-in-time comes from Japan. When you listen to someone like Taiichi Ohno, who is given credit for developing just-in-time, and ask him where he got the idea for just-in-time, he'll tell you he got it from the United States. He got it from two things: (1) Henry Ford and (2) the American supermarket. In fact, I think the better way of describing just-in-time is "just makes sense." And I think the simplicity of the system and the "just makes sense" aspects of it really can't have boundaries of Japan or the United States. It's a question of whether companies are willing to seize the opportunity and do something about it. It turned out that the Japanese, following Ohno's lead at Toyota, were the only ones that had the discipline required to stay with this program. It took Ohno about 30 years to develop a good just-in-time system. In the United States and perhaps in Europe, we tend not to have the patience to stay with something like this. So just-in-time, as far as I'm concerned, is not a Japanese thing; it's something that anybody can do without a lot of investment.

These conversations with Art Byrne and Frank Giannattasio took place in 1992. Wiremold has continued to work on JIT/*gemba kaizen* ever since. In a recent letter from Frank Giannattasio, he summarized the situation at Wiremold as follows:

> We at Wiremold have been operating our method of just-in-time manufacturing for four years. We call our just-in-time process the *Wiremold Production System*. We have reduced our delivery lead time of 67 percent in four years—from shipping our customer orders in 72 hours of receiving the order to shipping within 24 hours. At the same time we have increased our inventory turnover rate 367 percent—from 3 times per year to 14 times per year—while nearly doubling the size of our business. We have done this by practicing the three basic principles of our production system:
>
> - Build to *takt* time.
> - Incorporate one-piece flow.
> - Use a pull system.
>
> We implement the principles with a strategy of *kaizen* and teamwork. We try to sustain all improvements in standard work while evaluating our progress with various measurements.
>
> We have changed our production system from that of material requirements planning (MRP)–driven batch production to one of customer pull-driven one-piece flow in a series of phases:
>
> 1. *Phase 1* consisted of first reorganizing the factory both physically and organizationally. We organized our plant into business units based on product similarity. We then cleaned up our factory, following the guidance of the Japanese 5S model. This made waste very evident.
> 2. *Phase 2* focused on setup time reduction and disciplined adherence to production schedules. The schedules are now based on recent sales instead of MRP forecasts. Our efforts in this phase were to develop flexibility and responsiveness. We knew that we had to shorten the lead times with our production process and to improve operating dependability of those processes. The gains we made with each *kaizen* generally were to reduce the existing setup time by at least 50 percent and in many cases by as much as 90 percent. We repeated

> these gains with successive *kaizens*, improving our setup times from hours to minutes.
>
> 3. *Phase 3* capitalized on our quick response as a result of reduced setup times and improved process dependability by fully implementing our pull system. We now build today what was sold yesterday for all moderate- to high-volume products. We use some inventory cushion on the level of customer demand from day to day.
>
> These three phases, as conducted over the past four years, have resulted in 20 percent productivity gains each year, over a 40 percent reduction in defects each year (we now realize only 1.4 defects per 100 in-process audits), an increase in inventory turnover of 367 percent, and a 67 percent reduction in delivery lead times.
>
> We continue to execute a hundred or more *kaizen* exercises every year, and we try to involve 100 percent of the organization. Our next phase will focus on associate development and teamwork. We feel that our future gains will come from the creative problem-solving efforts of our associates. Our associates have to be able to identify waste even in what seems to be an acceptable operation by most traditional means of evaluation.

Frank Giannattasio also writes about Wiremold's activities in improving flexibility in the past four years as follows:

> Our firm has been involved with pursuing continuous improvement and just-in-time delivery now for about four years. We've experienced large degrees of success in many areas. One example of this success is the Plugmold area. In the Plugmold area, we make over 200 different end products.
>
> We originally produced all these 200 products in a batch mode, producing large quantities of each product at random and infrequent intervals. Our first approach to improving customer service and reducing inventory levels was to develop a daily product-mix schedule. To do this, we reviewed the daily consumption levels of all 200 products and developed a schedule that

systematically scheduled each product to be assembled. We then ran most of those 200 products each week, many of them every single day. As a result of this production mix, we were able to assign workers to specific products, and we uncovered a multitude of problems. We found a variety of reasons why assemblers could not build products at the rate we anticipated each day. Problems such as bad vendor parts, cracked receptacles, cut wires, and miscut or miscrimped wires that didn't assemble properly were seen, and the list went on and on. We found in excess of 60 problems on just one assembly operation. In our efforts to resolve these problems, it was important to have daily follow-up and intensive measurements. After only 20 days, we achieved our schedule on a daily basis in all operations. It was a culmination of efforts of many different team members.

One specific operation in the Plugmold area was a three-wire table assembly operation. This was an operation where four workers independently assemble a variety of three-wire products every day. As we embarked on a daily schedule on the table assembly operation, the associates targeted their average cycle time, or their best available cycle time, as the basis for our schedule. We found out during the process of assembling each part that different problems arose that made our assembly time vary. Some of these problems arose from the performance of a tool known as a *squasher*, which is a necessary instrument to assemble the receptacle front to the receptacle back that makes our Plugmold strip. We found that the squashers were constructed in a different fashion, one to another over all four tables, and they gave varying results, some good and some less good. We first moved to make all squashers identical, in which case we were able to, on a daily basis, repeat the best available performance.

Another problem that we found was various centerline loops on our grounding wire. This is a long, stiff wire with crimps that was a necessary component of our Plugmold strip. We found that if the centerline was not exactly the same for each assembled component, extra assembly time was required to customize the fit to the receptacle. We improved the centerline crimping device so that we were consistent from wire to wire.

Another problem was our vendor cord quality. We had problems with dirty cords directly out of the box that required assembler attention to inspect and clean. We worked with our vendor to improve its quality and began to get clean wires, requiring no more inspection time or operator rework.

Another example of a problem was varying lengths of Plugmold strip or metal base. We went back and reviewed our specification, and we were able to develop standards and train our operators at the mill to sort these parts within the limit samples and, as a result, deliver no variability to the assembly operators.

Again, eliminating unnecessary inspection time and sorting at the assembly, all contributing to our efforts to produce a daily schedule. Now, four years later, following numerous projects, *kaizen* activities, and short-duration *kaizen* events, we have increased our volume on these assembly tables 67 percent. We can make every product every day, taking no time to change over the table from part to part. We have increased our turnover on this product line from 3 to 14 times per year, and we rarely identify any poor quality issues.

Byrne also had his side of the story of Wiremold's achievements, which were published in an article entitled, "How Wiremold Reinvested Itself with *Kaizen*" [*Target Magazine* 2(1), January–February 1995]*:

At the old Wiremold, a product might take as much as six weeks to make its way from raw material to finished product. We'd make huge quantities of a single component because our changeover took so long. Often a batch of components would sit gathering dust in our large work-in-process inventory areas before products could be assembled because the other parts weren't scheduled to be run that week. Finished goods were sent to our 70,000-square-foot warehouse down the road to wait until needed for shipping to a customer. We were cash poor, yet had such large in-process and finished goods inventories that we were shopping for more warehouse space.

---

*This excerpt is printed with permission from the Association for Manufacturing Excellence, 380 West Palatine Road, Wheeling, IL, USA.

We've come a long way since then, reinventing ourselves into a vibrant, growing firm. In just three years, our sales have doubled and our profits have tripled. We've grown our base business by more than 50 percent and supplemented that internally generated growth with six acquisitions—five of which we were able to make without borrowing because we had freed up so much cash from inventory reductions.

Our success is not the result of any complex business strategy. Nor is it the fruit of some intensive program of capital investment. Rather, we turned our company around by turning our manufacturing operation on its head: We adopted *kaizen*.

We began to implement our *kaizen* program of continuous improvement in late 1991. In slightly less than three years, here are some of the changes we've made:

- Productivity has improved 20 percent in each of the last two years.
- Throughput time on products has dropped from four to six weeks to two days or less.
- The defect rate on our products fell by 42 percent in 1993 and by 50 percent in the first half of 1994.
- Inventories have been slashed by 80 percent, resulting in our space needs being cut in half.
- Profit sharing layout for our employees has more than tripled.
- Equipment changeovers have been reduced dramatically—in some cases from as much as 10 hours to less than 10 minutes.
- New-product development time has been slashed from almost three years to under six months.
- Vendors have been cut from more than 400 to fewer than 100.

One ingredient essential to our success has been the way we look at *kaizen*. At Wiremold, we believe it's a fundamental part of our business strategy. After all, our business delivery systems are what the customer sees.

If we fall behind in quality or on lead times, we disappoint our customer, and we won't succeed no matter how good our strategy. On the other hand, if our systems can outperform the competition, then we can outrun them.

We've actually made *kaizen* part of our business strategy—to continually "fix" our base business. We believe that the minute we stop doing so, we'll fall behind.

In summary, here are three "tips" that anyone getting started needs to keep in mind:

1. Changing people's mind-set is a critical part of the job. People are naturally skeptical, and you have to take dramatic and sustained action to overcome objections. In the long run, you must change the culture of the organization. The "concrete heads" must go.
2. Senior management must lead the "change." That means not only at the beginning but throughout, continually putting pressure on the organization. Lack of leadership attention is one of the major reasons that improvement programs die within a year to 18 months.
3. This is a long-term commitment. You have to acknowledge upfront that there's no end point. Be prepared for your people to ask, "Are we finished yet?" And be equally prepared to answer, "It's not good yet," even when you think it is.

## CHAPTER THIRTEEN

# The CEO's Role in *Kaizen*

My next conversation with Art Byrne took place in October 2011 in Tokyo. Byrne had long since retired from Wiremold but had, in his words, "failed retirement" and joined Boston-based J.W. Childs Associates, LP, a private equity firm, as an operating partner. The firm buys companies and subsequently sells them at a profit. Byrne's role is to improve their operations to increase their value, and *kaizen* is his secret weapon.

Byrne drives change from the top as he did at Wiremold, sitting as chairman of the board of some of the acquired companies and making *kaizen* an absolute requirement. This role has allowed him to build a unique track record in using *kaizen* to build stakeholder value in successive companies. It also has given him a very special understanding of the role of top management in introducing *kaizen.*

Byrne was in Japan leading a tour of Japanese factories for his portfolio managers so that they could see some good examples of *kaizen.* This kind of benchmarking tour to Japan is very popular, and at the Kaizen Institute, we have offered this service to our clients for many years. Byrne has a house in Tokyo, and I had the pleasure of visiting him there to have this conversation.

I wanted to know how things had progressed at Wiremold after our last conversation. Byrne gave me a summary of overall improvements from the time he came in as CEO in September 1991 to the sale of the company in July 2000.

- Lead time dropped from four to six weeks down to one to two days.
- Productivity improved by approximately 162 percent.
- Gross profit went from 38 to 51 percent.
- Machines changeovers went from three times a week to 20 to 30 times a day.

- Inventory turns improved from 3 to 18 times.
- Customer service improved from 50 to 98 percent.
- Sales grew from $100 million to more than $400 million.
- Working capital as a percentage of sales fell from 21.8 to 6.7 percent.
- Operating income improved by 14.4 times.
- Enterprise value increased from $30 million to $770 million.

**Imai:** How would you explain the increase in sales volume?

**Byrne:** During the period I was at Wiremold, we acquired about 21 different companies. And about half our sales growth came through acquisitions and about half through internal growth.

**Imai:** Did you use *kaizen* in the companies you acquired?

**Byrne:** Every company that we acquired—a lot of them were pretty small companies by the way, and some of them were just product lines—every one of them we did *kaizen* right away. The first week that we got there we started doing *kaizen*. We wanted to make sure that the people in the new companies that we bought understood that our whole culture was all about removing waste through *kaizen*, and they were going to be involved in it whether they liked it or not. It was really helpful to do it that way. There was no room for gray areas that might make the new associates nervous.

**Imai:** As a top manager, what was your role with *kaizen*?

**Byrne:** I led all the first *kaizen* sessions. Every time we bought one of those companies, we had a standard procedure, which was that I would go to the company as soon as possible after we bought it. I would bring all our new associates together, I would introduce everybody to Wiremold, I would give them a couple hours of lean training, and then I would start the first *kaizen* after lunch.

So people were kind of shocked by this. But, if you get the CEO of your new parent company down on the shop floor for the rest of the week doing *kaizen*, helping you to move the equipment around, and improving things, it leaves quite an impression on most people.

Most companies will buy something and then kind of analyze it for six months before they decide what they want to do. The way we looked at it, using lean and *kaizen*, we could come in the first day and start doing things. We already knew what we wanted to do. That would get us tremendous gains, and it allowed us to get our money back very, very quickly from an acquisition. In fact, in many cases, after three years, it had already paid back all the money we had paid for it. From then on it could generate additional cash flow that we could use to buy the next company. The process just repeated sort of like "the gift that keeps on giving."

**Imai:** That sounds like good preparation for the work you do now with J.W. Childs. Could you describe your role as an operating partner?

**Byrne:** What this means is that I stick to my area of expertise, which is manufacturing. When we are looking at a new company to buy with this type of characteristic, I get involved with due diligence, go visit the company, go visit the factory, and decide whether it makes sense to us and how much we're willing to pay for it. And then, if we are successful in buying it, I become the chairman of the company.

Then my role really is to represent Childs and to work with the management teams to improve the company. We only buy these companies to sell them—we don't hold onto them—so we're looking for some pretty big improvements in a fairly short period of time.

**Imai:** Can you give an example?

**Byrne:** One of the earlier companies that I was involved in was called American Safety Razor. When I got involved, the company wasn't doing *kaizen* at all and didn't really have any interest in it. I really had to sort of force it from the board of directors to get the company to do it, and I was able to get a good lean guy in to be the head of operations.

The combination of the board pushing and the lean operating guy in the business allowed me to make a lot of good progress in that company over the course of about three years—the three years where I was involved. We generated something like $65 million in cash just from freeing up working capital, we increased the gross margin by six points, we increased our on-time deliveries and service rate from the low 80s to 99 percent, and we had everything on time by the end. We gained a lot of productivity. I can't remember all the exact numbers, but when we sold the business, we achieved a return of 3.5 times our investment.

**Imai:** You've repeatedly been able to show that *kaizen* can produce financial results. Has anybody tried to follow your example?

**Byrne:** At Wiremold, a lot of people had heard of us from articles and books, including your *Gemba Kaizen*, and a lot were saying, "We want to do this." So we started hosting plant tours. We thought we should help those people because a lot of how we learned these things was going around with Shingijutsu looking at what other people were doing.

The tours were starting to take a lot of time—we still needed to run the company and make a lot of improvements ourselves. We wanted to help, but we also knew that unless the CEO is involved, no company is going to go home and do it. So I said, "Let's make a simple rule. You can come and visit us, but only if you bring your CEO. If you don't bring your CEO, you can't come to Wiremold." All the tours stopped, right away. It's true—that's exactly what happened.

**Imai:** This is the same problem we have been seeing—top management never seems ready to implement *kaizen*, even when the benefits are well known. Why won't they do it?

**Byrne:** Well, lean is something that's really easy to understand conceptually but hard to do. Most CEOs don't understand the basics, number one. In order to do this, you have to make some pretty radical changes in the way you do things. All their internal people are going to argue against them.

Let's say that you're the CEO and you have a staff of eight, and one day you say, "All right, we're going to change from doing everything in large batches at high speed to making it a one-piece flow. We're going to get 15 to 20 percent annual productivity gains, we are going to free up a lot of space, we're going to cut a lot of inventory, and we are going to drastically shorten our lead time and gain market share."

This may sound great, but it will be very threatening and scary for your staff. For example, they have inventory for a reason, because without it, they can't serve the customer very well. They have been doing all these things a certain way for many years, so they're convinced that it's the only way to do it. And they are convinced that if they drop the inventory, they are going to have a big problem, so they are going to fight this. The manufacturing person is going to tell the CEO that it can't be done. The sales and marketing person doesn't want to hear about cutting the inventory. In fact, he has always been arguing for even more inventory. The finance person is running a standard cost accounting system, and the idea of not making big batches to make the absorption hours every month is horrifying to her.

Even when you show them an example, they won't believe it. Let's say that you're the CEO of some company, and you say to your manufacturing people, "You know, I was just over at Art's factory and saw them change over the same machines that we have in one minute, so I need you to do the same." They'll think that you're nuts, but then they're going to tell you all the reasons why your factory can't do that. Even though you're the CEO, you won't win that discussion because now you're telling the people who do the work to do something that, after 20 years of experience, they believe is impossible. The only alternative you have is to show them how to do it. Unfortunately, most CEOs don't know how to do this, so they back down, and the setup stays at two hours.

**Imai:** A lot of companies say that they are lean, but they haven't embraced lean. The people at General Motors like to say that they are lean and mean. I say that they are mean, but they're not lean.

**Byrne:** I believe that when General Motors and Chrysler went through their restructuring, they did only financial restructuring but didn't do much to change their operations. They kept the good parts and got rid of the bad parts, but they haven't done much to the good parts to change the way that work gets done so that waste is eliminated. Their mentality and their focus are much more along the lines of, "What's the cost of labor?" "Can I get an agreement with the union to give me a two-tier wage system?" "Can I take the work offshore?" and other things like that.

The company has a new CEO, but otherwise, a lot of the management is the same. It's the same mentality, and it's the same way of looking at things, so I don't understand how the company all of a sudden can be terrific.

Perhaps one of the biggest difficulties for any organization trying to embrace lean has been a lack of objective criteria of what to measure in order to improve their chances of getting better in the future. You can't just manage the past; that already happened. Manage the things that will make you better next quarter and next year. If you can compare what Toyota is doing against what General Motors is doing, it's very clear. Toyota is all about removing waste and creating a *kaizen* culture, whereas General Motors seems more focused on financial engineering and appeasing its unions. Another reason is top management commitment. How many *kaizens*, for example, do you think the CEO of General Motors has ever been on? For a company to be successful and more profitable, implementing lean should be the number one priority for the shareholders. By simply ignoring this tremendous opportunity to make the company far more profitable, top management has failed.

**Imai:** When the twentieth-year anniversary of the joint venture between Toyota and GM was held at NUUMI in California, a senior GM executive was invited to go to the shop floors after the party but he gently turned down the offer because he had no time. This happened even after the plant had gained the highest reputation among GM plants from the previous dubious ones.

**Byrne:** I have a theory: Most top managers, as they are promoted up the ladder, become more and more dumb. Why does that occur? Organizations are a pyramid. The value-added work is at the bottom, so the further you remove somebody from the value-added work, the less they know what's going on, because they're getting further away from what it is the business really does. And then the people down below try to protect them from bad news, so the bad news doesn't go up to the top very often, only the good news. And now the CEO is saying, "Well, everything is okay."

I was very interested in Byrne's comment that companies lack objective criteria for embracing lean and asked him to name the criteria that he used as a CEO. He explained that while these obviously would be different in other industries such as banking or health care, the criteria he used at Wiremold should work for any manufacturing company.

Aside from meeting legal reporting requirements, he explained, the company was essentially run according to five lean measurements. These were all forward-looking measures aimed at improving where the company was going as opposed to the more traditional approach of spending a lot of time reviewing where they had been. Most companies try and "manage the results." Thus they spend a lot of time looking backwards. Wiremold never wasted any time on monthly financial reviews. The business was divided into six value streams, each representing a family of products and headed by a value-stream manager.

Byrne and his staff met weekly with the value-stream managers, and each reported his or her progress for each of the five key measurement categories. Emphasis was on keeping the discussion around issues that were immediately relevant to the tasks ahead—dwelling on past issues was considered a waste of time. In fact, the meeting room had a large digital clock, and each presentation was limited to 10 minutes.

Here are the five measurement criteria that they used:

*Customer service percentage.* This was the on-time percentage based on the time period between the receipt of an order and shipment to the customer. For stock orders, the benchmark was 24 hours. For custom orders, it was based on the time frame that the customer had agreed to at the time of the order. If an item was routed through a central warehouse, the value-stream manager was responsible for customer service up to final shipment to the customer.

*Productivity.* Byrne prefers to use units per worker-hour, but because of the large product mix at Wiremold, this was impractical, and instead, sales per employee was used. The target was aggressive—an improvement of 20 percent per year was expected from each value-stream manager.

*Quality.* Defects were defined across the whole company, and once again, the target was aggressive—a reduction of 50 percent in the number of defects each year. The strategy was to always be working on the top five defects at any given time. When one of these was solved, the company would move down the list. Byrne points out that the companies usually fell short of the 50 percent, but their quality performance, a reduction of about 40 percent each year, far exceeded that of companies with modest targets.

*Inventory turns.* This was a long-term target—improve inventory turns from 3 to 20 times—that many managers initially believed was impossible. As noted earlier, this increased to 18 times during Byrne's tenure.

*5S and visual control.* A 5S inspection team would conduct biweekly visits to all areas, including the offices, and report the results to all employees. Winning teams received a congratulatory banner and free coffee and doughnuts for a week. Visual control on the shop floor was constant, and quality and productivity charts were publicly displayed in prominent locations.

Byrne summed up his comments as follows:

**Byrne:** So those were the five measurements. If you just use those measurements as your lean measurements, and all you're doing is focusing on those all the time, and improving those, and getting to those stretch targets, you have to get better. If somebody says, "Well, five is too many things to measure. I need less," I would suggest that you focus on two measurements, customer service percentage and inventory turns. If those two things are both going up at the same time, everything else will fall in line. Your productivity will be better, your quality will be better, and your cleanliness and everything else will have to be better in order to improve those two measures simultaneously.

**Imai:** How would you adapt these criteria to a large player such as General Motors?

**Byrne:** I wouldn't change those measurements much at all. In General Motors' case, the company can measure productivity as vehicles per worker-hour much better than we could with Wiremold, say. So I think that the same measurement should apply to General Motors. There's no difference in the measurement—the company is making something, so I don't see much that needs to change as opposed to if you were running a life insurance company or a hospital. In those cases, the measures would have to change slightly.

The big thing for General Motors is that the company has to manage to get the union to go along to do this stuff. Management and the union are so far apart most of the time that getting this to happen is very difficult. And yet it's interesting, at Wiremold, we had a union, and my first real *kaizen* with Shingijutsu was at Jacob's Engine Brake, and we had the United Auto Workers (UAW) there. We just said, "Look, we're just going to treat the union like they're people; we're not going to treat them like they're the union." And I think that's the big mistake; the automotive people seem to treat their union employees like they're different creatures altogether. Both parties are at fault. The union's pretty tough there, but management is also pretty weak. So I think you need to treat the union like they're people and get the whole company to understand that everybody has a stake in this. That really helps. I think those five measurements can apply to any manufacturing company very well, and if you really focus on them and you set aggressive targets, guess what, you're going to get better.

Art Byrne's repeated success with *kaizen* in different companies confirms my belief that with the right approach from top management, *kaizen* is a reliable system for improving the overall performance and value of a company. Art's experience also suggests some guidelines for CEOs following a similar path. I would summarize these as follows: To ensure success with *kaizen*, a CEO must

- ▲ Make lean the company strategy.
- ▲ Be the hands-on leader and lean "zealot."
- ▲ Make it clear that all employees must adopt the *kaizen* culture.
- ▲ Act quickly and decisively to get improvement initiatives under way immediately.
- ▲ Personally go to the *gemba* and show employees at all levels what has to be done.
- ▲ Adopt measurements based on stretch lean targets, even if they are not achievable in the short term.
- ▲ Be ready to change everything a company has done in the past, even if there is considerable resistance and many reasons are given why this can't work "in our business."
- ▲ Avoid measurement systems that overanalyze the past, and instead, focus on simple, objective criteria that define a path toward better performance.
- ▲ Engage the best lean expertise available, and continually learn from others.

Beyond anything else, CEOs should make companies better at what they do. The value adding occurs at the *gemba*, not in the CEO's office. Anyone in the company who is not actively trying to improve the way value is added is waste. One can only hope that more leaders will be inspired to act accordingly.

CHAPTER FOURTEEN

# Going to the *Gemba*

## *Gemba Kaizen* and Overall Corporate *Kaizen*

The expression, "Necessity is the mother of invention," captures the *kaizen* spirit very well. This means that challenging situations help to inspire innovative solutions. Yet, when we speak of the need for dissatisfaction and the need to go to the *gemba* and do *kaizen* with the attitude that "it's never good enough," I am told that this is not appealing to Western culture. However, in my experience, this "necessity" as the mother is a universal notion. You must go see to be dissatisfied, and the dissatisfaction must be sufficiently strong to provoke action.

When you go to the *gemba* and observe the way your people work, the way materials are moved, and the way equipment is laid out, do you take everything for granted and accept it as satisfactory, or do you regard what you see as a starting point for *kaizen* and continuously look for opportunities to improve? Some Japanese managers go so far as to say to their subordinates, "Regard whatever you do now as the 'worst' way to do your job!" Your attitude could make a big difference in the *gemba* over the years.

The worst thing a leader can do is live in a world isolated from the *gemba*, making all decisions from a comfortable office. Even managers who *do* visit the *gemba*, though, cannot make improvements if they fail to see problems. A manager's ability to recognize problems brings success in *gemba kaizen*. What does "going to the *gemba*" really mean? Many people think that they know the *gemba* because they work there. But being physically present in the *gemba* is not the same as knowing the *gemba*.

Akio Takahashi, author of a Japanese book whose English title would be *Line Kaizen Textbook*, has been engaged for many years in assisting Nissan's major suppliers in *gemba kaizen*. He says that simply going and looking at the *gemba* is not enough. Truly knowing the *gemba* means expressing oneself in terms of actual nouns and numbers. Dissatisfaction with the status quo must be expressed in terms of a *problem statement.*

Statements along the lines of "This plant is not operating well enough" do not lead anywhere. A manager should say, "The operating ratio of line A is 65 percent, but it must be brought up to 85 percent." In Takahashi's view, expressing ourselves in specific nouns and numbers enables us to reach common ground for discussions, makes it easier to solve problems, and helps to fix in our minds an accurate picture of the *gemba.* It also allows everybody to engage in a *kaizen* project with a common purpose. After nouns and numbers come questions such as who, how, and when. Once a target has been agreed on, the person in charge of the *kaizen* project—namely, a person who has a stake in solving the problem—must be designated, the solution to the problem must be determined, and the deadline of the project must be specified.

Actually, the ability to identify problems in the *gemba* requires no sophisticated technology. To start, the manager must understand some fundamental *gemba kaizen* principles such as *muda*, 5S housekeeping, visual management, standardization, and the plan-do-check-act (PDCA) cycle of experimentation and learning.

*Gemba kaizen* means going to the *gemba*—observing, identifying, and solving any problems right on the spot in real time. Not long ago it was mostly Japanese manufacturers who practiced *gemba kaizen* to excel at delivering products of good quality at reasonable prices under favorable terms. Today, companies worldwide compete by using *gemba kaizen* practices to improve continuously.

After Toyota Motor Company had developed just-in-time (JIT) practices within its premises, Taiichi Ohno extended those practices to Toyota's primary suppliers. Ohno organized the autonomous study (*jishuken*) group as a vehicle to spread his philosophy. The group consisted of several employees of the company's suppliers. One of Ohno's "disciples" at Toyota who had been implementing JIT practices led the group. Each month, the group visited a *gemba* of a different supplier and conducted *gemba kaizen* there for three or four days. As members of Taiichi Ohno's autonomous study group began teaching the Toyota Production System internationally, the week-long *jishuken* became the model for the so-called *kaizen* event that has become widely popular in the West.

The *jishuken* activities invariably improved the productivity of the targeted process, cut inventory, and shortened lead time. Layout changes, such as eliminating conveyers and forming U-shaped lines, often took place as well. *Jishuken* proved to be such an effective way of spreading the Toyota Production System know-how and practices among its suppliers that the primary suppliers soon began involving their second-tier suppliers in the activities as well. Even to this day, *jishuken* activities take place regularly within Toyota and its group of companies.

## Two-Day *Kaizen*

In 1977, Nissan Motor Company and its suppliers introduced a process called *two-day kaizen*, in which a particular production line is targeted for improvements that must be completed within two days. Two-day *gemba kaizen* starts with a clear objective. For instance, a plant manager expects a 20 percent increase in demand next month, but he wishes to achieve the necessary 20 percent productivity increase with just one hour of overtime per day per employee. He discusses the subject with the line managers and agrees to conduct a two-day *gemba kaizen* on line A, a bottleneck process, and to improve the layout and jigs of line A as a means of achieving the target.

Thus two-day *gemba kaizen* starts with a clear target. Sometimes, depending on the circumstances, the project may take three days instead of two. In order not to stop the line, the layout changes usually take place during the night between shifts. Two-day *gemba kaizen* usually involves key players in the plant: the plant manager, line managers, supervisors, staff, team leaders, and operators.

A typical two-day *kaizen* project is carried out in the following manner: By the time the team members arrive at the *gemba*, they have had several meetings to study how to approach *kaizen* in that *gemba*, so they begin the morning session by explaining to operators what is going to happen. Then the team members take about an hour to observe and make notes on the operations. Afterward, they meet to discuss their observations, come up with *kaizen* ideas, and devise ways of implementing them. They record on designated sheets the data they have gathered and work out *kaizen* plans for each process of the line.

During the discussions, the team members go back to the *gemba* whenever they need to confirm something. The team leader must select from among several *kaizen* plans the items to implement the next day; this decision must be reached before 4 p.m. on the first day. Once the decision has been made, the team holds another meeting with line operators and explains the schedule for the next day. Another purpose of this meeting is to encourage operators to speak up about any difficulties they encounter in their work. Based on such input from the operators, the team finalizes the *kaizen* plans for implementation the next day. Then the team works with the maintenance people to explain the kind of tools, jigs, and equipment repairs that will be needed.

Since *gemba kaizen* necessitates equipment change, maintenance people and/or personnel capable of making the necessary jigs and tools stand ready to assist during the two-day project. This session will be finished by 6 p.m. Based on the instructions, new jigs and tools are prepared, brought to the *gemba*, and installed on the line. This phase usually lasts until 10 p.m. or even midnight. After installing the devices, the *kaizen* team and the supervisor start the line, process the work-pieces, and confirm any difficulties, such as operational or quality-related problems. Only after confirming that the line is functioning properly do the project team members go home.

Work on the second day starts half an hour earlier than usual. The *kaizen* team explains to the operators the changes on the line and the new work procedures. For instance, the team leader may say, "Up to now, six people have been working on this process, but we made changes so that the same work now can be done by five people. So, may I ask Mr.___ to stand back and watch while the other five people do the work?" The operation gets started at 8 a.m. as usual. Since the operators need coaching, *kaizen* team members stay with them until 10 a.m. to allow them time to get used to the new procedure. Between 10 a.m. and noon, the operators continue work on their own, and the team members make up a list of all the problems encountered during this period. If the tools and jigs need further adjustments, they are sent back for modification before noon.

As soon as any necessary modifications have been completed, the operators start working on the line, and the team observes and measures the effects of the *kaizen* project. The team prepares the summary of the two-day activities by 4 p.m., at which time it begins the wrap-up session.

Sometimes several teams may be involved in the activities; in this case, teams compete with one another at the wrap-up session. Often, the senior managers from the plant as well as those from the corporate office attend the wrap-up session. The session closes at 5 p.m., completing the two-day *gemba kaizen.*

It often happens that team members have no time to sleep during the first night, particularly when the line has to undergo substantial changes. Much can be achieved in such a two-day workshop because the people involved use various worksheets during the two days and have prepared themselves by attending several study meetings beforehand. Even after the session, many activities must follow, such as confirmation of the effects, revision of work standards, and sometimes revision of engineering rules and standards.

According to Takahashi, the following six items will help to achieve the target more easily during the two-day session:

1. Build a line that can produce according to the *takt* time.
2. Build a line flexible enough to meet deviations from the *takt* time.
3. Thoroughly eliminate *muri* (strain), *muda* (waste), and *mura* (variation) in operations.
4. Eliminate factors that disrupt a smooth rhythm of operations.
5. Develop work procedures that can be written into standardized work.
6. Minimize the number of operators on the line.

Takahashi suggests that the work standard (item 5 above) should include the following:

A. Conditions of work:
    1. How to place parts and jigs
    2. Where to place parts and jigs
B. Handling of parts and jigs:
    1. How to hold parts and jigs
    2. Locating where operator holds parts and jigs
    3. Body parts to be used
C. Combination of motion:
    1. Sequence of work
    2. Routing of work

Other key items to be included are safety considerations, inspection, cycle time, and standard work-in-process.

## Checklists as a *Kaizen* Tool

As a tool for carrying out *gemba kaizen*, Nissan has developed detailed checklist sheets for use during projects. For instance, when the team members observe the operator's movement, they use a checklist of economy of motion that includes such points as the following:

A. Eliminate unnecessary movement:
   1. Can we eliminate the movement involved in looking for or selecting something?
   2. Can we eliminate the need for judgments and extra attention?
   3. Can we eliminate transferring the work-piece from one hand to another (e.g., picking up a work-piece with the right hand and then transferring it to the left hand)?

B. Reduce eye movement:
   1. Can we confirm what we need to know by listening instead of looking?
   2. Can we use lamps?
   3. Can we place items within the relevant operator's field of vision?
   4. Can we use different coloring?
   5. Can we use transparent containers?

C. Combine operations:
   1. Can we process while carrying the work-piece?
   2. Can we inspect while carrying the work-piece?

D. Improve the workplace:
   1. Can we place materials and tools in a given area in front of the operator?
   2. Can we place materials and tools in the same sequence as the work?

E. Improve tools, jigs, and machines:
   1. Can we use containers that are easier to pick parts from?
   2. Can we combine two or more tools into one?
   3. Can we replace levers and handles with a button to operate a machine in one motion?

In addition to its checklist for economy of motion, Nissan provides guidelines for two-day *gemba kaizen* activities. The guidelines include the aims of the project, the schedule, and the major activities. The guidelines for major activities cover the following:

1. How to set the target
2. How to select the leader
3. How to check the line in question beforehand
4. How to confirm the inventory
5. How to explain the purpose of the project
6. What tools are to be prepared
7. How to select *kaizen* plans
8. How to instruct the operators
9. How to prepare standards
10. How to prepare the summary report

Specific individuals and departments are put in charge of each item on the list and given checklists to follow. For item 3—how to check the line in question beforehand—for instance, the following factors are included:

A. Person/people in charge
B. Items to be checked, as follows:
    1. Name of the line
    2. Product type
    3. Volume of production during this month
    4. Hourly production volume (for one week) (This item is particularly important for purposes of confirming the effects of *kaizen* and follow-up activities.)
    5. Number of operators on the line
    6. Does the line have a second shift?
    7. Percentage of overtime
    8. Rate of operation (previous month's record)
    9. Failure rate
    10. Required *takt* time
    11. Layout

## *Gemba Kaizen* Workshops

The Kaizen Institute has been conducting *gemba kaizen* sessions worldwide for the past 27 years in nearly every size and type of organization. The format and duration of these *kaizen* workshops have been adapted to serve a wide range of customers. We have found a few keys to successful *gemba kaizen* workshops:

- Set challenging but well-defined targets.
- Form cross-functional teams to solve problems.
- Take action with speed, on-the-spot at the *gemba*. The 60 percent solution right now is better than 100 percent next week.
- Invest time in preparation, communication, and planning the workshop.
- Make learning and skill transfer a *kaizen* objective, not just QCD results.

Management should plan on a long-term schedule of education and *gemba kaizen* workshops, often covering a period of several years. At the Kaizen Institute, such consultations typically begin with a two-day lecture on *kaizen* basics to all managers, including top management, followed by a best-practices benchmarking trip to Japan or nearby excellent companies. Then *gemba kaizen* activities are planned and conducted at one of the client's *gembas*. The type of consultation selected depends on the requirements and objectives. In many cases it is best to select a *gemba kaizen* target area that will demonstrate a dramatic change within a few days, creating excitement and belief in *kaizen*. It is also important that the initial *gemba kaizen* areas show management how much room there is for improvement.

Often, different locations within the same *gemba* are targeted for different kinds of *kaizen* efforts; the *gemba kaizen* sessions are held repeatedly to transfer know-how to the client's management. Engaging in *gemba kaizen* also identifies cross-functional (interdepartmental) problems in the company. For instance, *gemba kaizen* often shows that customers' quality requirements are not being properly communicated to the *gemba* by the sales department because there is no formal communication channel between the sales staff and the *gemba*. Identifying such inadequate internal procedures makes it possible for top management to address these problems and build better internal systems. For this reason, it is important that *gemba kaizen* workshops be planned across a series of linked processes rather than as unconnected islands; otherwise, the overall flow of material and information across the enterprise will not improve as quickly.

Another common risk of companies starting out with *gemba kaizen* that must be avoided is to do only *gemba kaizen* workshops without establishing the people foundation of the house of *gemba* (see Chapters 7, 9, 10, and 13). For the results-driven manager, it can be exciting to plan many *gemba kaizen* workshops and plan ahead for the many wonderful savings, but without the foundation, there is no long-term sustainability of those results.

Table 14.1 shows an average of improvements by type among Kaizen Institute clients that have engaged in a sustained and balanced program of *gemba kaizen* and development of human capability.

One of the reasons the Kaizen Institute starts with *gemba kaizen* is that it helps to identify many inadequate upstream management systems in the company. The *gemba* is like a mirror that reflects the real capabilities of the company: The problems encountered in the *gemba* are often the result of poor support by various departments. Some examples include the following:

A. Engineering Department:
   1. Poor layout design
   2. Inadequate equipment
   3. Inadequate preparation for production

B. Inspection and Quality Department:
   1. Not enough failure mode and effects analysis (FMEA) studies before production
   2. Insufficiently detailed analysis of rejects
   3. Poorly prepared inspection criteria
   4. Lack of feedback

**Table 14.1** Average of improvements by type.

| | |
|---|---|
| Setup time | –66.4% |
| Lead time | –55.7% |
| Cycle time | –17.9% |
| Downtime | –52.1% |
| Operators required | –32.0% |
| Work-in-process | –59.3% |
| Finished goods inventory | –43.5% |
| Distance-traveled/part | –54.1% |
| Floor space | –29.4% |
| Parts required/unit | –57.0% |
| Cost quality rejects | –95.0% |
| Rework | –71.7% |
| Scrap | –45.9% |
| Equipment required | –34.0% |

C. Production Control Department:
   1. Failure to understand process capabilities of the line
   2. No grasp of inventory level
   3. Changing plans, ignoring *gemba* conditions
   4. Insufficiently precise production plan
D. Purchasing Department:
   1. Ignorance of supplier capacity
   2. Inability to provide technical guidance to suppliers
   3. Insufficient quality audit to suppliers
   4. Inadequate management of incoming supply
E. Sales Department:
   1. Failure to understand capabilities of the *gemba*
   2. Failure to provide vital customer information to the *gemba*
   3. Insufficient liaison with customers
F. Accounting Department:
   1. Requesting more information than actually needed
   2. Delayed monthly reports
   3. Inadequate cost analysis
G. Administrative Department:
   1. Introducing flavor-of-the-month programs that bear little relevance to the needs of the *gemba*
   2. Inadequate training programs
H. R&D and Product Development Department:
   1. Designing products that fail to take into account the capabilities of the *gemba*
   2. Failure to advise the *gemba* of anticipated changes in advance

Thus *gemba kaizen* becomes a starting point for highlighting inadequacies in other supporting departments and identifies internal systems and procedures that need to be improved. This must be done based on a shared feeling of dissatisfaction with the current state and also an environment of no blame for the current state. As many problems are exposed, these must be celebrated as opportunities to realize savings rather than embarrassments to hide.

Since 85 percent of the total cost of production is determined at design and planning stages upstream from the *gemba*, and since the conditions for quality and delivery are also determined in the design planning stages, improvement in upstream management is the key to achieving successful quality, cost, and delivery. *Gemba kaizen*, therefore, is but a starting point for making much more exciting, challenging, and beneficial changes by bringing *kaizen* to upstream processes such as design, planning, and marketing. However, unless the caliber of the *gemba* is first elevated to internationally competitive, world-class standards, no matter what improvements are made upstream, the *gemba* will not be able to reap the benefits. *Gemba kaizen* is truly a continuous improvement strategy that must encompass the total extended enterprise.

# CASE STUDIES

# Lessons from a 20-Year *Kaizen* Journey

I first heard of Masaaki Imai in the early 1990s, when I was an engineer working for Bosch in a plant in Portugal. My boss had been in Germany with other top executives to train with Mr. Imai, and when he came back, he told us about this fantastic philosophy that he had learned about. It was called *kaizen*, he said, and it would guide our company in the coming years.

Years later, I returned to Portugal after being abroad and took over the role of senior vice president of the plant. This was a fairly large plant; it employed about 1,000 people and had an annual turnover of about $300 million. The plant is a leading manufacturer of tankless water heaters, competing with Japanese companies on a worldwide basis.

Bosch had just begun to implement a worldwide initiative called the *Bosch Production System*, which was modeled on the Toyota Production System. Under the program, all plants must meet the same strict standards and are measured on a point system that applies equally to all countries.

I knew that there would be challenges. My plant is not an automotive plant, so the cost structures are different from others at Bosch. Portugal is also on the outskirts of Europe, so we have high logistics costs toward our markets in Europe. Finally, our blue-collar workers are not as well educated as they are in countries such as Germany. I knew that our people would have to work very hard to compete and would need strong management support.

When I arrived, I spent the first week walking around the plant and speaking to people. I learned that the machinery and solutions had been implemented correctly, but there was still a long way to go in achieving the proper mind-set among the people. The *kaizen* tools and processes were available, but the evidence for seeking continuous improvement every day at all times was not there.

For example, people kept saying that they had quick changeovers, that they had implemented single-minute exchange of die (SMED), but I could see that they were not applying the methodology of continuous improvement or setting new targets and constantly improving, as Masaaki Imai always stressed. I told my staff that we were missing a great opportunity.

## *Kaizen* Must Begin With a Vision

For me, it was very clear that we needed to define a certain vision. For one, we wanted to have a continuous-improvement mind-set not only on our shop floor but also for people who worked in the indirect areas of the company, including management, whom are often not involved in such matters. So we implemented activities such as value-stream mapping and value-stream design in all our areas.

This was not always easy. Early on, I had two middle managers who were sabotaging our company's *kaizen* efforts. They would come to meetings and say "Yes" to everything, and then they would turn around and tell their people to forget about *kaizen*. After several attempts to get them involved, I decided to put one of them on different tasks and dismiss the other. This gave a very clear signal to the rest of the company that we were not going to divert from our improvement efforts.

We developed our vision for the plant and were able to convince most of the team that this was our new direction. The vision embodied our goals of aspiring to world-class manufacturing standards and embraced the eight basic principles of the Bosch Production System. Much emphasis was placed on the involvement of our people.

## Use *Kaizen* to Develop the Workforce

The key for us was getting all employees to embrace the spirit of continuous improvement—something we could call *kaizen* culture. Bringing this kind of change involves many considerations, of course, and a very important one is that you have to know what country you are in. For example, if you want something to get done in my country, you have to ask every day. I worked for a couple of years in Germany, and if you ask your employees to do something several times there, then you are considered to be out of line because they are more independent.

In Portugal, we also have a very extreme class system in many companies. On the manufacturing level, you have low-skilled, uneducated people, and in the indirect areas, you have very skilled, educated people who may speak three, four, or five different languages.

*Kaizen* helped us to build a culture that valued our blue-collar people even more than our white-collar people. When we implemented total productive maintenance (TPM), for example, we realized that the operator has to have ownership of the machine.

I became very aware of this when I was visiting a plant in Japan and saw a shop floor worker screaming very angrily at the plant manager. I asked one of my translators what had happened, and he said that the shop floor worker had just gotten married. When you get married in Japan, you leave work for five days, and during those five days, nobody had taken care of his machine. In the *kaizen* culture, this machine was not the machine of the company or the owners—it belonged to the operator. This is very important.

I returned to Portugal and explained this principle to others, and that might have been a turning point. All of a sudden, people started believing in TPM, and I realized that *kaizen* didn't require so much energy. At the beginning it does, but once you have people on the path, it becomes much easier.

Our numbers began to improve, but this did not happen overnight. In 2003, we achieved 275 points out of 800—not the strong rating we had hoped for. By 2007, however, we had reached the 500-point level, which is the benchmark for a good plant at Bosch. Bosch held a contest that year, and we were nominated as one of the top five Bosch Production System plants in the world. In addition, some of our practices for involving and empowering our people were considered for Bosch worldwide. This is not bad for a small country like Portugal.

## You Cannot Delegate *Kaizen*

None of this is possible without leadership. The first advice I give to CEOs is that you have to be ready to put yourself—as a manager and as a person—out of your comfort zone. You have to be willing to try new things and to work against difficulties; otherwise, don't bother.

In addition, you cannot delegate *kaizen*. I always tell this to people whenever we implement something within the company. *Kaizen* is a job for everybody, but especially for top management. You cannot create a *kaizen* department because *kaizen* must be within everyone's spirit.

As a top manager, you have to set an example. When we implemented 5S throughout our company, this included the offices of senior managers. I told people who were auditing our offices not to tell senior management when they were coming. It is very important that people see that you are committed to the same process as they are.

## Never Stop Learning

Self-improvement is also very important. We told our blue-collar workers that no matter what they did, everything they did to make themselves a better *kaizen* employee was for their own good. If they got another job, they could take that education with them. This also made them keen to do their training, to learn more.

With *kaizen*, management has to commit to self-improvement as well. I try to learn everywhere I go. It's not easy, but every time our people go to another country or another plant, I tell them to look for things that others are doing that are different.

For example, some weeks ago I was visiting a company that had many problems. It had almost no *kaizen* activities, but it did one thing very well—better than we did. It had a packaging system in which it used a *pokayoke*, so it couldn't make a mistake. Immediately after I left the company, I called our plant manager and told him to go and learn from this company.

## *Kaizen* Builds Our Future

*Kaizen* offers a lot of hope for today's economy. It's more and more about cutting down waste these days, doing things right the first time, and becoming as efficient as possible. Especially today, when we are so strongly affected by competition, especially from Far East products, we need to stay competitive. Here in Portugal, if we hadn't continued to do *kaizen* since 2000, our clients would have moved to a lower-cost supplier somewhere else in the world.

Above all, it is the people in the company who make the difference. Very often I read books that say human resources are the most important asset. It's a very nice sentence, but it's very difficult to live by. *Kaizen* gives you the chance, the tools, and the methodologies to give your people a feeling that they are all part of one organization. It brings different levels of the organization together. I believe that companies that have implemented *kaizen* have more open discussions, are more transparent, and have no or fewer taboos. Workers can ask anything, and by doing this, they create a very open atmosphere that leads to improvement. Here in Portugal, our numbers show this.

JOAO-PAULO OLIVEIRA
*Senior Vice President*
*Bosch Termotecnologia Portugal SA*

# Changing the IT Culture at Achmea

Quality experts like to remind us that you can't manage what you can't measure. However, as this case study shows, it is often more important to remember that you can't manage what you can't understand. Achmea, a large European insurance firm employing 20,000 people, had all the tools it needed to transform its 7,000-employee information technology (IT) department but was initially unable to get to the bottom of the cultural issues that were holding the company back.

The company had been so prosperous in the past that cost was never an object, and waste and nonaccountability were embedded in the corporate culture. When the financial industry started facing problems in 2008, efficiency became a top priority almost overnight. Concerned about the future, the company brought in a major consulting firm to implement lean-related improvements.

While success was achieved in other departments, IT remained a stumbling block, and it became clear that the lean consultants didn't have the IT expertise to understand the department's difficulties. To address this, the company engaged the Kaizen Institute Netherlands, which has experience with both lean and IT.

The initial target was a subdivision of the IT department's 600-employee software development division, which is responsible for designing, building, and testing the software applications that support the firm's core business. This diverse group includes engineers, builders, application managers, testers, business analysts, and other software professionals. Because these workers are highly paid and their output is strategic to the organization, the potential for gain was considered to be substantial.

As in most IT organizations, the division had employees trained in a variety of IT management methodologies. The best known of these is a process framework called *information technology infrastructure library* (ITIL), which creates criteria for standardization and establishes the output of an IT organization as a collection or "catalogue" of services that are delivered to the business. This latter aspect is similar in some ways to value-stream mapping.

The problem was that people weren't using available tools and were under no pressure to do so. Project planning was haphazard at best, employee performance was virtually unmeasured, and project stakeholders had little sense of the costs or time frames of their projects. Workflow was highly disjointed—it was considered "normal" for an engineer or builder to complete a task and then have to wait six months to complete the next step in the project.

"The culture of not applying available tools arises from the often-correct assertion that they don't align with the work," says Wijbrand Medendorp, managing partner with the Kaizen Institute Netherlands, who consulted on the project. "To address this, management has to accept feedback from workers, acknowledge the shortcomings, and lead the way to step-by-step improvement."

Even when it is obvious to management that there are problems, it is very difficult to identify broken processes in a software-development group because much of the work is invisible code that is only comprehensible to very few. Even within the development group, there are many technical "languages" that are not universal.

Consequently, training in lean methods had to be supplemented by high-level instruction in various IT principles. For example, people who wrote flowcharts had to learn more about the business of writing code, and business analysts had to learn about the limitations of the existing SAP system. This information was conveyed through facilitated sessions where representatives from functional groups were able to share their objectives and issues with their counterparts.

One process that brought impressive results was a type of facilitated session called the *customer arena*. Here, key stakeholders seated around an inner table would air their key issues, and members of the delivery team would listen to the conversation from outside the circle. This helped many to hear, for the first time, how their actions or inactions were affecting others in the value chain.

Games also were used to better understand the roles of other team members and to learn about the benefits and dynamics of cross-functional teams. Traditional lean games, such as the Airplane Game, were adapted for IT.

Once some basic understandings were established, cross-functional improvement teams were able to proceed with value-stream mapping. In a typical software-development project, the invisible product moves from the establishment of business requirements to flowcharting, code building, troubleshooting, and user testing. An example at Achmea was adaptation of the retirement savings plans application to accommodate new tax rules.

Seeing their work represented in value-stream maps helped employees to uncover some major organizational problems. Project managers, who were accountable for seeing their projects completed on time and on budget, had no authority over team members who were creating the product. Instead, team members reported to line managers, whose work was aligned functionally—engineers reported to an engineering manager, testers to a testing manager, etc. There were team meetings to discuss problems, but these were only within the same functional groups.

The lack of coordination between these functional groups meant that each project was, essentially, a free-for-all. Project managers often had to compete for resources to complete their projects and frequently were usurped by others with more influence. To protect themselves, they typically overallocated. And while line managers understood the skill sets of their workers, nobody knew how to build and manage the cross-functional teams that were counted on to complete projects.

The establishment of value streams also identified non-value-adding waste, making it possible to measure worker productivity. This figure is characteristically low in software-development groups because of the complexity of interactions—25 percent is considered normal.

Workers at Achmea were highly resistant to surveys measuring their productive versus nonproductive time—many said that they felt they were being spied on. Surveyors were able to make headway, however, when workers were asked to log their ideal hours—time where they were able to work on projects uninterrupted.

"When management shows that it is not taking a hostile approach, then it becomes much easier to get the facts on the table," says Medendorp.

For creative workers such as software designers, it is widely accepted that every interruption—say, a phone call—costs 15 minutes of productivity. The surveys showed that often workers were being interrupted four times or more per hour and, consequently, had productivity of zero during those periods. In addition, the chaotic schedule resulting from the lack of coordination between teams was causing significant wastes such as work-in-process, rework, and wait time.

Measurements eventually showed a productivity rate of between 2 and 4 percent. In other words, a sum of approximately 20 million euros a month was going out the window.

Through a series of *kaizen* sessions, the teams were able to identify some simple improvements that took aim at this waste and, when implemented, raised the productivity of the group substantially. They included

- The establishment of a kickoff meeting for each project, where requirements and expectations would be tabled in the presence of all team members. This gave the opportunity for early warning of some of the challenges and a chance to develop team rapport.
- The establishment of daily team *kaizen* sessions to allow problems and issues to be tabled without delay.
- Adaptation of access rights to make systems more easily available for application testing. Previously, testers had to wait weeks to be able to continue a project.
- The public posting of the ASAP quality-checking procedures that the organization had adopted. Previously, workers had no idea how the quality department was checking their work.
- The creation of performance standards for workers, which included participation in *kaizen* activities. Previously, there had been no standards, and workers were almost automatically given a high rating.

After the improvements were implemented, an assessment was conducted for the four business units within the software-development group. In each area, the measurements showed a cost reduction of 30 percent, pointing to 20 million euros of cost savings annually.

The success in the software-development group is now being replicated in other parts of the IT organization. The CEO and the CIO have reinforced the mandate for continuous improvement, and this is now reflected in the employee performance-evaluation process maintained by human resources. Senior management now understands the cultural issues around creating change within IT, and all IT employees are now aware that improvement always must be part of their work and that there will never be a return to the old ways.

# Daily *Kaizen* at Tork Ledervin

The success of daily *kaizen* at Tork Ledervin, a weaving plant in Brazil, is a wonderful illustration of my definition of *kaizen* as "everyday improvement, everybody improvement, and every way improvement." It's also interesting to remember that Taiichi Ohno's first experiments with lean began in the early days when Toyota was still the Toyota Automated Loom Company, making the kind of machinery that can be seen on Tork Ledervin's shop floor.

The 30,000-square-meter plant is located in the city of Osasco and employs 350 people in the production of high-performance yarns and fabrics for the automotive industry, conveyors, machinery and equipment, and other areas.

Tork Ledervin's *kaizen* journey began in late 2009 when the Kaizen Institute conducted a value-stream design (VSD) workshop. This five-day event, coordinated by project manager Fernando Andrade, included managers and coordinators from the sales, factory, quality management, maintenance, logistics, and production planning and control departments. The industrial director, Irineu Bergamo, also participated, and Ledervin's two vice presidents, Laerte Serrano and Frederico Lima, were present the opening and closing days of the event as project sponsors.

Kaizen Institute consultants began the workshops by showing team members how to look at their company from a lean perspective. Soon, managers began to see how their current processes created waste, including frequent loom stoppages, a cluttered factory, long changeover times, and quality defects. Team members also learned how *kaizen* tools and concepts could help to streamline these processes and create a desirable future state that would better serve customers through increased productivity, improved quality, and less inventory. These improvements, in turn, would help the business achieve its goals of reducing lead times and increasing flexibility to meet customer demand.

After the initial training, team members established a focused plan that would help them to implement *kaizen*. The first step was to create two present-state maps of Ledervin. The first of these analyzed the flow of materials from the receipt of raw components to shipping of the finished product. The second followed the flow of information from the client's order, to production planning, to materials planning, to customer payment.

Once the maps were completed, employees were assigned to their respective processes and, using the *gemba kaizen* workshop format, held a series of regular *kaizen* sessions to brainstorm improvements that would help them reach the future state. Reaching their goal would mean reducing setup times, increasing availability, reducing product defects, and improving layout and line design by establishing *kaizen* methodologies such as supermarkets and *mizusumashi* routes.

Tork Ledervin knew that culture change doesn't happen overnight, so the company focused its energy on the weaving department. Starting in February 2010, everyone from managers, to maintenance personnel, to team leaders began a series of workshops lasting between 7 and 10 days that aimed to reduce the number of defects, increase the efficiency of the looms, and reduce changeover times. Each workshop applied one *kaizen* approach to help improve an existing process. In one session, for example, workers constructed a supermarket that would supply yarn more efficiently.

Despite achieving significant positive results within three months, *kaizen* champions at Ledervin knew that they needed to ingrain a culture of continuous improvement in the company that went beyond the workshops. A daily *kaizen* approach would ensure that these standards would be maintained and that deviations would be handled properly for years to come.

Working with the Kaizen Institute, Tork Ledervin's managing director, Irineu Bergamo, built a model with several key objectives in mind. These were

- Improve area management by creating a visual system to display key performance indicators (KPIs) and share it with the work teams.
- Establish an effective way to report deviations, analyze causes, and implement corrective actions.
- Ensure compliance with operational standards and updates.
- Establish various levels of leadership within the *gemba* team.
- Evenly distribute tasks among workers, ensuring less downtime and less excess workload.
- Educate all workers about *kaizen*.

To achieve these objectives, the following measures were put in place:

- A daily *kaizen* board was established that aided in the visual management of KPIs and provided a space for recording deviations, root-cause analyses, and descriptions of actions and deadlines for completion.
- Daily *kaizen* meetings were established to ensure that problems were discussed by operators and leadership and that solutions were defined and completed.
- A leadership checklist was created, a daily routine that shop floor leaders would follow to ensure compliance with standards and the consistent attainment of expected results.
- A *kamishibai* system was put in place to help provide scheduled audits for various levels of leadership in the *gemba.*
- An activities leveling box helped to distribute tasks to each employee at specified times.
- *Kaizen* tools manuals were distributed to help explain the approaches, methods, and tools being used for continuous improvement.

In May 2010, workers from across the weaving department spent 15 days fine-tuning the tools and methodologies just mentioned, ensuring that they were well adapted to Tork Ledervin. Training materials were prepared, and workers were instructed on their new routines. An important tool that was used widely was the one-point lesson. This provides a structure for 5- to 10-minute instructional sessions, each of which addresses one particular learning point in a worker's environment.

When the workshops ended, Ledervin had what it called a "K Day." In the presence of their vice presidents, workers held an official "Turn Key" ceremony that marked a total changeover to the *kaizen* approach. From now on, the entire plant would follow a lean philosophy.

Several weeks later, results began to appear that not only sustained previous gains but further increased the efficiency of the looms. Results included

- A 40 percent reduction in the number of loom stoppages
- A 60 percent reduction in loom changeover times
- Increased loom efficiency, from 60 to 86 percent
- A 25 percent reduction in defects

A daily *kaizen* approach has provided increased motivation and more enthusiastic participation from people working in the *gemba*. Better management of day-to-day tasks has ensured a superior working environment for all and, ultimately, better service and a better product for the customer.

"Daily *kaizen* gave a better definition of responsibilities and greater efficiency in reaction to deviations," said Irineu Bergamo. Based on the success in the weaving department, more *kaizen* activities are now being planned in other areas of the organization.

# *Kaizen* in Public Spaces: Transforming Rome's Airports

Improving processes that are highly visible to the customer presents a special kind of challenge. Whereas manufacturers, for example, can openly experiment with different *kaizen* solutions in the seclusion of their factories, processes that are plainly visible to the public require caution and precision.

This was one of the main challenges faced by Aeroporti di Roma (ADR) when it began its *kaizen* journey in 2008. ADR—which manages Rome's two international airports, Leonardo da Vinci–Fiumicino and Ciampino Airport—decided to adopt *kaizen* because it was confident that it could help ADR set the highest airport service standards in Europe.

Reaching this goal would not be easy. ADR would have to undergo a deep organizational change if it wanted to attain a culture of continuous improvement. This would mean not only changing its behavior but also reassessing how it thought about and approached problems and solutions.

## Streamlining the Security Check

ADR's first improvement project centered on the passenger and hand baggage security check. In their first workshop, workers focused on defining and identifying lean concepts such as value, flow, and *muda* and how they related to the passengers, ADR, and the airport's stakeholders.

"The main challenge was getting the security people engaged in order to improve their work processes," says Bruno Fabiano, Kaizen Institute *sensei* who managed the project. "This had to start with a clear identification of 'value' from the airport customer's point of view."

From this perspective, the security check was about ensuring the well-being of passengers while respecting an international array of social norms and customs, all the while complying with strict international regulations. Real value lay in making passengers feel safe while not making them feel scrutinized or embarrassed by intrusive security procedures.

Three *kaizen* teams were selected, each consisting of 10 to 12 members representing all of the subareas within the "passenger and hand luggage control" category, including passenger preparation, screening, scanning gate operations, control room, and supervision. The teams were trained in basic *kaizen* concepts and lean tools.

Once they had a grasp of the concepts, the teams began a "walk-through" to help them identify waste, as well as make suggestions for improvement. Each team member followed the path the passenger took from being issued a ticket, to the security lineup, to the actual security check, to exiting the security area. During their observations, staff members were surprised at what they saw, and many experienced an epiphany that helped them to internalize the customer-centered *kaizen* perspective.

Next, the team used value stream mapping to identify and visualize processes and sort value from waste. One particular source of *muda* was long lines because this resulted in lost time for passengers and ultimately less revenue for the airport (because those passengers spent less time shopping). The walk-through helped to identify some of the causes. For one, there weren't enough trays for passengers to put their pocket items into for the baggage scan, forcing them to wait for trays to become available. Also, the screen that displayed instructions for passengers was too high up, meaning that people often didn't see it and were left confused about what to do.

Teams issued a plan-do-check-act (PDCA) list to help them eliminate problems and sources of waste. Guided by the results from the value-stream mapping exercise, workers began, through a series of *kaizen* sessions, to generate ideas to realize a vision where passengers could flow smoothly and comfortably through the process.

Pilot tests were implemented in the *gemba* with employees standing in for passengers. In the spirit of encouraging creativity, a wide variety of ideas was tested. In one scenario, passengers were allowed to keep certain items such as belts and cell phones on them, for example, and a chair was placed next to the baggage scanner where people could place their shoes, which was much more desirable than having them walk through the scanner in their bare feet.

The teams also used visual *kaizen* tools to rearrange benches and monitors to improve passenger flow. While an ideal visual configuration was not possible in the existing facilities owing to space constraints, suggestions are being incorporated into plans for new construction.

In the spirit of continuous improvement, regular *kaizen* meetings also were held, whereby workers could discuss opportunities to improve quality and efficiency, as well as record any other important observations or happenings in a book.

As the improvements progressed, it became clear that security clearance is not in island but must work in conjunction with other functions within the airport. This is a complex matter because of the huge numbers of people involved, as well as the way work shifts are structured.

To understand these interactions better, the teams sought input from other groups within the airport community on how they managed their own processes. An airplane pilot, for example, was invited to one workshop, where he outlined the checklist that all pilots perform before every flight, regardless of how experienced they are. At the end of this meeting, the team prepared a draft checklist that they soon applied to their own *gemba*.

## Welcoming Passengers to Rome

Behind the desk of Fabrizio Mariotti, human resources development manager at ADR, one can see a board covered in colored Post-it notes showing the two airports' training schedules for the entire year. On another wall of his office is a giant picture of Rome, which was posted there after the award-winning Archimede improvement plan was realized.

Archimede came to life in 2010 while a *kaizen* team in the security department was brainstorming ideas on how to improve the quality of processes that had the biggest, most direct impact on customers. The team found that when international passengers stepped off the plane and into the airport, there was nothing to communicate the historical atmosphere that is Rome.

*Kaizen* had taught ADR that the energy that drives continuous improvement comes from *gemba* workers who see thousands of customers every day, and for this reason, these workers were asked to propose simple, concrete actions that could make passengers feel the significance of where they had landed. In order to encourage teamwork, projects had to be presented by groups with a minimum of five people. Roughly 200 people submitted a total of 111 project ideas, which were evaluated by a special committee. After considering both the impact and practicality of each idea, 14 were selected, each of which was then assigned to a team charged with implementing it. Some of these were as follows:

- Thirty different posters were set up by the staff in the airport terminals to welcome passengers to Rome. Images included pictures of the city, as well as pictures of Leonardo da Vinci and Ciampino Airports.
- Two designated areas were constructed where parents could keep their children while they waited for their baggage.
- Courtesy strollers were placed in two terminal departure halls, and personalized assistance was provided at the arrival and departure terminals.
- Waiting lounges were created for passengers with reduced mobility.
- Nine hundred phrase books were distributed to terminal operators, each containing the most common English expressions. Three hundred more were distributed with the most common Russian, Chinese, Japanese, and German sentences, as well as a pronunciation guide.
- A manual was placed in every shop that detailed communications procedures, new products, and proper general procedures.

The Archimede project also contributed to the airport security function by developing behavioral standards that would help to make passengers more comfortable during the potentially tense examination process. Some of the issues that were covered in the booklets developed by the team were eye contact with passengers, handing physical objects that had been through the baggage scan directly to their owners, and having two officers stand apart at a specific distance in order to make it clear that each of them could be expected to conduct respective security checks. One hundred and twelve multilanguage phrase books also were distributed to security checkpoint officers and are now being used to train new employees.

Behavioral standards also were reinforced through the training within industry (TWI) method, which helped to provide instruction on the job. This approach followed a simple four-step method:

1. I explain; you listen and observe.
2. I do; you observe.
3. You do; I correct.
4. You do autonomously.

The Archimede project confirmed the willingness of managers at ADR to build a policy around actively listening to their workers, as well as building improvements based on research and results. Above all, though, it demonstrated that involving people is the best way to motivate them to work together to improve the customer experience, one step at a time. In Michelangelo's words, "Perfection is in the details."

## Pushing Ahead

An award ceremony was held in 2011 to recognize the hard work and achievements of ADR's 200 *kaizen* workers. All present understood, however, that this was not an end but a beginning. *Kaizen* seminars, which continue to this day, are open to all workers and encourage people to bring their own ideas to the table, as well as practice what they have learned on a daily basis.

Thanks to *kaizen*, there is a renewed sense of energy at ADR. Results have been so positive that management now considers it a distinctive competency and encourages workers to continuously pursue improvement actions on a daily basis. Results such as this ensure that ADR—in the true spirit of *kaizen*—will continue to improve the passenger experience for years to come.

# Sonae MC: The Silent Revolution

This story about *kaizen* in a large retail chain is another good example of how well the *gemba kaizen* principles can be applied to industries very different from manufacturing. Also, it demonstrates that *kaizen* is not a prescriptive method but that it evolves and adapts based on the unique needs of each organization. In this case, what started as a training program turned into a comprehensive transformation process involving all 25,000 employees.

Managing a retail operation requires a delicate balance. Subtle changes in the retail environment can have a major impact on customer behavior. This means that improvements have to be implemented in ways that are not visible or disruptive to the retail floor in order to avoid the risk of driving customers away. Consequently, when Sonae MC implemented *kaizen* in its 171 food supermarkets, it referred to this work as its "silent revolution."

Sonae MC is a subsidiary of Sonae, the largest private company in Portugal. Sonae MC is the longest-standing business unit within the company, with revenues of 3.27 billion euros in 2010. In business for 25 years, this experience has helped Sonae MC become a leader in its industry but also had created a culture where many people believed that the old way was the right way.

*Kaizen* was first considered at Sonae MC as a way to comply with a new government requirement, effective beginning in 2006, that all companies provide a minimum of 35 hours of vocational training annually for each employee. This, in turn, has been part of a national strategy to make Portugal more competitive within the European Union and beyond.

Because of the large number of workers—Sonae MC is Portugal's largest employer—it was clear from the outset that this would be no ordinary training project. The scope spanned classroom training for managers complemented by training at the *gemba* for employees working in the stores and warehouses. The approach was called the *Team Development Program*

(TDP), and the Kaizen Institute was asked to present a training plan to make up a part of the TDP.

*Kaizen* training, of course, is not just about changing the worker—it is about changing the workplace—so it became clear that the expectations were far greater than simply meeting the government's training mandate. When Manuel Fontoura, manager of Sonae MC, was shown the results of a study made by the Kaizen Institute about the *gemba* at Sonae using photographs to illustrate the opportunities for improvement, his immediate comment was, "I've seen these pictures for the last 25 years. I want solutions, not problems."

Fontoura did not accept the assumption that the methods that had made the company successful in the past were the only way to run the business, and he agreed to test *kaizen* training first for managers. If successful, he agreed to expand the training to frontline workers in the following year.

## Old Ways and New Ways

The *kaizen* training was introduced in a cascade fashion, where concepts were learned by leaders and quickly applied in the *gemba*. The program began with two-day seminars for the leaders that covered basic *kaizen* principles and tools. On the first day, trainees learned many concepts such as *muda*, 5S, visual management, just-in-time, and standardization. On the second day, they observed a work shift in the *gemba* that had begun five to six hours previously, allowing them to observe the processes in full operation and then suggest improvements at the end of the shift.

There was a lot of resistance at the beginning. Comments such as "We have always worked this way" or "I have no time for this type of thing" were typical. However, as time went by and results started to appear, the most vocal critics turned into the most enthusiastic supporters.

As the teams began to test improvements, it became clear that it would be impossible to attack all areas of the retail environment at the same time, so the managers used the *kaizen* process to identify where the quickest and most visible results could be achieved. Selecting an area that would help to establish a culture of continuous improvement was an underlying objective because they realized that this would be the common denominator behind any visible change.

The areas on which the team members decided to focus were restocking, multi-restocking, and backward management (warehouse-area logistics) in selected areas. Their task was to seek out improvements to these three areas wherever possible. The *kaizen* tools used were 5S, visual management, and standardization.

## A Place for Everything and Everything in Its Place

As the teams became more confident in their use of the *kaizen* tools, a dramatic transformation began to take place. The 5S initiatives ranged from the basics of cleaning, painting, and labeling to the comprehensive reorganization of floor layouts and the redesign of racks to facilitate selection and restocking of merchandise. The result was a degree of visual order rarely seen in logistics areas of the retail industry. This included a complete system that prevented items from being left out of place.

In less than three months, the transformation was complete in all the stores' warehouses, and the company began to notice improvements in the work processes themselves. Management had identified the key sources of *muda* during its training sessions, and now managers were watching the waste disappear. For example, a participant had noted that approximately 80 percent of the time spent by restocking operators was due to such tasks as moving articles from one place to another or folding cardboard—activities that brought no value to the process. The transformation that had taken place in the *gemba* was making a noticeable reduction of this type of waste in particular.

In keeping with the theme of a silent revolution, the transformation remained invisible to store customers. All that was evident was improved employee morale, a lack of merchandise in the aisles and corridors, and the elimination of items being "out of stock" on the shelves.

## Shared Success

While the training and support for *kaizen* concepts came from management, the changes were designed and implemented by the workers themselves. Increasingly, these workers gained a comfort level with the *kaizen* process. The management teams had developed the eyes to suggest improvements but also knew that the people who performed the tasks every day had the most knowledge of how to improve them. Management also knew that its belief in and commitment to *kaizen* were essential to making the project believable to employees.

As a result of the first year of *kaizen* activity, the level of employee satisfaction had increased greatly. Better planning and a friendlier visual environment helped workers to accomplish their daily tasks with less effort, less stress, and in a shorter period of time. All this was aided by a visual management method that created a new level of transparency in which all employees could see how well their team was doing. One of the examples was the application of dynamic work plans that allowed a better management of human resources and tasks.

The improvements in the stocking and restocking processes also were remarkable; the productivity of these tasks increased by 35 percent. The quality of the restocking improved, and the store shelves started to be more complete and the products better organized. Besides, there was more time left at the end of the day to organize the backstage areas and fulfill the cleaning plans.

By the end of the first year, there were no doubts that *kaizen* was the path that Sonae MC needed to improve its productivity and efficiency and consolidate its leadership in the market.

## Replicating *Kaizen* Success

With a *kaizen* culture firmly in place, the field of initiative was broadened to include all the areas of the shop, namely, fresh foods and points of sale, embracing all the workers involved in the operations, including administrative clerks. The people who had already been trained with *kaizen* in the previous year were introduced to new tools in order to consolidate their knowledge. From that point onward, the whole company was speaking the same language and was rowing in the same direction.

The team of internal trainers, supported by the Kaizen Institute, continued to develop new skills, but as the scope and number of *kaizen* initiatives increased, the importance of leadership increased. At the same time, there was a need to give workers some autonomy to make improvements on their own and to help leverage the role of supervisors and leaders. The *kaizen* methods and tools applied to these ends included:

- *Kobetsu kaizen* to encourage autonomous maintenance of automation equipment.
- Standardized job instructions so that workers did not need to check with management all the time.
- 3C (contain, control, correct) problem solving to help workers solve problems on the spot.
- Standardize-do-check-act (SDCA) cycle instruction to give workers a tool to improve their daily processes.
- Training Within Industry (TWI) to help workers integrate *kaizen* into their daily routine.

These tools were reinforced through daily meetings of 15 to 30 minutes that included operational briefings and the TDP agenda for that day.

The opening of pilot stores, working as showrooms of the improvements achieved, was another important action that helped spread the positive solutions already implemented. Several leaders were called to pay a visit to these stores, and after observing what had already been done, they tried to follow the same procedure in their own commercial units. This type of positive infection created some healthy competition, all in the spirit of continuous improvement.

A system of implementing organization-wide *kaizen* improvements called *Sistema de Implementação de Melhorias kaizen* (SIMk) was introduced to develop a unifying strategy for all the *kaizen* initiatives.

## Stabilizing the *Kaizen* Culture

As the initiatives broadened in scope, management needed to collect real data on the adoption of the new practices in order to ensure consistent execution of improvements. Consequently, an audit based on samples was made in order to evaluate how well the *kaizen* culture was being assimilated. The main lesson learned was that assimilation was not consistent at all locations. This was not a surprising outcome given the number of employees and locations. To address this, management decided to institute a companywide certification program.

In 2009, all stores prepared for two audits. The first was diagnosis; the second was certification. The diagnosis audit included an evaluation and an improvement plan outlining the requirements that the particular store would have to meet to achieve certification. For the second audit, the stores were expected to have prepared their employees and made the required corrections.

The first round of audits showed that there was significant need for improvement at many stores. However, many were surprised to find that the store that scored lowest in the initial evaluation scored the highest in the certification audit at the end of the year.

## New Leadership

While the teams were busy stabilizing the *kaizen* culture, there was a major change in management. A new executive, Mário Pereira, took over the role of manager of Sonae MC. Some feared initially that this might jeopardize the newly developing *kaizen* culture. However, the transition was carefully managed, and Pereira immediately understood the importance of the *kaizen* initiative, which had been considered the biggest innovation project of the previous year.

Pereira agreed not only to continue with *kaizen* but also strongly reinforced it. He personally attended training sessions, embraced the *kaizen* culture, and assembled an SIMk steering committee of senior managers charged with high-level planning for continuous improvement.

## The Internal Logistics Projects

This new management committee, working with the Kaizen Institute, decided to take on a large and aggressive target: internal logistics. Plans were made to divide the work into three initiatives:

- *Internal food logistics (IFL).* The primary objective of IFL was to open each store with a full shop and restock only after the shop was closed. Using selective multi-restock, the planners hoped for zero returned goods to the warehouse and increased productivity.
- *Internal nonfood logistics (INFL).* Here, the goals were somewhat different. The underlying priority was optimization of the stocking levels in the warehouse, reduction of stocks, and zero return in restocking. In addition, priority would be assigned based on potential sales volume. This transition would require a revised layout, selective

sorting stations inside the warehouses, revised timetables, a logistics train, and enhanced informatics tools.

- *Manufacturing logistics model (MLM).* This began in the bakeries, with the objective of having quality hot bread throughout the whole day and freeing workers to spend more time with customers. This required creating internal supermarket systems for raw materials, as well as schemes for push production before the opening time to ensure the availability of fresh bread when the store opened and pull flow during the day in response to demand. This called for new production equipment and processes.

Several stores were selected for pilot testing of the new logistics systems. This provided a valuable test environment to see what worked and what needed improving. These stores became showcases within the company to generate momentum and enthusiasm for the project.

The following early results were achieved from the pilots:

- The IFL project yielded a productivity increase of 17 percent, a reduction of medium stocks of 14 percent, and a reduction in losses of 11 percent.
- The INFL project saw a productivity increase of 31 percent and a sales increase (with the help of supplementary campaigns) of 20 percent.
- The MLM project reduced bread-making losses by 4 percent and increased sales by 2 percent.
- Taken together, these projects allowed the optimization of work schedules, resulting in a 52 percent reduction in the cost of night-shift premiums.

## Aligning *Kaizen* Progress with Corporate Performance Measurements

The year 2010 saw the addition of the balanced scorecard (BSC) methodology at Sonae MC. The balanced scorecard articulates monetary and nonmonetary objectives for corporate performance. Management felt that this was essential to ensure that the *kaizen* improvements that were being achieved in the *gemba* were aligned with the strategy of the organization. The BSC helped to articulate the *kaizen* strategies developed with the SIMk process so that they could be understood within a strategic context.

The implementation of these high-level measurements helped to ensure that measurement of *kaizen* progress—an essential part of *kaizen*—would not be neglected. The balanced scorecard made the business management indicators much clearer and standardized the way those results were presented. It also allowed the monitoring of financial performance along with customer satisfaction, worker competency, and other important metrics according to established benchmarks. The system noted—in red—the areas that were below ideal performance, indicating that areas of priority action.

Management also recognized the need to manage the many improvements that were being proposed in the *gemba* and consequently established an internal process called *method of managing the improvement* (MMI). Between the first of March when it began to operate and January 2011, the system had registered 1,089 improvements. In order to objectively measure the value of these contributions, an evaluation committee was set up, and a ranking system based on impact and ease of implementation was established. In addition to speeding the process for approvals, this also made it easier to assign credit to those who had innovated improvements.

The year 2011 was marked by the rollout of the new internal logistics model in all of the group's 172 food retail stores. This was a massive undertaking that required global training for all managers as well as training specific to the IFL, INFL, and MLM projects.

António Costa, director of the Kaizen Institute Iberia, reflected, "Everything started as a training program, then it became a project, and now it is a companywide system that keeps on improving."

## *Kaizen* Continues

Sonae MC accomplished a great deal through *kaizen* in five years and become the first and only major retail chain to achieve this level of *kaizen* proficiency. Through a joint effort that eventually involved over 25,000 people, the company had seen an overall productivity increase of 35 percent, a reduction in medium stocks of 14 percent, a reduction in losses, a sharp decrease in returns and leftovers, and substantially lower expenses from night-shift premiums by the end of 2011. Even more important, the improvements have led to a higher level of customer service, ensuring Sonae MC's future competitiveness.

The *gemba kaizen* initiative has been marked by a comprehensive culture change. All employees from senior managers to stocking clerks have firmly embraced the *kaizen* culture. There will be challenges in the future, but Sonae MC has the *kaizen* culture in place to stay the course on a journey that never ends—the journey of continuous improvement.

For Manuel Fontoura, Sonae MC COO, five years later, there is no doubt that the company made the right decision. "Had we not followed this path, we would be far from where we are today. More than a methodology, this is a way of life, a mutation of our DNA, that goes beyond the frontiers of operations, extending to other areas of the company," says Fontoura.

And Sonae MC is now prepared to face the future, whatever it may be, because the *kaizen* methodology has fertilized the ground where the company will lay the seeds of new business opportunities. "People are now more receptive to change, to disruption, and accept easily the changes imposed by an increasingly competitive market," concludes Luís Moutinho, Sonae MC CEO.

# Surpassing Expectations through *Kaizen* at Embraco

The pursuit of lean production, with high-performance, motivated employees and without waste, has always guided Embraco initiatives. Some years ago, while engaged in mirroring itself in the excellence described by Richard Schonberger, in his *World Class Manufacturing*, the company went beyond that. The area leaders of engineering, production, and quality then had a new mission. To take the Embraco unit in Joinville, Santa Catarina, which already worked with advanced production systems, to a new level of efficacy based on lean manufacturing practices.

Researching examples in the market, Silvio D'Aquino, one of the managers of the industrial area, had the opportunity to observe different cases of lean manufacturing applications and the range of this methodology, comparing the presented reality with what already was being practiced at Embraco. His conclusion was that, albeit advanced—since the nineties, the company had already been engaged in implementing the total quality concepts and their requirements—Embraco still had room to grow in terms of quality and efficiency.

Focusing on promoting quick changes in its manufacturing, Embraco identified the importance of total productive management (TPM) integration with the lean philosophy. Thus, in 2005, after selecting consultancy companies in the market that were fully aligned with the objectives of Embraco's operational strategy, the first steps of *lean thinking*—as the *kaizen* project was baptized at the company—began.

With experience in diagnosing industrial scenarios and implementing, together with the client, *gemba kaizen* and lean manufacturing projects, the Kaizen Institute Consulting Group–Brazil was the provider chosen to share with the Embraco professional team the challenge of achieving better productivity indexes at the Joinville plant. In addition, the Kaizen Institute Consulting Group–Brazil masters TPM practices capable of renewing the organizational culture and promoting positive changes on the factory floor with agility, competence, and flexibility to respect the peculiarities of each business.

## Agility and Compatible Cost

Meeting with efficacy the market demands with a maximum of quality and the best cost and providing quality for the delivered product, with an intelligent cost, and respecting the environment and the health and safety of the employees, all with a high dependability throughout the process and agility in the response time to the client—this was the goal. A chain this complex inspires full-time dedication, where each link needs to give feedback to the other so that this entire dynamic is exemplary and the objective is fulfilled. Transforming this mission into reality for manufacturing is a challenge and then some. Embraco did it. Its operational excellence is a reference and has contributed to strengthening its leadership in the global hermetic compressor market (see Figure CS-1).

## Commitment to Quality

Embraco, whose headquarters is in Joinville, South Carolina, is a company committed to quality. The good performance in its industrial area is the

**Operational Excellence OPEX**

- **Lean thinking**
- Supply chain management
- Quality
- Demand management
- P3M
- Embraco management model
- Competitive intelligence

**Figure CS-1** Operational excellence at Embraco.

fruit, among other efforts, of a culture constantly searching for excellence and which has been part of the company since the seventies, when it began its activities. This vocation was confirmed and still exists to this day.

Since the beginning, the company objective was to expand its competitiveness by increasing productivity and reducing operational costs. *Kaizen* arrived at the company with the task of yielding good results and "paying for itself" in the short or middle term. Challenge taken, the project began in Joinville. The starting point was opportunity identification through the value stream mapping (VSM) and definition of the objectives to be met, followed by the development of an initial plan capable of transforming one of its units and guiding the first steps of its implementation.

Without losing sight of the Embraco operational strategy objectives, the Kaizen Institute and the Embraco project managers focused efforts on developing and implementing the actions identified in the *kaizen* vision outlined in the future value stream. Promoting workshops directly in the *gemba*, which gave good results in just one week, was a determining factor that motivated the team to present these preliminary results to the company's administration and to receive approval to carry on with the project. To structure this first step, the team organized some sensitizing events for the leadership, bringing together about 70 professionals, among them directors, managers, and leaders. All were involved in a workshop that had as its objective to present the *kaizen*-lean concepts and tools in a practical way, laying the foundation for one of the principles of the renowned Japanese methodology—learning by doing (see Figure CS-2).

The games conducted during the *gemba kaizen* workshops consolidated the bases that the teams needed to execute the proposed project successfully. The result was a collective adherence. Everyone was convinced that if well conducted, the *kaizen* not only could make an analysis in the Joinville plant's processes but also could indicate the improvement points and lead the unit toward better results. Convinced of the theory and affected by the workshops conducted, the team was faced with the hardest task: to actually guide the change and to achieve the established objectives.

Each week, new workshops of *gemba kaizen* were conducted, and more employees were sensitized to understand their important role in the transformation process that the company intended to implement. One more step brought another positive result as new achievements and new rounds of *kaizen* took place. Each week the operations director at the time

**Figure CS-2** Simulating lean introduction.

personally checked results and validated everyone's engagement—as a way of supporting the entire transformation process.

As important as it was to deepen the analysis of each existing value stream and to align each of those transformations with a group of *gemba kaizen* workshops, it was equally important to create a strategic plan for each of the value streams that was perfectly aligned with the Embraco strategic objectives. This activity was extremely important for breaking down organizational goals into clear goals for each of these transformations, as well as creating and laying the foundation so that a true cultural change in the organization could begin and continue (see Figures CS-3 and CS-4).

Another important point was to design structured actions focused on people development within the concept of "learning by doing and learning by teaching." To sustain uniform communication capable of guiding everyone, Embraco created a group of communication actions with the great slogan "Continuous Improvement Is Our Goal" that aligned all dissemination and communication actions with all the results obtained by the *kaizen* teams (see Figure CS-5).

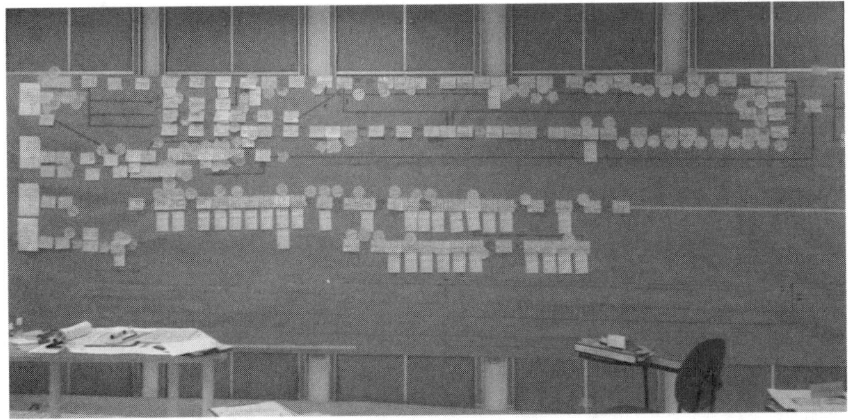

**Figure CS-3** Value-stream design at Embraco.

**Figure CS-4** Creating the strategic plan for transforming the value stream.

## Learning by Example

Goals were met at the Joinville plant, surpassing expectations. The increase in productivity was around 30 percent, which meant avoiding investments of approximately $45 million. Permanent advances and improvement were achieved and maintained by an engaged and totally committed team.

The positive scenario motivated a new step: international benchmarking. The *kaizen* consultants selected some projects from their inter -

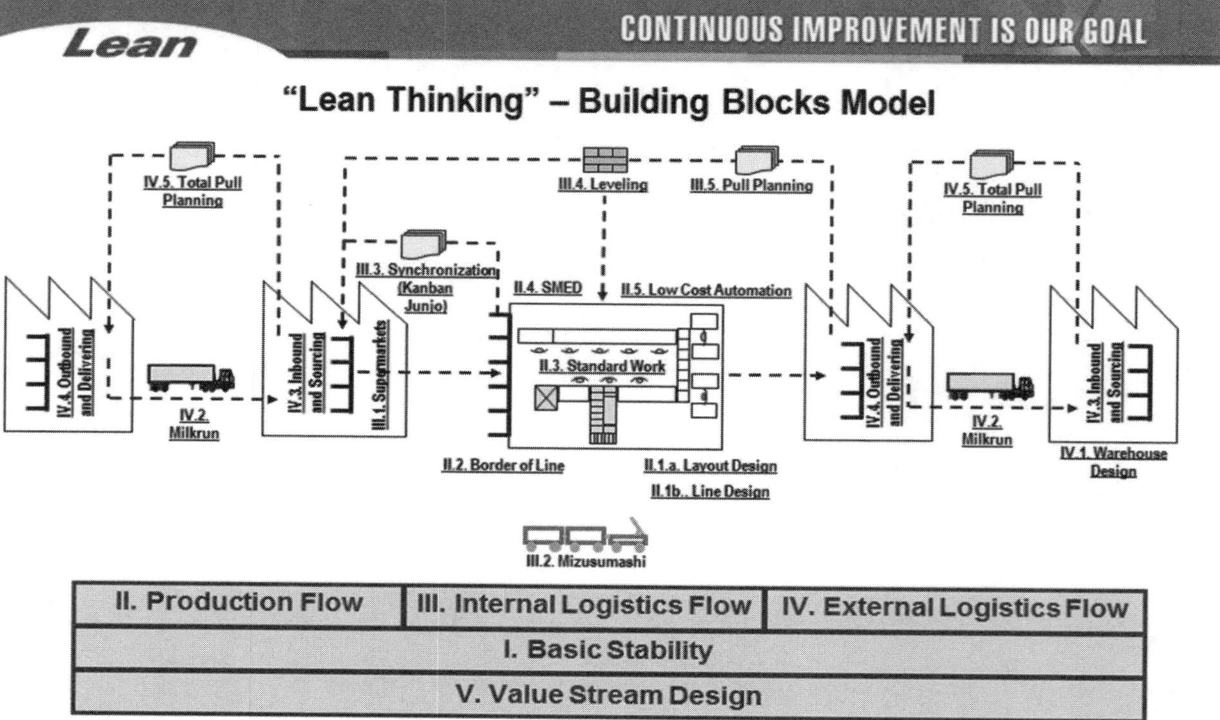


**Figure CS-5** Building blocks of operational excellence.

national clients and coordinated the visit of a group of executives and leaders from Embraco. Once again, to venture beyond the walls of their reality was a rich experience. There were companies doing even more and achieving excellent results. It was necessary to seek more results!

Soon another very interesting practice also had the chance to occur: an action of internal benchmarking with other plants from Embraco outside Brazil. Leaders of plants in Slovakia, Italy, China, and the United States visited the headquarters in the south of the country to check for themselves the results obtained by the Brazilian plant (see Figure CS-6).

In 2007, the *kaizen* projects adopted in Brazil started to take their first steps outside Brazil. Considering the needs and characteristics of each plant in each country with its own culture and specific productivity challenges, a new stage was started. The results obtained in Brazil inspired Embraco's other plant managers to review their productivity and performance indexes and seek a continuous-improvement project (see Figure CS-7).

To sustain the project in its entirety, leaders were designated from each plant in each country, with Kaizen Institute international teams responsible for closely monitoring this work. China, for instance, which did not yet have a Kaizen Institute in the country, was supported by a team from the Kaizen Institute of Portugal, whereas the Embraco North America Project had the direct support of the team from the Kaizen Institute Brazil.

The objectives were the same for all: to achieve the best performance at each operation and to implement the *kaizen* concepts in all units in order to provide continuity to the *kaizen*-lean culture in their respective plants.

Adjusting the approach in each country, mapping its needs from the industrial point of view, and reconciling the cultural and behavior differences of each country were great exercises that brought maturity to all those involved, such was the magnitude of the projects and challenges to be faced daily. The transfer of know-how and application of the processes adopted in Brazil in each of the plants abroad also needed to be adjusted to the characteristics of each plant.

In Slovakia, for instance, the great challenge was to create a climate of empathy among all so as to involve the employees in the project. Holding workshops was fundamental to demonstrate to the local team how much the results were directly connected to people's participation and how the interactivity of all made a difference in the tasks' execution. Today, the

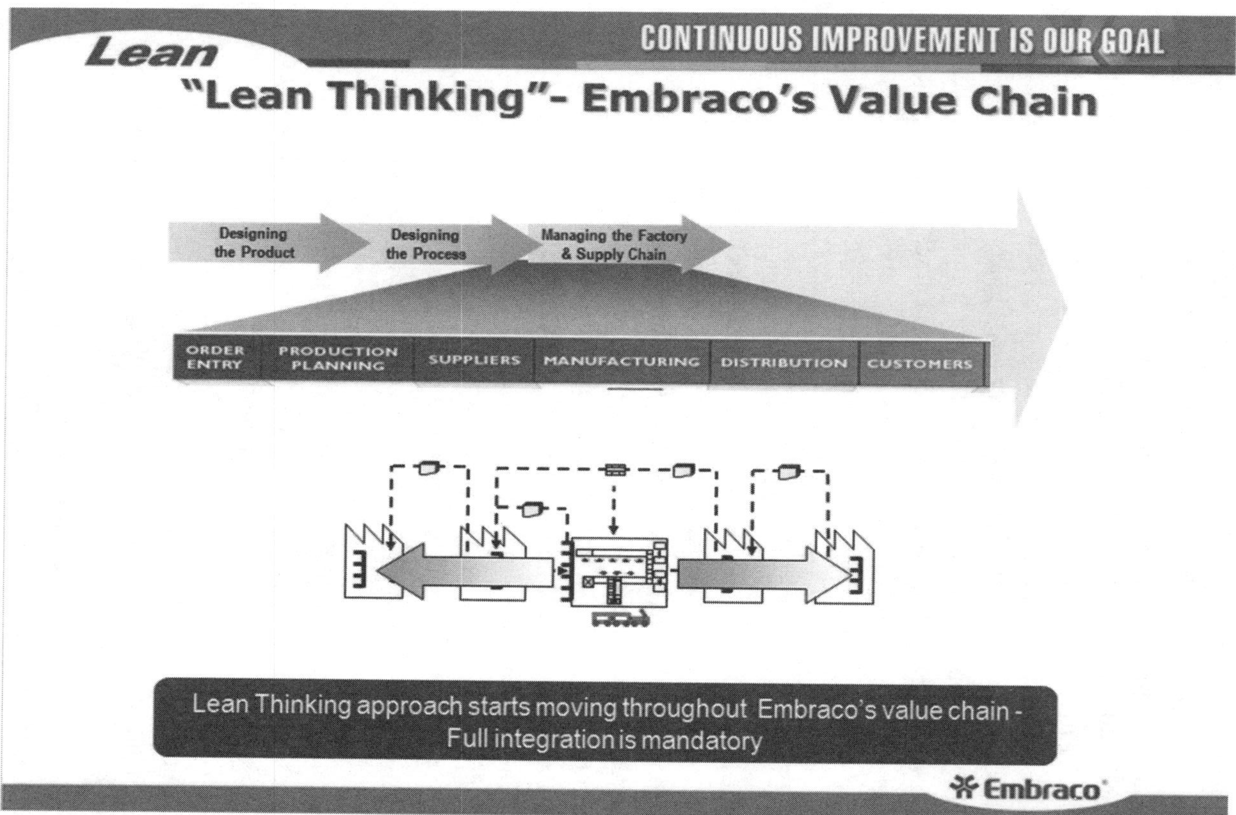


**Figure CS-6** Lean thinking and Embraco's value-chain model.

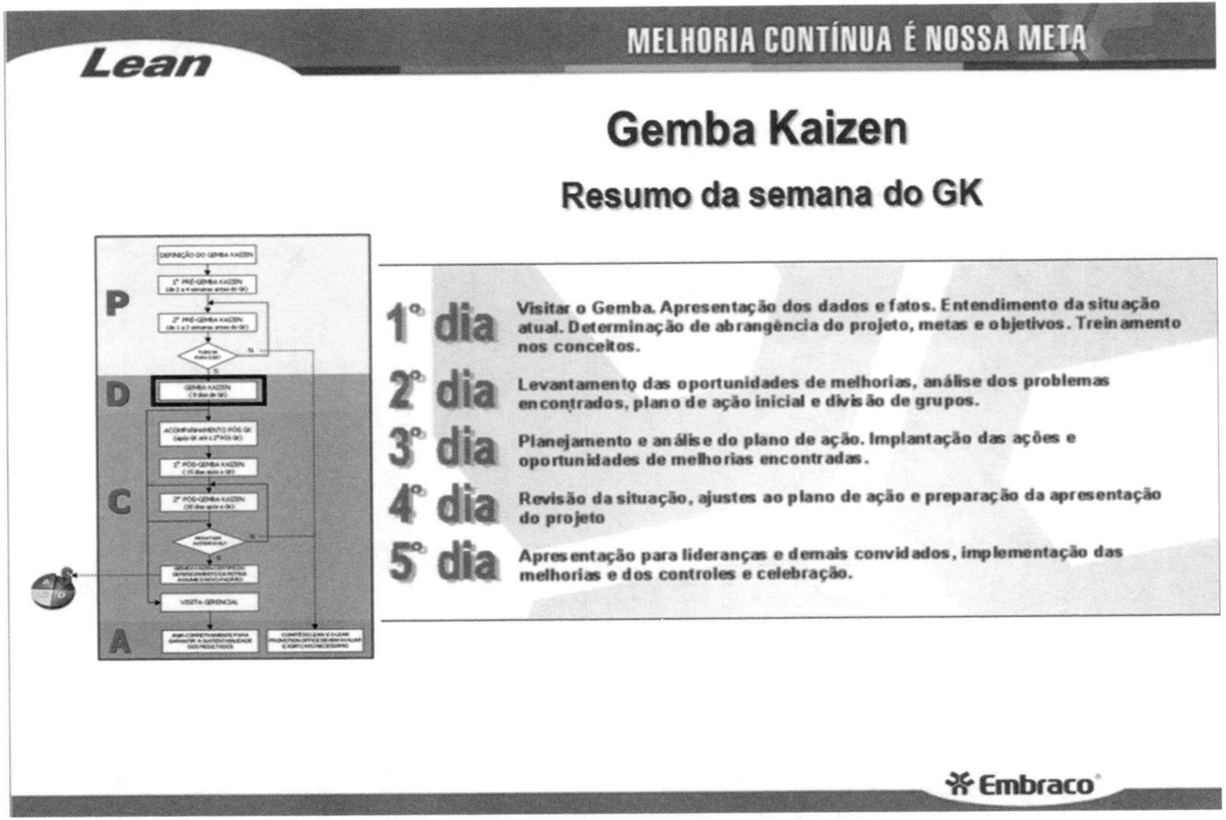

Lean

MELHORIA CONTÍNUA É NOSSA META

# Gemba Kaizen

## Resumo da semana do GK

P

D

C

A

1° dia Visitar o Gemba. Apresentação dos dados e fatos. Entendimento da situação atual. Determinação de abrangência do projeto, metas e objetivos. Treinamento nos conceitos.

2° dia Levantamento das oportunidades de melhorias, análise dos problemas encontrados, plano de ação inicial e divisão de grupos.

3° dia Planejamento e análise do plano de ação. Implantação das ações e oportunidades de melhorias encontradas.

4° dia Revisão da situação, ajustes ao plano de ação e preparação da apresentação do projeto

5° dia Apresentação para lideranças e demais convidados, implementação das melhorias e dos controles e celebração.

Embraco

**Figure CS-7** *Gemba kaizen* workshops at Embraco.

Slovakia plant has highly superior performance, and the leaders mirror themselves in the *kaizen*-lean concepts to guide the processes there.

Already very much dedicated to the optimization of processes, the Embraco operation in Italy responded quickly to the new methods as it advanced each week and the results of the improvements began appearing. Culturally, the local employees have the need to visualize the benefits in order to be motivated and then promote the changes quickly. With each *gemba kaizen* workshop, the teams focused efforts on the solution of a problem, and so the project went on week by week, achieving the desired levels of improvement (see Figure CS-8).

In China, the initial step was to overcome the cultural differences that precluded greater interactivity among teams. With this challenge met, the operation responded very well to the new proposal. The China unit advanced very quickly to the best technical qualifications and the systems that needed to be implemented. On account of a local characteristic that results in high turnover in companies, the workshop routine for standardizing new methodologies and maintaining the improvements being achieved is still intense in China.

Overall, each unit ultimately found its equilibrium axis to attain the goals established for the unit facing lean-thinking implementation. With a different response time and with peculiarities from one plant to another, the plants in the United States, China, Italy, and Slovakia have developed a productive

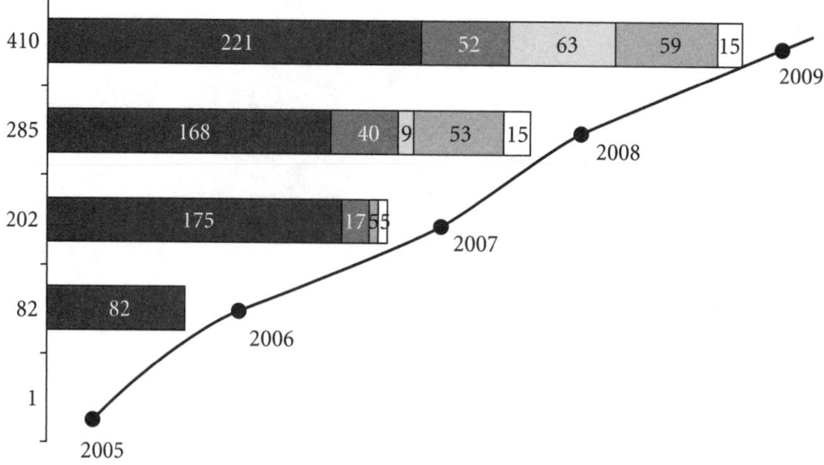


**Figure CS-8** *Gemba kaizen* workshops evolution—all plants.

force that, alongside Brazil, contributes to strengthening Embraco's global leadership in the hermetic compressor market for refrigeration.

## Valuing Human Capital

Creativity, innovation, and technology. Worldwide respect for employees is an Embraco value. Some years ago, in parallel with adopting continuous-improvement and greater productivity systems, the company also started adopting procedures even more oriented toward the inclusion of disabled people. In this phase, *kaizen* also had a strong participation.

At each time point, besides operational changes, parts of the Joinville plant underwent small structural changes to adapt the environment for the arrival of employees with difficulties in locomotion. More than a small construction project, the workers themselves in a given area got engaged, and all, jointly, enlarged a space, built a ramp, and made wheelchair user access easier, for instance. In this way, in an engaged manner—a key element of *kaizen*-lean projects—Embraco marched to a new level of excellence—of productivity and respect for its employees (see Figure CS-9).

The arrival of lean thinking always had this same orientation. Initially, the *kaizen* concept frightened the employees owing to a false association of the *kaizen* methodology with substituting workers with machines. To clarify the real foundations of *kaizen* and their alignment with the Embraco values was one of the tasks of the team heading the program.

Embraco at that time made a commitment to all its employees that the people involved in the process would not be fired. At each stage and at each improvement, the professionals with a higher profile and greater collaboration in that achievement became the multipliers of the knowledge acquired and the *kaizen*-lean transformation agents. And this dynamic, which gave priority to valuing talent and recognizing individual and collective efforts, sustained lean thinking in a way that permits the benefits obtained to go beyond the numbers. The company achieved differential quality standards with a high productivity index and a professional team committed to sustainability of this success.

**Figure CS-9** Respecting people—one of the Embraco values.

# *Kaizen* at Oporto Hospital Centre: Making Patient-Centric Care A Reality

One of the most exciting developments of the past decade has been the adoption of *kaizen* by the health care industry. Today, demographics and economics are such that many hospitals and clinics face increasing patient loads with no funding to add staff.

Toyota faced a similar situation when it saw growing demand for its trucks at the outset of the Korean War, yet it was forbidden by its banks to bring on more people because of past financial problems. Taiichi Ohno was forced to use *kaizen* instead of money, and the rest, as they say, is history.

A distinctive feature of health care is that the lean value stream revolves around the customer, who is generally the patient. At the input, the system sees a person with a particular health-related complaint. At the output, instead of a manufactured product, we see a healthier person whose complaint has been resolved satisfactorily.

Because this relationship is so intimate, problems in health care are very visible. Few people will ever see the *muda* in an automotive plant, but most people have experienced long wait times in hospitals, crowded emergency areas, and situations where "the right hand doesn't know what the left hand is doing." These are the health care problems we read about in the newspapers.

The use of *kaizen* tools at Oporto Hospital Centre (OHC), located in the north of Portugal, clearly showed that it is possible to eliminate these kinds of inefficiencies through simple solutions and minimal financial investment. The center is made up of three health units: Santo António Hospital, Maria Pia Hospital, and Maternity Júlio Dinis. Santo António Hospital alone has 3,200 workers, about 600 beds for inpatient units, and 50 different medical specialties. When this hospital became an independent institution in 2004, management decided to combat the inefficiencies that were preventing the OHC from attaining better results.

This case distinguishes itself by the improvements made in lead times and service quality. In care, there was a significant reduction in costs and an improvement in customer service, particularly in areas such as the hospital center purchasing and supply center.

OHC also accomplished a significant cultural change. Now all the hospital personnel are concerned about reducing waste with the purpose of reaching the strategic goals proposed by the organization.

## First Steps: Improving the Hospital Logistics Systems (HLS)

The first *kaizen* project at OHC took place in the hospital logistics and supply areas and was called the *Hospital Logistics System* (HLS) (see Figure CS-10). The goal was to improve the efficiency of supply management processes and to take an important first step in the cultural changes that would be made at the institution. Despite not having a direct effect on patients, the adequate distribution of products is crucial in the provision of essential medical care.

After receiving training in 5S, visual management, and standardization from the Kaizen Institute, a designated lean team that included managers,

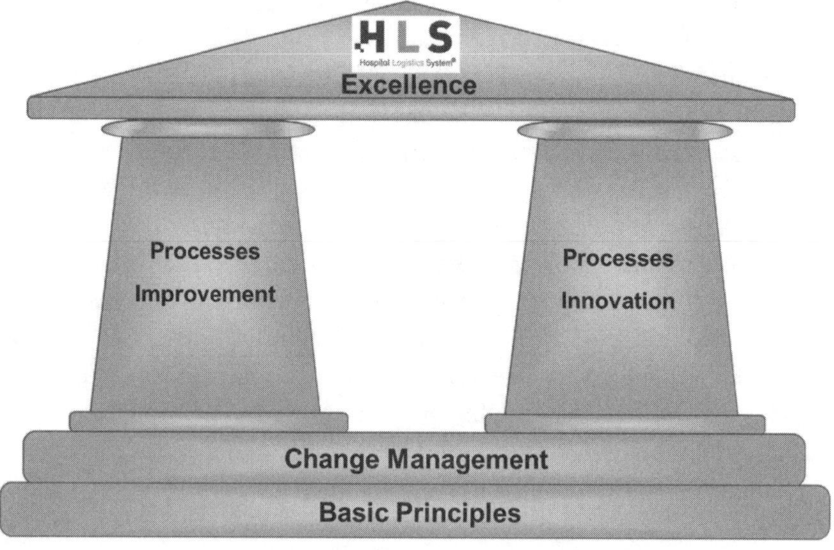


**Figure CS-10** HLS organization.

care workers, and logistics personnel took on the task of establishing priorities for change and improvement. Teamwork was emphasized from the outset, which is an essential ingredient of *kaizen* efforts. The team attended regular sessions, documented their workflows, and encouraged workers to search for more efficient ways of performing their daily tasks with the help of *kaizen* tools. Each daily meeting focused on a particular group of indicators that pertained to factors such as stocking levels, defects, space requirements, workload balance, urgent requests, and overtime hours.

The team decided to target the replenishment system for clinical and pharmaceutical supplies. The previous system, which used preestablished stocking levels that had to be monitored continually, was replaced with a two-bin *kanban* system that was updated twice a day. A simplified picking system, achieved by reconfiguration of the central warehouse using 5S, and the introduction of *mizushumashi* (a logistics person, also called a "water spider") were measures that made the transition possible.

The key to executing the change was broad participation among workers, and it took persistent effort to overcome the usual cultural challenges that *kaizen* efforts confront. "Changing paradigms is not an easy task," says Vitor Herdeiro, hospital administrator at OHC, "and when one changes patterns of behavior, it is far from being peaceful. Therefore, it is essential to involve all people, for these are the ones who ensure the success of cultural change. I believe in the amazing transformation that *kaizen* brings to the teams, but the success is only achieved if all are part of the project, if there is commitment of management and greater involvement of all. Therefore, I must emphasize that nothing should be imposed, the solutions must come from everyone, or otherwise the process of continuous improvement may not achieve the proposed objectives" (see Figure CS-11).

The results were particularly motivating for staff. Because the new system responded directly to demand, a number of nurses no longer had to manage their units' stocks and were able to spend more time with patients. In the administrative area, processes were standardized and simplified, and unnecessary tasks were eliminated.

Numerical results (see Table CS-1) included the following:

- Stock in the medicines warehouse were reduced from 5 million to 3 million euros.
- Space requirements were reduced by up to 70 percent.

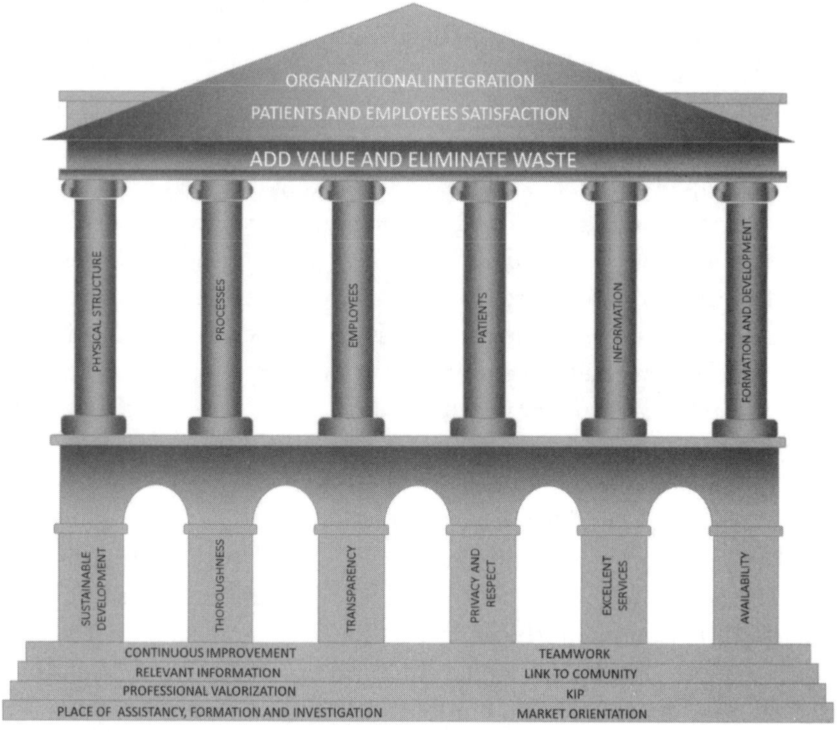


**Figure CS-11** An integrated, patient and employee-centric system.

- Overtime, urgent requests, and defective orders were reduced to zero.
- Productivity in clinical services replenishment increased by 75 percent.

## SAH Reduces Waiting Time in Care

Encouraged by the results obtained in logistics, the OCH decided to apply lean practices to the care division of Santo António Hospital (SAH). The significance here is that the *kaizen* initiative was now moving into areas that touched the patient directly.

The obvious target was wait times. In Portugal, it is traditionally accepted that one has to wait months for an outpatient appointment in a public hospital, and therefore this has become accepted as normal. The OCH management team knew from its experience in logistics that such wait times are unnecessary symptoms of faulty processes. According to Luís Matos, manager of the care section at SAH, "We just had to look around us.

**Table CS-1** HLS results.

| Scope | Objective | Results |
|---|---|---|
| • Internal clinical services replenishment<br>• External clinical units replenishment<br>• Warehouse reorganization | • Stock reduction in the internal clinical services<br>• Stock reduction in the external clinical units<br>• 0 ruptures<br>• Increasing logistics productivity<br>• Space reduction<br>• Reducing extra hours<br>• Reducing urgent requests | • Stock:<br>  ◦ 40% stock in the internal clinical services<br>  ◦ 70% stock in the external clinical units<br>• 0 ruptures<br>• Productivity: +75% increasing in the clinical services replenishment<br>• Space: –50% space used (internal clinical services)<br>  ◦ –70% space used (external clinical units)<br>• 0 extra hours<br>• 0 urgent requests |
| • Medications purchase<br>• Unit dose pharmacy preparation | • Stock reduction in the warehouse<br>• Planning fulfillment<br>• Balanced workload | • **40% stock reduction (2 M€)**<br>• 0 ruptures<br>• 20% productivity increase<br>• Service level increase<br>• 14% space reduction in the general medications area<br>• 50% space reduction in the nutrition area<br>• 65% space reduction in the antiseptics and disinfectants area |

We had papers everywhere, and doctors' offices were littered with huge piles of clinical files. We knew we had to improve, but we didn't know how."

The team set to work improving the various workflows. In order to speed up the referral process, delivery of appointment requests to appointment managers was changed from weekly to daily. Envelopes containing requests were color-coded by category so that they could be quickly recognized, and available appointment vacancies, as well as the waiting time, were published so all staff were aware of work status.

The work and dedication of staff that was required to implement these changes were, as they had been for logistics, significant, but it paid off. Without investing in information systems, the team was able to continuously improve the hospital's response to the increasing number of appointment requests.

The care section of SAH had an average of 50,000 patients per month, and with lean, the waiting time for the first outpatient appointment was drastically reduced from 70 to 46 days (see Table CS-2). "The patient is already aware that the hospital is responding much faster in the outpatient appointment's booking," says Matos.

The project also was able to reduce the waiting time of the first outpatient appointment from 38 to 7 days by reorganizing the process of appointment referrals.

## Lean Operating Room

Before implementing *kaizen*, there was an atmosphere of constant dissatisfaction among the teams in the operating room. "It was terribly tiring," says Laura Galego, the operating room's chief nurse. "Processes were constantly backed up, and we felt that they had to be standardized, that the flow had to be improved." Staff, especially nurses, were anxious for improvements.

**Table CS-2** Lean in the outpatient appointment results at SAH.

| Indicator | Before | After |
|---|---|---|
| Number of outpatient appointments (daily average) | 1500 | 1800 |
| Outpatient appointments' sorting plus reference time | 38 days | 7 days |
| Waiting time for first outpatient appointment | 70 days | 46 days |
| Replenishment frequency | weekly | daily |

Sometimes, health care professionals were spending more time on the phone dealing with requests than taking care of patients.

Again, it was fundamentally important that all the staff were involved in designing and implementing improvements. The initial lean team had 12 people, all of whom were directly or indirectly connected to the operating room, and they studied each process individually. Little by little, more were called to join, until a team composed of people from all levels was involved in the improvement process.

Team members used the same lean tools as in previous projects, but these were adjusted to the requirements of the operating room. In the new, 5S-transformed environment, material and medicines were stored and organized properly, there was more floor space to help facilitate the transportation of patients, unnecessary tasks had been eliminated, and paperwork had been reduced.

Procedures involving patient flow also were improved. Methods of transferring patients and communicating with orderlies were improved to prevent delays at the beginning of surgery. The booking of new surgeries was rationalized, standardized, and simplified. Surgical material kits were reviewed and rethought; of the 145 existing items, 23 were eliminated, and 6 new ones were created. The need for paper forms to reorder items was reduced by the introduction of *kanban* cards.

The improvements were small, but they added up to an environment that was more conducive to caregiving and less stressful for staff. "I think that the work done with *kaizen* has a big influence in the satisfaction shown by the hospital's collaborators, especially in the case of nurses, whose pressure has been reduced," says Simão Esteves, anaesthetic doctor and director of the operation room. "Little changes such as the use of mobile phones to call the next patient were very useful. Visual management has clearly reduced workplace stress among surgeons and their assistants. With the dynamic work plan in place, it is now much easier to know where the workers are and to have a fairer distribution of tasks."

The results also included a surprise—the critical waiting list was reduced by 9 percent and the outpatient waiting list by 75 percent (see Table CS-3). "We never wanted to do more surgeries, but we did want to increase the quality of surgeries by having more time available for the patients," says Luís Matos. "If waiting lists are reduced, that's great, but that was never our main goal."

**Table CS-3** Lean in the operating room results.

| Indicator | Result |
|---|---|
| Efficiency | +5% |
| Nonconformities | –60% |
| Waiting list | –9% |
| Waiting list (verify normal priority) | –75% |

## Cultural Impact at the Hospital

Despite the fact that this case involves three different projects in three different hospital areas, all have seen similar increases in productivity and improvements in patient care. In all three instances, doctors and nurses are spending more time taking care of patients and less on disorganized bureaucratic processes. The OHC *kaizen* experience shows that better-organized work reduces stress and creates more relaxed professionals who make fewer mistakes. Organizations that reduce costs by eliminating waste ultimately have more caring and motivated professionals.

"This way of doing things has to be seen as a transformation of organizational culture," says Manuel Valenta, head nurse at the SAH operating room, "valuing people, innovative ideas and simplification of processes in the environment where everything happens, in the original Japanese *gemba*, where value is created."

# *Kaizen* Enables Innovation and Customer Intimacy at Densho Engineering

When I speak about *kaizen* at conferences, I frequently begin by pointing out that *kaizen* means "every day improvement, everywhere improvement, and everybody improvement." This case study is a wonderful example of how such an approach can help a small high-tech company innovate rapidly enough to keep pace with much larger companies while maintaining its status as a trusted supplier of highly critical components.

Densho Engineering is located in Saitama, Japan, and processes glass screens for mobile phones, tablets, and other electronic devices. Since the industry evolves at lightning speed, these glass panels must improve in quality continually by becoming thinner and stronger. To succeed, Densho must conform to the ever-increasing technical requirements of market leaders such as Sony, who make up its customer base.

Densho is privately owned and employs 110 people. Although many companies in this industry are attached to large corporations, Densho has managed to stay independent, even though its customers are many times larger and have multimillion-dollar research and development (R&D) departments.

The business challenges for Densho are significant. The Japanese high-tech industry, especially the semiconductor and liquid-crystal display (LCD) industry, is in a constant state of drastic price reduction, technology improvement, and changing work content. Mr. Iwao Sumoge, president and owner of Densho, says, "If you want to play in this industry, you need to be able to move both people and money fast! It's not for the faint-hearted."

At Densho, *kaizen* does not just reduce costs and ensure the best quality—it also ensures a culture of rapid innovation. Densho employees have to learn to think and act like leaders.

Mr. Sumoge is a respected expert on processing glass. He has published widely read academic papers on the subject and holds a number of patents.

As *shacho* ("leader") of a Japanese company, Mr. Sumoge takes an active role in imparting knowledge to employees. In fact, another meaning of the word *densho* is to "pass on to the next generation."

## Daily *Monozukuri* Class

Every day, Mr. Sumoge leads a 30-minute *monozukuri* class. In Japanese, *monozukuri* means "making things," but it has a deeper feeling of craftsmanship and pride in manufacturing. The objectives of this class are to help participants see problems and opportunities and to learn to work outside their comfort zone.

### *Seeing Problems and Opportunities*

At Densho, priority is given to solving problems that have occurred in the *gemba.* If there was a defect during the previous shift, this is investigated using the 4M approach (i.e., man, material, machine, and method) followed by asking the five whys to find the root causes. Once the root causes are found, the company can take action (see Figure CS-12).

After the class, each small group gets together to look at the schedule and to take actions based on the outcome of the class. Through this process, the company is able to minimize defects much more quickly than the competition (see Figure CS-13). Mr. Takakura, head of operations, said, "In the end, we just eliminate one reason after another, and every *kaizen* is an experiment."

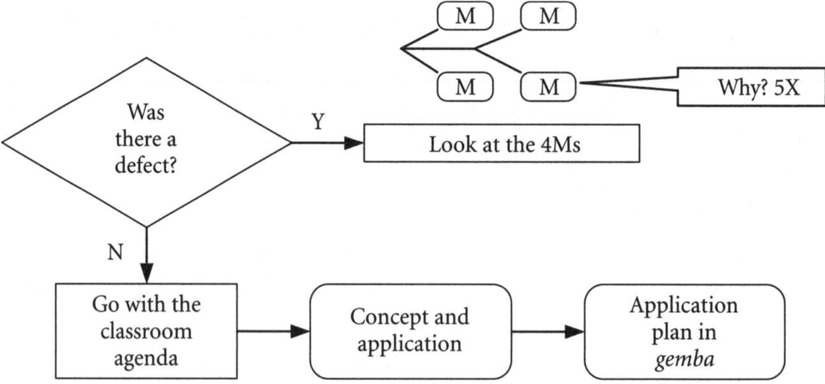


**Figure CS-12** The simple yet highly effective format for the *monozukuri* class.

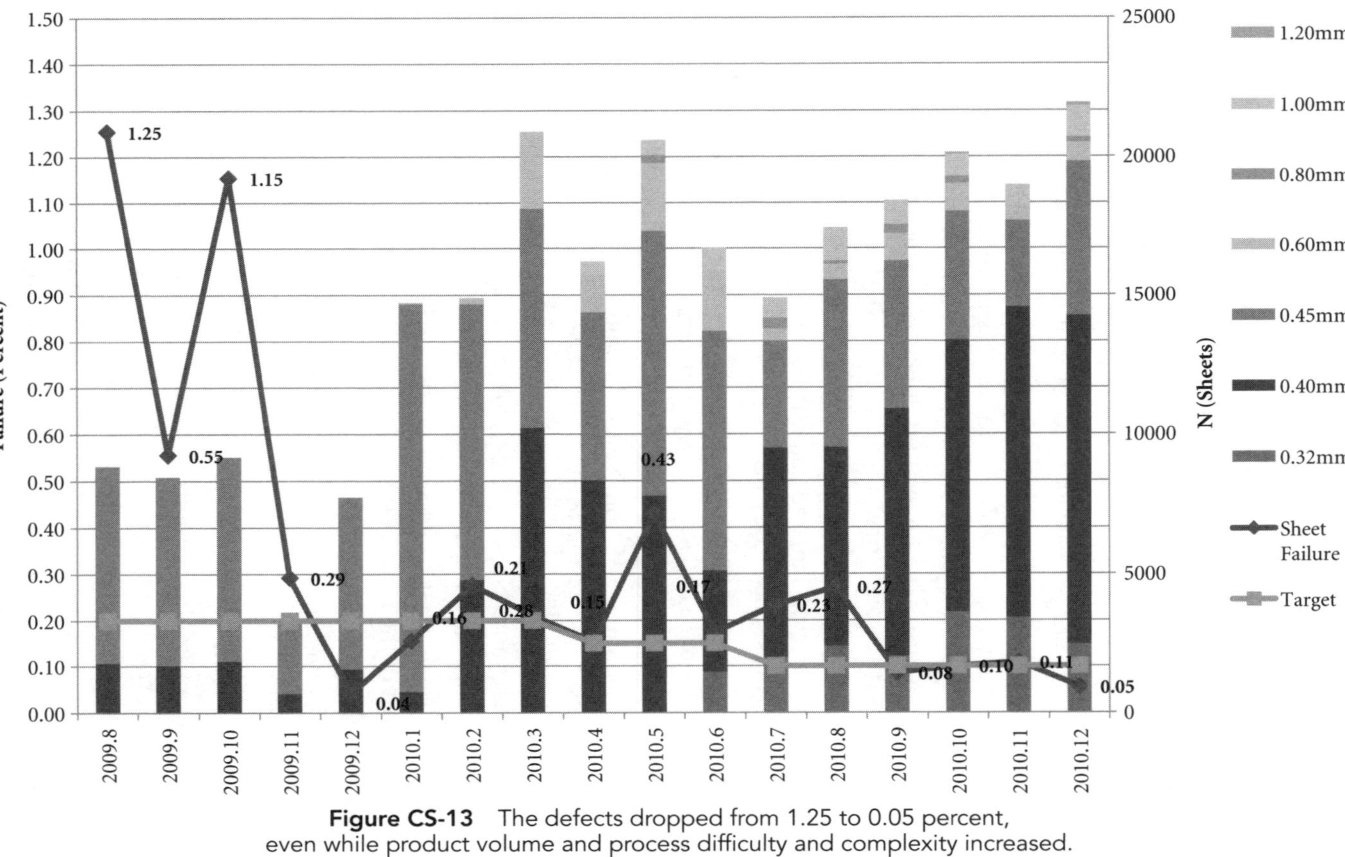


**Figure CS-13** The defects dropped from 1.25 to 0.05 percent, even while product volume and process difficulty and complexity increased.

## *Moving Beyond the Comfort Zone*

Getting workers to think and act outside their everyday comfort zone is essential to the company's culture of rapid innovation. Mr. Goto, Densho's production manager, recalls an incident where Mr. Sumoge taught employees to overcome their fear of maintaining highly sophisticated equipment:

> We had a lot of issues with robotics, but since none of us were robot experts, we were reluctant to go ahead and take apart the robot. One day, during our *monozukuri* class, the *shacho* brought in an antique rifle used by the *samurai*. He asked us to take it apart. All of us refused, but he insisted. It was very scary since this was such a valuable piece of work, but in the end, after a lot of nervousness and sweat, we did so and then put it back together again. The *shacho* then said, "I trusted you with the antique gun, so I trust you with the robot. Just go ahead and try and maintain it yourselves." That was a huge turning point for us. Now we pretty much do all the maintenance and even heavy construction ourselves.

The development of confidence in autonomous maintenance of machines as a result of the *monozukuri* class extended to other indirect facilities management areas. Over a period of two years, the company evolved from contracting out all external construction and factory layout work to handling such work in-house. Recently, Densho purchased a brand-new machine, the first of its kind in the world, and Densho employees handled most of the installation of this machine in the factory. As a result, they were able to reduce the installation cost by 96 percent.

The daily *monozukuri* class is complemented with two additional regular activities:

1. *Team* kaizen. The team *kaizen* happens once every three months. Employees are divided into groups and work on issues that concern them. Normally, people are looking for ways to make the job easier or safer and to improve quality. The "before" and "after" conditions for each *kaizen* are summarized in a report, and prizes are awarded in a formal ceremony. Not only are the *kaizen* project teams good for the company, but they also provide great evidence to help the sales

department promote Densho's qualifications to customers. Every three months the sales force can demonstrate Densho's commitment to continually making products better, safer, and quicker than the current situation through kaizen.

2. *5S and visual management.* Densho's layout is always changing because the company continually takes on bigger jobs and adopts new processes. The practice of 5S is essential to ensure that when the work environment changes, order and cleanliness are not left behind. The following steps are used:
   a. Clean the area and throw away the garbage. This is easier said than done because there are many rules and restrictions around the proper disposal of chemicals and waste materials used in Densho's processes.
   b. Put everything in place, and label it.
   c. Keep the environment clean.

Mr. Hinosugi, who leads the 5S effort, said

> The spirit of 5S in Densho comes from an old Chinese saying: "Even if you are not rich, if you clean your house, it will still be respectable." We at Densho are not a rich company and are much smaller than our customers, but we can still be impressive if we have good 5S.

### *Achieving Densho's Corporate Goals*

The overall objectives of Densho Engineering are to continually innovate and to build open, trusted relationships with customers. By doing both, Densho truly can become partners in responding to the market.

## *Kaizen* Enables Innovation

One of the biggest innovations for Densho in the past two years has been the recycling of chemicals. Densho uses hydrofluoric and sulfuric acids to etch the glass. These substances are costly to dispose of and potentially damaging to the environment. Working with Kansai University, Densho has developed and patented a method to recycle these acids. Based on the success of the project, Densho received the very prestigious New Energy Development Organization (NEDO) Award from the Japanese government. The 5S and *kaizen* activities provided Densho with the open space in the factory for the recycling plant, eliminating the need to invest in additional space.

## Becoming a Trusted Partner

A key to building trust is sharing information openly to ensure that everybody is working from the same data. Some companies find it difficult to have this kind of candor with the customers they are trying to impress. Densho makes a special effort to keep this priority in perspective for all customers and employees.

As a supplier with unique processes and capabilities, a quality problem or line shutdown at Densho can put 10,000 to 20,000 people out of work at Densho's customers' factories. Solving quality problems at Densho therefore is not just an internal matter—it is about "protecting our company, the customer employees, and even the country of Japan." Keeping this in perspective ensures that both Densho and the customer can solve problems without trying to find out who is to blame.

The resulting level of trust has allowed a strong partnership. Densho and its customers in Japan and the United States jointly own many patents. Densho uses *kaizen* to continually show customers that the company is

- ▲ Always innovating and doing *kaizen* to reduce cost and reduce lead time
- ▲ Not hiding problems
- ▲ Maintaining a spirit of cooperation that makes problem solving transparent and easy

As a result of *kaizen*-fueled innovation and partnership, Densho is assuming an expanded role for customers in the LCD manufacturing process. This is making Densho a one-stop shop for the Japanese electronics giants, a true supply-chain partner that will drastically reduce lead times and make its customers more competitive in the global market. This huge responsibility shows that the people at Densho are trusted to maintain perfect delivery and quality in order to keep the LCD industry running in Japan. *Gemba kaizen* is a key enabler of the long-term strategy of Densho.

# Cutting Red Tape at a Public Utility: Enexis

When an individual does business with a large organization such as a utility, a phone company, or a government organization, it is common for such interactions to be slow and unnecessarily complex. The cause is a particular type of *muda*—characterized by unnecessary steps and poor coordination of different players—that is frequently described as "red tape."

The people who best understand this kind of waste are the workers in the *gemba* who interact with customers every day. When senior management is willing to step back and allow such employees to take ownership of their processes, organizations can, with the help of some simple lean tools, improve their customer service significantly in a very short time with very little investment.

A *kaizen* project implemented by Enexis, an energy distribution company based in the Netherlands, provides an excellent example of this. The company provides electricity to 2.5 million customers and gas to about 2 million, collaborating with a variety of third-party providers to ensure that customer requirements are met in a timely and qualitative manner. Enexis operates within 10 different regions in the Netherlands.

Before implementing *kaizen*, Enexis was receiving frequent complaints about the time required for power and gas connections. Their process has four steps:

1. *Intake.* A customer sends a request for services via the Internet. An administrator receives the request and sends an acknowledgment. A price quotation is sent several days later to the customer. The customer signs and returns the document.
2. *Technical preparation.* A request is sent out to a third-party installer.
3. *Realization.* The power or gas service is installed.
4. *Invoicing.* An invoice is completed and sent to the customer.

These four steps were typically taking two to three months to complete, which senior management deemed unacceptable. To speed things up, management decided to implement *kaizen* to streamline the intake phase.

Working with the Kaizen Institute Netherlands, management began by coordinating *kaizen* workshops, but only the workers—from all 10 regions of Enexis—actually were involved. While some managers were a little nervous about taking a hands-off approach, this allowed workers to take control of their processes and provide solutions to the problems they were intimately aware of.

In their first *kaizen* session, employees measured the length of the intake phase, which varied widely by region. In one region, for example, the customer typically would receive a price quotation within 5 days 35 percent of the time and within 10 days 85 percent of the time. Another, smaller region, though, was achieving 68 percent within 5 days and 86 percent within 10 days.

In the next three workshops, employees used value-stream mapping (VSM) to complete a current-state map for each of the 10 regions. This uncovered a number of problems, including the fact that procedures were different for each region. To address these issues, the team then drew up a future-state map that would standardize a single, streamlined intake process across the company. Using the *kaizen* cycle methodology (plan-do-check-act), workers developed several key changes.

One problem area, for example, was the acknowledgment letter that administrators had always sent out to customers. These letters took up workers' time while adding very little value for the customer. The letters were eliminated for routine cases and used only in rare cases where it would take more than 10 days to complete the quotation.

Intake time was further reduced by modifying the review process. Before, administrators would draft a quotation and then send it off to the technical division for review. Nearly every time, though, the technical people wouldn't make any changes, resulting in a lot of non-value-added time. In the future state, administrators would send offers to the technical people only when they had doubts about the offer.

When the future state was finished, a pilot program was developed and then implemented in one of the regions. The results were positive, so the pilot program was expanded, culminating in great success throughout

Enexis. One region's response times, for example, jumped to 95 percent in 5 days and 99 percent in 10 days.

Enexis owes this success with *kaizen* to enthusiastic engagement of frontline workers and constructive support from management. Rather than try to control every detail, senior management formulated some general goals, initiated the *kaizen* workshops, and then let the workers take over the process. As a result, people felt responsible for the work they were doing and fought to see the changes they wanted implemented. It was their solution they were working toward, not something that was dictated to them from outside the *gemba.*

"I now have more fun in doing my work because I have to make important decisions on my own," said an Enexis administrator. "In the past, other people made those decisions for me, and my work was just the administrative part."

Management's initial fears about giving up control were assuaged by consistent feedback from frontline workers, who were very positive about the changes they were seeing. Regular progress reports also were provided.

"A great result was achieved by giving the people from the *gemba* the responsibility they deserve," said Eric van de Laar of the Kaizen Institute Netherlands, who consulted on the project. "It was a joy to see them taking this responsibility and feel them committed to realize the change."

Today, plans are in the works to expand *kaizen* to the technical preparation phase. Lean thinking, after all, is a never-ending journey.

# People Power: Participation Makes the Difference for Electrical Manufacturer in China

This case demonstrates that every *kaizen* effort depends on people. From senior management, to middle management, to the shop floor worker, all must collaborate with strong focus and enthusiasm to ensure a true lean transformation.

For many years, China has relied on low-cost labor to compete in the manufacturing sector. This, however, is changing—as living standards in China improve, labor costs are rising, and manufacturing companies there are needing to find other advantages than low cost to compete in global markets.

People in factories, however, are used to the old way, where it didn't really matter if they weren't as productive as they could be. One of the main challenges for manufacturing companies in China is changing that attitude.

Xuji Group Corporation is a leading manufacturer of electrical components and systems headquartered in Xuchang City, China, with multiple factories employing a total of 5,000 workers. Senior management decided to implement *kaizen* because they were concerned about improving lead times and productivity, so they appointed a lean team to implement practices that they had seen succeed in other companies.

After two years, the lean team was making very little progress. The problem was that the small team was trying to do everything on its own and needed more support. Rather than abandoning lean at this stage, senior management saw what needed to be done and acted on it.

"Initially, we believed that lean tools could be deployed by just the lean group in the company," says Dr. Zhen, Xuji Group's lean leader, a highly respected expert in human resources and performance management, "but we finally realized that was wrong. It should be the people. Without them, there is nothing, there is no way for continuous improvement."

In 2011, Xuji Group engaged the Chinese team of the Kaizen Institute Consulting Group to help it move forward in a more comprehensive way. The first step was to provide lean training for senior managers so that they could understand their role as sponsors of lean. Training ranged from the organizational methods of visual management to the inventory-shunning approach of the pull system.

With top management support now firmly in place, the focus of the training moved to middle management. An initially cool reception gradually thawed as people began to understand how Xuji Group could benefit, and soon the group was able to form several teams of enthusiastic workers who were willing to champion the lean process in their areas.

The next stage was the implementation of these processes. Since Xuji Group is a large company, however, senior managers knew that the transformation would have to take place in stages. "You can't try to boil the whole ocean," says Zhen.

Xuji Group began with four of its subsidiaries. Those companies concentrated their resources on a limited number of lean initiatives—four to five for some and seven to eight for others. All these initiatives, however, were aimed at two principal targets that senior management had established. The first was lead times. "There were lots of customer complaints about the delay of the delivery," says Zhen, "because it took so long to manufacture the product." The second was low worker efficiency and poor productivity, which the managers now understood was due to the waste in the production processes.

The now-energized lean teams started with the basics. They implemented 5S to create a visual environment where work cells could easily communicate and changed the layout of machinery to minimize distances that shop floor workers had to walk, as well as create a more organized, visually pleasing workplace.

Gradually, workers began to fully grasp the meaning of *flow*. One subsidiary company, for example, had been manufacturing electric power instruments by creating them in large batches, mistakenly believing that this would maximize productivity while minimizing costs. Instead, this was forcing unneeded parts down the line to the next work cell. In order to curb the congestion this caused, inventory then had to be created to store the extra parts, resulting in wasted floor space and lengthened lead times.

The factory's *kaizen* team eliminated this waste by introducing 5S to help different work cells communicate and a supermarket to help the cells respond faster to customer demand. Now, instead of creating the product in large batches and pushing the units down the line, the factory responded to demand only when customers made their orders. The results of these changes were significant; in addition to eliminating inventory, lead time dropped from 6.7 days to just over an hour, a decrease of 97 percent. Workers' productivity also jumped 30 percent.

Another subsidiary of Xuji Group found success by transforming its assembly process. "Originally, one operator would build the product from the first process to the end process on his or her own," says Zhen. "There was a lot of motion, movement, and transportation, all considered waste." Using lean thinking, workers broke production down into five distinct stages and then assigned one worker to each step of the process. Each person would complete his or her job and then pass the product down the line to the next stage, thus creating a smooth flow. "This was all done within six months," says Zhen. "Efficiency improved by 45 percent."

Results like this would have been impossible without the enthusiasm of Xuji Group's workers. "The people are the driving force behind lean," says Zhen. "Because of our achievements, we are giving back to them." Salary increases, better workplace ergonomics, and improved work environments are senior management's way of recognizing the workers' key role in improvement, as well as driving greater enthusiasm about lean.

It now looks easy, but Xuji Group's lean journey required considerable patience and a concerted effort by people at all levels of the organization. People tend to shy away from change, and it takes consistent leadership to show workers that managers really mean what they say. Indeed, lean is now so important to senior management at Xuji Group that they have tied it to their key performance indicators. Progress on these goals is measured and updated every month.

Zhen's advice for others thinking of embarking on their own lean journey is to start with a small group of devoted people. "Look for team members who are very enthusiastic about lean," he says. "They need to be willing to spend most of their time with their team members and must be willing to sacrifice."

While the early successes are impressive, Zhen acknowledges that the journey is only just beginning. These days at Xuji Group there is enough enthusiasm among the workers to take aim at some more ambitious goals. The organization now has developed a short-term plan and a three-to-five year plan that will implement lean across the entire corporation.

The long-term plan, in the true spirit of *kaizen,* is a never-ending one. "We want to change the thinking and behavior of all people," says Zhen, "and improve the quality of all goods within China."

# Rossimoda: *Kaizen* and Creative Product Development

One of the most common misunderstandings about *kaizen* is that it is only suited for predefined assembly-line tasks. This *kaizen* journey of a high-end shoe manufacturer in Italy shows that, quite to the contrary, *kaizen* can help to establish an environment where creative ideas can flow more freely and "right brain" thinking can flourish.

Rossimoda operates a 260-person shoe factory in the Brenta Riviera, a region between Padua and Venice renowned for production of high-quality shoes. Founded in 1947, the company has manufactured and distributed under license for many prestige brands, including Yves Saint-Laurent, Ungaro, Calvin Klein, and Celine. The key to the competitive success of Rossimoda and other similar companies in the region is the uninterrupted flow of creative ideas between the designers and the factories.

Over the years, a strategic positioning shift was achieved, with the average reference price for shoes growing from 150 to 450 euros. Today, all the world's important footwear brands prefer the Riviera companies' savoir faire to produce their premier collections.

When the world economy began to decline in the first decade of the millennium, the major brands began to outsource their higher-volume work to countries where labor costs are lower. In response, Rossimoda and others in the region changed their strategy to focus exclusively on low-volume, high-end shoes, which had been their differentiator from other parts of the world. While it was fortunate that this option was available, the shift to lower-volume production put more emphasis on the challenging process of implementing new products.

To realize new designs in this high-fashion market, skilled workers must come to terms with the intentions of very ambitious and finicky designers who are constantly trying to outdo each other and often expect the impossible. This is not just a matter of following orders from the designer—this is a highly creative process that requires visual sensitivity, close communication, and the ability to think "with the right brain" in order to facilitate the flow of creative ideas. Furthermore, each situation is unique—there's no predicting what will come in the door next.

A diverse set of functional skills is also involved in this process. The Rossimoda facility actually houses a "minidistrict" of smaller factories, which include a shoe-shaping factory, a shoe-insole factory, a shoe-soling factory, and a shoe-heel factory. These functions are highly interdependent—an issue in one area often affects others.

Each time a new design is received, prototypes and samples must be built on a very tight timeline and sent back to the designer for approval. The process has three phases (see Figure CS-14):

- The initial prototype is first created during the research stage, which permits progressive refinement of the designer's ideas. Here, compromises must be made between the designer's ideal look and the practical realities of production. This requires cooperation between the product manager, who interfaces with the designer, the structure technicians, pattern makers, and representatives from all stages of shoe production.
- Once the prototype has been accepted by the designer, the production of samples begins. This process also includes preliminary production

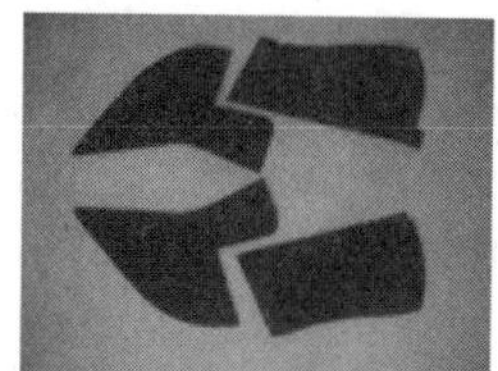
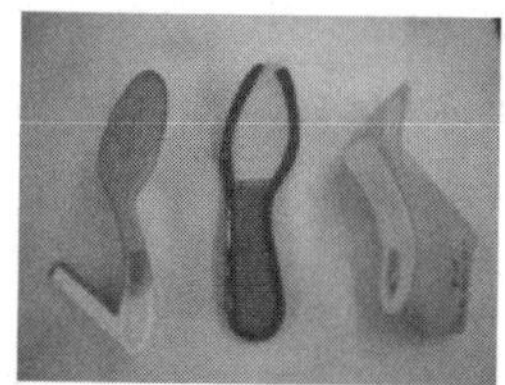
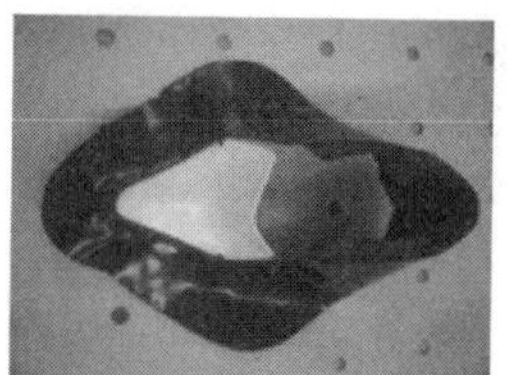

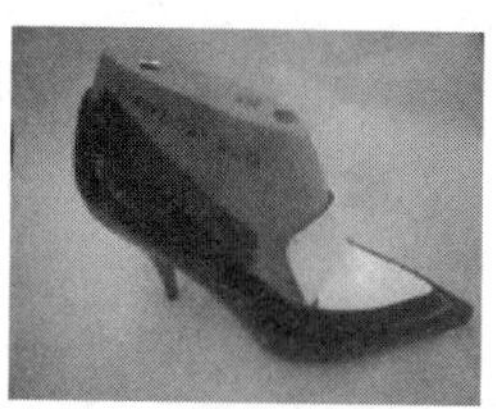
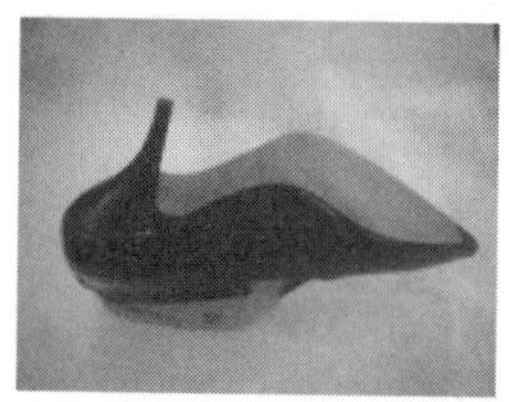

**Figure CS-14** From idea to prototype to sample.

setup and defines in detail how the production will be executed. Samples are also subject to approval and feedback from buyers.

- Production receives the information gathered during the development process, and production schedules and delivery targets are established, along with commitments to buyers.

The tension between designers and the factory during these steps is notorious. It is common for designers to complain about the lack of accuracy in the prototypes and samples, whereas the technicians responsible for realizing the designs complain about "absurd and impossible requests."

When such differences are not resolved satisfactorily, this can have two effects. On the one hand, the designer's creative capacity could be impeded, perhaps diminishing the brand's success in the market during a crucial period. On the other, unworkable solutions to designs can lead to quality problems such as high levels of customer returns, low production efficiency, delays in deliveries, poor turnover, and increased costs.

## Setting a New Course

It was clear to management that Rossimoda needed a new mind-set to strengthen the creative development process. CEO Frederic Munoz had past experience with *kaizen* and decided to engage the Kaizen Institute Italy to help the company realize this change.

Rossimoda's *kaizen* journey began with the training of managers and then supervisors in lean basics. Techniques that were emphasized were 5S, visual management, and value stream mapping. Once the training was in place, a cross-functional improvement team representing all the subfactories within Rossimoda was created.

The first major exercise for the team was to visualize the entire existing process through value-stream mapping. Team members, whose job functions were very different, had very diverse perspectives on the problems, so it was important to create an atmosphere where they could both listen to each other and speak openly. The following team process was used to create such an environment:

*Step 1:* Observe the process as a team—no judgment, just observe.

*Step 2:* Generate ideas outside the box. Use "right brain" thinking—avoid paradigms and rules.

*Step 3:* Select the best practicable idea, and sketch out a quick, rough implementation plan.
*Step 4:* Create a new work standard.

The team first mapped the prototype-creation stage, which revealed many complexities in the existing process (see Figure CS-15). For example, the entire team had to climb the stairs connecting the product office, the pattern-makers office, and the structure and production office, which were situated on different floors. This was particularly troublesome because the flow of a workpiece from one section to the next was not steady and predictable. Instead, the product frequently moved back and forth while questions were answered and issues were resolved.

This chaotic process was not consistent with the requirement of delivering a prototype to the designer within several days. It also was very difficult to verify the status of particular items in response to customer inquiries.

Similar problems, with some differences, were noted in the mapping of the development process. With the prototype approved, commitments now were being made to many players in the outside world, and with them, there were a variety of pressures. Product managers struggled to get answers to their questions while *gemba* workers scrambled to solve the practical issues around production.

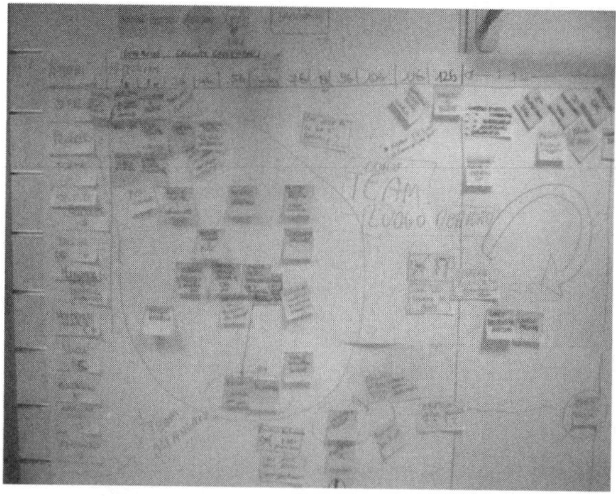

**Figure CS-15** Mapping the prototype-creation process.

## Rethinking the Process

It became clear that what was supposed to be a linear process was not linear at all. Essentially, it was not possible to predict exactly how a particular prototype or sample would take shape. Instead of orderly steps, the development process was, the team found, a bit like a random walk, with the product moving back and forth in a trial-and-error fashion until all the problems were ironed out and compromises, where necessary, had been reached (see Figure CS-16).

Through the value stream mapping process, it became clear that this "random walk" was normal and couldn't be fixed. Instead, the team had to find a way to accommodate this natural process.

What was needed was a better way to interact when problems were encountered. A heel problem, for example, could create difficulties for assembly down the road, and better communication was needed to minimize the number of times a problem had to be passed back and forth between different functional groups.

Discussions of this dilemma led to a paradigm shift. The team decided that the best solution was to have all of the product-development people in the same room and "breathing the same air." This called for creating a

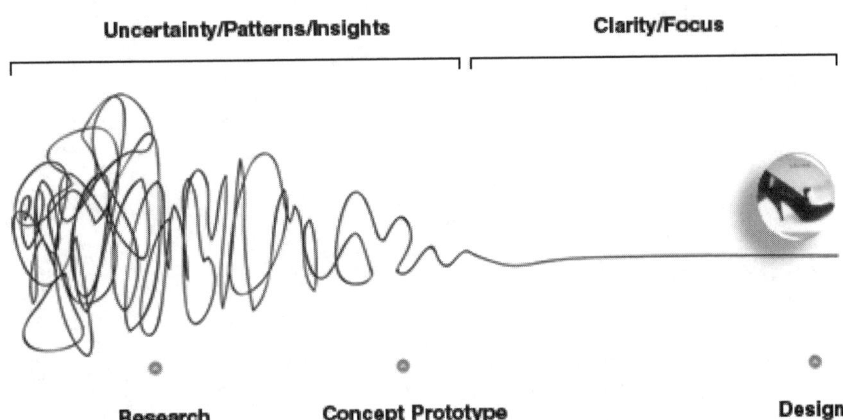


**Figure CS-16** From uncertainty to clarity.

dedicated team of pattern makers, hemmers, assemblers, cutters, and finishers to work together to, as the founder said, "author prototypes for fashion maisons." In this way, the creative spirit could be preserved through team involvement.

To make this possible, a development team of 10 workers adapted the largest available space, a former hides hall, and equipped it with all the amenities needed to produce prototypes, including tools, desks, technical positions, and producing machines. Workers used their knowledge of 5S and visual management to create a workable space that would allow the development process to flow.

The layout emphasized a section in the middle of the room where workers could access drawings and view the visual project management tools. Some initial problems colocating computer-aided design (CAD) stations and sewing machines, a combination nobody had seen before, were quickly overcome.

The first meeting, hosted by the managing director, saw people with business suits, blue overalls, and even without shirts altogether in the same room for the first time, with a great deal of curiosity and also some skepticism. Many doubted that something this unusual could work. However, by focusing on the benefits of a high percentage of "passes" and the prospect of fewer problems and less running around, team members set their minds to the task (see Figure CS-17).

**Figure CS-17** The design-process *kaizen* team at work.

Two basic ground rules of interaction were established:

- Don't just talk about a shoe—show it on a model or on your hand.
- Never say "impossible"—always make another prototype that attempts to solve an issue.

As the teams got comfortable with this new arrangement, the results began to show. With all the functional groups together, issues such as hemming problems could be resolved on the spot instead of being sent down the line with defects. Clearer communication made it much easier for the product manager to obtain information about possible options to be conveyed to the designer. Soon the factory stopped sending the product manager off to the designer with a list of what couldn't be done. Instead, there were useful recommendations for how the designer's intentions could be realized.

An example of this occurred when the factory received a challenging request for a shoe with a resin heel, a feature normally produced using injection molding technology. One of the issues with this technique is that the product shrinks when it hardens, causing wrinkles that are slightly visible on the prototype's heel and even more visible after leather dressing has been applied. The time available for the prototyping stage was insufficient to address the wrinkle issue for this model using the traditional method.

Working together, the team came up with an alternate solution that met the required time frame. In a follow-up meeting, the team compared their new approach with what would have happened the "old way."

- *What would have happened:* E-mails would have been sent informing the designer of the operational difficulty, followed by a delayed sample with wrinkles. This would have been accompanied by a detailed justification for being unable to proceed with the design. Even if the factory was not at fault, this traditional way would have obstructed the designer's work during one of the most crucial moments.
- *How the team succeeded:* After a heated discussion on the techniques and technology available, the team quickly realized that there was no possibility of executing what was requested with the traditional system. Then they changed the system. After having made a brief verification, the team decided to abandon injection molding and execute the heel by milling it in balsa wood. This broke the old rules but was much more suitable to the moment's requirements.

This experience taught a powerful lesson about the *gemba*—that the job is not to build samples and prototypes but to create a successful collection of shoes. This broader perspective has enabled the workers to think outside the box and come up with creative solutions that, according to the traditional "rules," might have been considered unworkable. By coming through these situations together, the workers began to build confidence in their ability to face challenges as a team.

The team also learned that special expertise or additional capacity sometimes is required. To accommodate extra team members, the work-table in the *gemba* was enlarged. This has given the core team additional flexibility in handling volume peaks and solving special problems the moment they occur.

## Advanced Visual Management

With multiple prototypes and samples on the go, the team also needed a visual process to make communication easier and to diminish the number of interruptions owing to unnecessary one-to-one contact. To accomplish this, the team implemented a visual management system based on simple, centrally displayed charts. These show the progress of each prototype and sample through each working phase, giving an overview of work status for all team members and product managers. This also makes all problems readily visible, allowing countermeasures to manage proactively using a plan-do-check-act (PDCA) actions list (see Figure CS-18).

Visual management made it possible to control an extremely rapid process where much must happen within the two to three days from the prototype's request to its delivery. For product managers, it meant no longer having to nervously wait with their suitcases open, hoping that a prototype would be ready in time for them to catch the last available flight for their meeting with the designer.

## Conclusion

Rossimoda's experience with *kaizen* shows the power of cross-functional teamwork. If the information being passed to the customer is coming from 10 separate groups in the *gemba*, then the process is highly complex and

**Figure CS-18** Visual management to manage the development process.

difficult. If the 10 groups are brought together as a single team in the *gemba*, the process becomes very simple.

By agreeing to some ground rules and establishing a core team and an extended team when reinforcement is needed, the workers created an improved *gemba* that is now delivering on time and fostering creative dialogue in a constructive way. Creativity can't be turned on like a machine, but *kaizen* can help teams to get the distractions out of the way so that workers can pursue what they are best at—serving customers by creating the best shoe collections money can buy.

# Finsa Uses *Kaizen* to Emerge Stronger from a Crisis

Finsa was founded in 1931 by Manuel Garcia Cambon. It began as a small sawmill in the village of Santiago, Spain. In 1965, Finsa began the manufacture of wood chipboard in Cesures' factory, which had been completely renovated. Over the years, other factories were added, and while the factories accommodated between 100 and 200 workers and most of the production facilities were in Galicia, the leadership model worked perfectly, with the owner as a role model for teams of managers. The leadership model achieved such a commitment from workers that one could safely say that it was better than that of any other factory or any other industry.

From 1995 to 2005, the company expanded significantly and quickly through technological renovation of emblematic Finsa plants and the acquisition of other companies, which, in most cases, were facing serious problems of survival. The fact is that in 1998 the company had 1,760 workers, and fewer than 10 years later, there were over 3,300; in other words, the number of workers doubled.

This made management rethink the issues. The owner's leadership was being conveyed to workers by personal example, and it was becoming clear that this was no longer possible with plants employing 600 people when 300 people, or more than half, had three or fewer years with company. However, it was clear that the leadership and its know-how were the main key to the relative success of the company.

Based on these observations, it was clear to our team from the Kaizen Institute that there was a need to write this philosophy out because hearing about it just once was no longer enough. The third generation of the owner's family transcribed, with the help of their parents, the language of the old philosophy into a clear list of Finsa's values today. These values were based on two principles: mutual respect and effective effort. The best way to apply this philosophy to the operational model was through the use of continuous improvement. Thus the company sent three executives to Japan in 2007 so that they might see how Toyota worked and later apply the model to Finsa.

Shortly after making this decision, an event occurred, at the beginning of 2008, that would help to align the staff with the new era: the start of a violent financial and economic crisis such as this country had never known before. Finsa's CEO, Mr. Carballo, said in a communication to all employees: "If the ways and forms of work that brought us here do not change in each and all of us, we will not remain as a company within 25 years. We need to change the way we work."

Two factors had driven the company to a firm commitment to continuous improvement: (1) the desire of the third-generation owners to recover and renew the basic principles of Finsa's foundation, which were just the first bricks of what would become the Finsa Management Model or MGF (*Modelo de Gestão Finsa*), and (2) a crisis that caused a drop in domestic consumption by 40 percent. All this put together helped the entire organization to have an open mind toward change in order to ensure the company's survival.

Finsa's activity was in the technical world of wood production. The company manufactured chipboard made of wood particles and medium-density fiberboard (MDF). These products were covered with decorative paper impregnated with melamine resins. Furthermore, the company also transformed the melanin into furniture for different uses, such as in kitchens, bathrooms, and offices. The boards were covered with noble wood veneers. The company manufactured MDF laminated flooring impregnated with coated paper that was resistant to abrasion. The company also had sawmill production-oriented packaging and a timber garden, which was a complement to the production of boards by employing waste wood from the sawing process.

## *Kaizen* Culture in Finsa

With the help of the Kaizen Institute, in January 2008, the company directed its first project with the business unit closest to the final consumer—furniture and components, which supplied kitchen units, stewards, and countertops with large surfaces to more than 40 stores in the Iberian Peninsula and which represented more than 50 percent of billing and, with a delay of less than 24 hours, allowed us to know which range of Finsa's products consumers took home.

The selection and management of this project marked the path of many of the decisions that were taken afterward. First, the team decided that each step to take should be based on a pilot study that always began with the client's vision, in this case external, and was supplied by the commercial network.

Before beginning the project, the company indicated that despite having a stock of finished product equivalent to three to four weeks of manufacturing, service in 12 to 15 days was unacceptable. The challenge was to replace in those 40 stores, twice a week, what was taken daily to final customers and with a stock of finished goods in the factory that would not reach 1.5 days of manufacture.

And this led to the second step/milestone of this project: establishing in Finsa the concept of pull. It is the company that chooses what it wants at the pace it wants. In this way, the company was forced to pull all the members of the supply chain and used this concept internally.

This led to a physical realignment of production machines in a new layout and to a new organization in production planning based on new concepts for the enterprise: supermarkets, *kanban*, and *muda* that led to the major shift of the production paradigm.

Machines that were designed for the manufacture of bulk lots (3,000 to 5,000 pieces) had to be adjusted to more modest manufacturing batches (180 to 200 pieces) without a loss of overall equipment effectiveness (OEE) using tools such as the standard work and single-minute exchange of dies (SMED) in order to allow more and faster changeovers and at the same time reduce the number of defective parts.

The new paradigm allowed greater flexibility in production, which reduced intermediate stocks and lead time. This first phase was somewhat confusing to the eyes of the workers: Some time was spent explaining the project and training the workers in the basic tools, but the initial vortex of changes in the locations of the machines, which played out over a longer period of time than expected, kept them disoriented, wondering "When and how would this end?"

One fact appeared to be a fundamental change in perception: The conception, design, and manufacture of a new line of kitchen modules packaging, in which the operators' participation was fundamental to the selection of up to six different production line types, building models out of wood, testing them, and choosing one of the options to build with the help maintenance team of the section. Then a pilot was developed that was tested and tweaked until the go-ahead for the final version was given by an external supplier, with the drawings provided by the workshop team.

This marked the beginning of the third step, the active involvement of employees in actions that very positively affected their working conditions.

## Planting the Seed

Conceptually, this pioneering project addressed one more item: The line known in the factory as the "daisy line" because of its similarity in shape to the flower itself consisted of minifactories where batches of one to five pieces from modules B and C were manufactured, and these constituted 20 percent of turnover, thus accumulating most of the *muda* in this line and enabling a greater productivity of the lines that make module A (80 percent of turnover).

For those looking for tangible results, we can say that even in times of crisis, it was possible to improve service to the level that customers were looking for and allow the company to maintain the orders and therefore its level of activity. External claims halved, whereas internal defects declined from 1.77 to 0.69 percent despite reducing the production batches to one to five but without undermining the OEE, and this thanks to the fact that production changeover times dropped from 20 to 12 minutes. The material was supplied in frequent cycles inside the plant, with a lead time reduction of 40 percent.

All steps in the implementation of this improvement process were very carefully thought out and included the lessons learned from previous projects. The next priority therefore appeared quite clear—involving the maintenance staff.

The objective was very simple: Teamwork between production and maintenance in order to force production to perform some easy and repetitive maintenance tasks so that the maintenance personnel could be released to focus on making improvements and tasks that required greater expertise.

The project moved into the maintenance leadership with a coleader from production. Tasks emerging from the action plan that could be performed by the production staff were the first to be put in place. With this step, the production operator gradually moved from being the machine's worker to being its owner.

This project proceeded somewhat slower because there had been many years of antagonism between production and maintenance, and therefore, a higher degree of patience was required. However, in the pilot machines (the process always should start with pilots), the cost of maintenance had been reduced by 30 percent in relation to the initial value and this 2 years from starting to walk. Obviously, the company must check whether this good start will be confirmed in the coming years.

However, we must not look for shortcuts or immediate results; these are a consequence that sooner or later will be reached after applying the principles that drive any projects that are addressed in the company and that can be summarized as follows: observe the process, look at the people who work in the process, improve the working conditions of those people included in the previous and subsequent analysis in the chain (overview), and when problems arise, do not blame but look for the root cause of the problem.

## Maturing into a Daily Management System

The experience of four years showed that if this scheme was followed, people would get involved and would be encouraged to present ideas, and ultimately, the results not only would appear but also would be strong and durable because they came from the ideas of the workers, and the workers are the ones who will defend and fight for them.

Each year the company has been incorporating more productive areas in the continuous-improvement process. In 2010, we decided to apply *kaizen* in logistics in order that the good service level achieved in the unit of furniture and components could be extended to all the other products made in the company.

The next step was the personnel department, in which the team is currently mapping, improving, and simplifying office processes. A key tool in the implementation and consolidation of continuous improvement was the daily MGF.

The daily MGF had two main components:

- The first component of communication occurred initially from the top down so that each worker knew his or her specific daily goals and thus could assess them.

- The second component was the help chain that started at the next level of command and escalated if necessary to more resourceful control levels.

This led to a new way of exercising leadership called *servant leadership*, in which the work of leadership was to help operators achieve their goals because they were the ones who added value to the chain.

The first link in this chain was a five-minute meeting at the beginning of each shift because prevention was the first thing to take into account on the production lines, in the maintenance shops, and in the offices, where obstacles sometimes were so complex that they were unresolved at this level.

This was the key to preventing continuous improvement from being just a matter of having a few people doing some workshops on certain days and in certain places. With daily MGF, continuous improvement happened every day, everywhere, and with everybody involved.

## Lessons Learned

The five years of this experience provided much more information than the most optimistic person ever could have dreamed of in the beginning.

### *Patience*

We firmly believe that most businesses fail in the implementation of continuous improvement because of a lack of patience. We witnessed this in our trip to Japan, whose culture is used to waiting patiently for planting and harvesting. In Spain, the trend is to plant and ask for the fruit the next day. If it doesn't appear, people change the seed. This is the great obstacle of our culture. We are addicted to the principle that the right process will lead us to the expected result.

### *Culture*

We started in a state of freely running the business; meanwhile, the crisis came and changed the playing field. However, the initial situation allowed us to propose the project as a culture not looking for immediate results but seeking the involvement of our employees toward the company.

We have decided to take into account the opinion of each and every person instead of just considering the very best, who at that precise moment would be able to present the best solution. However, we have realized that if we really want to be the very best in the business, we need not only our workers' physical efforts but also, and mainly, our workers' willingness to work for the company.

## *The Challenge of Sustaining*

It is quite simple to get an improvement or to get perfect 5S, as well as to implement a system to solve quality problems or any other, either of a line or of a section, and finally, to achieve an SMED and substantially improve OEE. But do not overestimate the difficulty of preserving these achievements. We would say that it is much harder to sustain than to do. It is crucial to achieve sustainability, and therefore, the improvements we get come from the bottom to the top, and not vice versa, as happened in traditional leadership.

The higher we rise, the more humble we should be in order to uphold the principle that "your idea makes you more committed," and this is the only way of making sustainable and continuous improvement come true.

# Innovating with *Kaizen* at Group Health

Many people think that *kaizen* takes place only on the shop floor. When I mention lean to people who are unfamiliar with it, they often say, "Oh, you mean lean manufacturing." Lean, however, is a total management system. At Toyota, lean extends to all business processes, including the planning and development of new products.

Toyota uses a planning method called *production preparation.* This has been adapted and introduced to Western companies as the *three Ps* (3P), or the *production preparation process,* by Chihiro Nakao, a student of Taiichi Ohno, as a member of the autonomous study group. The method of 3P organizes activities for groups working on a complex plan or design. Toyota has used 3P successfully to shorten its development time for new vehicle models from five down to two years. This has helped Toyota to respond much faster than North American automakers to changing market conditions.

The 3P approach is different from lean processes on the shop floor because it involves the flow of ideas instead of the flow of materials, yet there are many similarities. When the *gemba* is a shop floor, lean brings related processes in close proximity so that workers literally can see how their work contributes to the value stream. When the *gemba* is a drawing board or planning table, 3P brings people together so that their ideas can flow back and forth easily, allowing them to develop within the context of each other.

An important part of 3P is the use of cross-functional teams, which bring very diverse ideas to the table. An idea can range from a simple fix to a transformative change; participants are encouraged to think without boundaries or, in other words, to innovate. Teams then develop quick and inexpensive prototypes for dealing with a designated problem, the best of which is selected through a testing procedure.

Group Health, a nonprofit health care provider based in Seattle, Washington, has shown that this same 3P process used to design cars can work very well in the health care field. The organization serves over 660,000 residents of Washington state and Idaho through its own clinics and those operated by an associate network. Two wholly-owned subsidiaries offer a variety of health plans to large and small employers. Group Health is one of the few U.S. health organizations that is consumer governed, resulting in a strong mandate for patient-centric care. The organization is also fully committed to following lean principles.

Group Health began using 3P in 2008 to improve the way it designs new health care products. Product development in health care is a balancing act—varying patient needs, a changing market, and a complex cost structure make health plans a multifaceted, moving target. Providers in the United States, in addition to being regulated by state governments, are now under increased federal scrutiny under the Affordable Care Act. This means that tolerances for error are small, and all aspects of products have to be defined precisely.

"We like to say that creating a product at Group Health is harder than building a car," says Melinda Hews, executive director of product management at Group Health, "because it's more complicated in some ways."

Like many other providers, Group Health had previously employed a methodology that breaks development into phases, passing the product sequentially from one business group to another. With this approach, deficiencies are often found after the fact, and products frequently have to be sent back for rework. This means delays in development and an environment that does not lend itself well to innovation.

The 3P approach is an advanced methodology that requires a better-than-average understanding of lean principles, techniques, and philosophy and how they must work within an organization. Without this background, it is difficult for participants to visualize the world-class outcomes they are aiming for. Group Health began using 3P after seven years of working with lean and based on a complete commitment from senior management to use lean as a total management system.

The design process began with the establishment of a product design team charter, sponsored by members of the senior management team. Team members were mandated by the charter to solve the following problem:

> Group Health derives its value from its uniqueness as a delivery system and its ability to directly affect the quality and cost of care. The challenge is both to create market competitive health

insurance products that leverage these unique capabilities and to demonstrate value to our customers through improved health, productivity, and cost outcomes.

Elements such as deliverables, measurements, evaluation criteria, strategic alignment factors, time frames, and the names and roles of team members all were included in this concise yet comprehensive document.

The 3P approach uses a scheme referred to as 7-3-1 as a roadmap for the design process. During the first phase, seven initial product ideas are developed. In the second, these are narrowed down to two or three. In the final phase, the remaining designs undergo a rigorous testing process, which includes market research, customer feedback, and feasibility testing.

The Group Health product design team, consisting of over 30 members, included, in various roles, such diverse people as nurses, doctors, executive vice presidents, sales and marketing people, a legal specialist, and human resources staff. The team process for the first two phases took place between December 2009 and February 2010. Team members committed to 24 hours over six meetings and, in addition, to 8 hours of homework over this period.

The mandate to come up with seven initial ideas is a stretch for team members and pulls them outside their comfort zone. These cannot be just "wild ideas"—members of the team have to create concepts that meet strict project criteria. Here are the criteria that were used at Group Health: The product concepts proposed should

- Meet the needs of the commercial middle-to large-market customers
- Leverage the group practice
- Reframe the health maintenance organization (HMO)/defined-network platform
- Improve the health of customer's employees and Group Health members
- Be affordable and offered at a significant discount relative to a comparable Group Health HMO product
- Be developed in time for the 2011 middle-to-large-market renewal cycle
- Benefit administration cost neutrality
- Be compliant with all legal and compliance regulations
- Produce a positive production margin

This approach makes innovative thinking and risk taking unavoidable, so it becomes virtually impossible for the group to follow a conventional path. Many corporations like to brag about how innovative they are, but it would be interesting to see how their executives would fare in this kind of planning environment where everyone's ideas are tested "by fire."

The process for Group Health was refreshing for participants who had rarely interacted with their peers in other areas of the business. "The first substantial benefit was the simplest one," says Chris Schrandt of the Kaizen Institute, who led the initial sessions, "Just getting everyone together, sitting down, and sharing ideas with each other."

Participants also were energized by the opportunity to get down to the core of their business. "People are thrilled to help," says Hews. "They're used to just dealing with the creation of the product, delivering on it, or fixing the problems when they arise. And they have great ideas."

The cross-functional nature of the teams led to instant feedback from subject-matter experts. "A product management team is usually accountable for that work, but obviously it's cross-functional in its nature," says Hews. "You need the input and the financial help and all sorts of different things. The 3P process really helped to bring the right people into the process as early as possible."

## New Directions

The 3P process moved Group Health team to take on one of the toughest challenges in health care product design—value-based insurance design (VBID). The approach is regarded as one of the most innovative ways to address rising costs while improving the overall health of the patient because, essentially, it motivates patients to look after their own health.

For instance, magnetic resonance imaging (MRI) scans are often overused, so VBID could charge a customer extra for this service while also providing a drug that lowers blood pressure for free. This places more emphasis on preventive care, which has been gaining prevalence. A value-based insurance package, for example, may provide discounts on gym memberships or on ergonomic office supplies in order to help curb chronic conditions.

The question is, How do you design such a product? VBID adds considerable complexity to the development process in that measuring value involves including elements that are difficult to quantify. For instance, a health insurance company may opt to include a discount on back rests for employees, which involves variables such as cost, product effectiveness, and market relevance. This is the challenge that Group Health took on with the 3P process, and by following the process through, the company became the only provider in Washington state to offer such a product.

Group Health's new 3P-based development process proved to be an ideal fit for handling the many variables of VBID. Since implementing 3P, Group Health has seen a reduction in product-to-market time from about 18 to 12 months. In addition, stakeholders have renewed confidence in the quality of their products and anticipate further innovation in the near future.

## Breaking Ground

Group Health is also using 3P in another important planning area—facility design. In 2012, construction will begin on a 43,000-square-foot clinic in Burien, Washington, in which staff will be able to care for up to 20,000 patients. The facility will support the concept of lean health care, which centers around the patient. Lean goals for the facility included 50 percent less walking around the clinic by patients and staff and 80 percent fewer "handoffs" during patient care.

Designing a lean facility is challenging because the planners have to fully understand the lean processes that the facility will support. In addition, the team has to include not only the owners of these processes but also the design team, consisting of architects, engineers, and builders.

The 3P process was similar to that used for the product planning group. The charter for the Integrating Care and Facility Design Project had executive sponsorship, followed the 7-3-1 format, and used cross-functional teams. The criteria for concepts related to how the design would support lean concepts such as less travel, particularly for patients, suitable sightlines, and a pleasant, orderly environment.

A unique aspect was that the team included a group that they called *keepers of the concept.* This group served as a steering committee, clarifying the criteria and holding all participants to the vision of a patient-centric lean facility.

Another unique aspect of the project was the use of 3D modeling to test the three selected designs. In collaboration with the architectural firm CollinsWoerman, who use this technique with other clients, the designs were mocked up to scale with cardboard in a large warehouse in Tukwila,

Washington. This allowed the teams to literally walk through the facilities, testing the lean concepts that were the design criteria for the facility.

"There were 60-plus value streams that the team looked at—scenarios that the team went through," says Kaizen Institute USA Director Mike Wroblewski, who led many of the sessions. "We have a mother who's pregnant coming in for a checkup. We have a person coming in for a wellness check. We have somebody coming in for a flu shot. They tested and measured all these scenarios."

The team also was able to test some of the fine points. With life-size modeling, the team actually could assess whether a 12- × 14-foot room or a 13- × 14-foot room was better—a comparison that never could be made with 2D drawings. Then there was the innovation. In one case, team members didn't like the idea of the patient having to go to the lab for the lab work. To address this deficiency, team members came up with a novel concept—a lab on wheels. "Now patients don't have to leave their rooms—the lab comes to them," says Wroblewski. "This has made a tremendous impact on the flow of patient care."

The finished design for the facility, which has been submitted to the architects for detailed drawings, has a phenomenal level of buy-in. The architects, builders, and engineers are thoroughly familiar with the requirements. They know every detail—how much lighting is needed in the room, where the light switches should be located, and where the sink and hand lotion dispensers should be. Senior managers know exactly what they are getting. And most important, the frontline care workers who will use the facility are confident that it is the best environment possible to support their lean processes and provide optimal care for their patients.

The adaptability of 3P to planning and design in a health care environment illustrates how powerful the method is. If the flow of ideas can be managed efficiently, as it has at Group Health, there is no limitation on the areas where the process can be applied.

# *Kaizen* Helps Caetano Bus Deliver on Schedule

One of the most immediate indicators of how well a factory is running is whether it can deliver on time. If lead times are too long or are not met, the customer can sense right away that there is a problem without even seeing the factory or the product.

When Caetano Bus, Portugal's largest passenger bus manufacturer, increased its focus on the highly competitive European market, the company began to face complaints about delivery. Through *kaizen*, the company was able to make some remarkable changes in its organization that helped the company not only meet these challenges but set an example for other companies as well.

Caetano Bus belongs to the industry business unit of Salvador Caetano Group, which employs over 6,500 workers in markets such as automotive, industrial machinery, renewable energy, and automotive retail. The first Caetano Bus factory in Portugal was built in 1946 and produced passenger buses made of wood and then steel a decade later. In 1966, a new factory was inaugurated in Vila Nova de Gaia in the north of Portugal, and exports to England began. Some years later it began production of the Cobus model, which today reaches all continents. In 2001, it established a joint venture with Daimler Group and started producing Mercedes models for the European market.

The Gaia factory has only been using *kaizen* for a short time, but it has already produced such excellent results that it is being used as a model to help other factories in the group face the global economic crisis. The *kaizen* transformation began immediately after Jorge Pinto became the company's new CEO in 2005. By then, he had noticed that the factory's 500 workers were unmotivated, that productivity was low, and that lead times were not being met. This last aspect was affecting customer relationships, and the company's business partner, Daimler Group—which owns Mercedes-Benz—was not at all satisfied with the variability in delivery times.

Variability and excessive lead times are often due to poor distribution of workloads, where some sections are unable to keep up and others are idle much of the time. The *kaizen* approach to this problem is to reconfigure the lines so that there can be a balanced workload where all stations are contributing equally and the output of the factory is constant and predictable.

Pinto was already aware of *kaizen* and knew that the approach could put the factory on the right track. In cooperation with the Kaizen Institute, he began a pilot project on the Tourino model's assembly line, which was having particular problems meeting deadlines.

With the help of training sessions and workshops involving multifunctional teams of workers, the first *kaizen* tools were introduced. These included 5S, visual management, standard work, *mizusumashi*, supermarkets, and line balancing. The objective was not just to apply a "quick fix" but to help make sustainable cultural changes.

## Standard Work on the Assembly Line

Armed with knowledge of *kaizen* tools, the improvement team set to redesigning the workflow patterns in order to balance the workload among all stations. Previously, the workers were organized according to their function and only did the task for which they were qualified. A painter, for example, did nothing else but paint, and a welder did nothing else but weld. The rebalancing of work required workers to adapt so that they could perform multiple tasks. Using tools such as standard work and *yamazumi* charts, the team broke down the paradigm of "division of labor by type" and replaced it with a lean model.

To implement this kind of change in the *gemba*, it was necessary to break several paradigms. Many employees didn't believe that it would be possible to produce more with the same number of people. Little by little, however, people's resistance decreased mainly because they realized that it was possible to increase productivity without increasing their effort.

"Increasing productivity does not mean increasing the workload but rather providing better working conditions with discipline and cleanliness," says Jorge Pinto. Cleaner, more organized workstations were achieved with the help of 5S. After its implementation, the workers knew the standard place for everything and didn't have to spend time searching for tools. In addition, a team responsible for 5S audits was established to ensure that the improvements were sustained.

## Delivery Time According to Consumption and Direct Supply

The same tools also were used to reorganize the supply of materials to each work area to ensure a more consistent workflow. The creation of a *mizusumashi* supply system also helped to keep the workflow consistent and uninterrupted. Reconfiguration of workstations also helped to enable a smooth supply without interrupting the operator.

It was also necessary to draw up a new production layout, so workers moved the assembly lines to the opposite end of the pavilion to widen the aisles and repositioned all preassembly, which included large items such as roofs, pipes, and dashboards. The facility also was remodeled and organized to support a supermarket system. Before, pallets were unloaded and transported to the line by the operators, who were wasting time separating items. Now they are unpacked and stowed in the racks of the warehouse according to their turnover. They are then distributed throughout the workplace in small boxes by *mizusumashi*.

Parts now are delivered according to consumption needs; the intermediate stock has almost disappeared. Local suppliers are expected to deliver high-volume parts directly to the assembly line on a daily basis. For example, empty boxes from the assembly line are taken to the warehouse and placed on a specific rack. Every day the supplier collects the boxes, fills them with the necessary parts, and returns them to Caetano. Suppliers of lower-volume items deliver parts to the warehouse, and these are then processed and distributed by internal logistics operators. The little stock that exists is only sufficient for a week of production, and shortages are identified by visual inspection.

## Cobus Line Project

In 2007, Caetano decided to replicate the Tourino line's successes on the Cobus line, maker of the company's star product. Cobus is a bus model used for transportation to airports, and it holds a 90 percent share of the global market for such vehicles. The effort achieved extraordinary results, with a reduction in production lead times of nearly half, and the model then was applied to all other lines.

The increase of productivity, which reached 40 percent, was crucial to meet the growing demands of the customer. Caetano Bus went from a production of about 400 to 700 vehicles per year during this period, productivity gains that were achieved by increasing output without hiring more workers. A notable improvement was that inspectors were no longer needed at the end of the line because lean methods were being used to detect defects earlier. This also meant less waste owing to rework.

## Lead-Time Reduction on the Assembly Line

Reduction in the number of workstations took place between 2007 and 2008 and was implemented in order to reduce delivery time to customers. The finishing line of bodies, for example, dropped from eight to four workstations. During this project, the team developed a new approach to assembly that was nicknamed "the cage." Previously, assemblers had attached the front and rear side panels to the chassis, which was used as a base. In the cage, the bus skeleton is preassembled, welded, and then lowered onto the chassis. With the help of a multitasking *kaizen* team, work that was done in three stages is now done in just one. Furthermore, quality has improved dramatically because there is now a template for joining multiple parts (i.e., front, side panels, roof, and rear) that helps to achieve a better fit.

To realize this approach, it was necessary to make some model changes, such as gluing in the rear bumper instead of welding it. This reduced problems associated with welding and increased the quality of the product. This concentration of jobs in a smaller area also freed up floor space that was being used for preassembly.

Overall, Caetano's *kaizen* efforts have brought significant benefits to the business. According to Jorge Pinto, the barriers between the company and its partners have disappeared, along with the associated administrative burdens; response times have been reduced; transportation costs and packaging have been reduced; productivity is up 35 percent; and space has been freed up by reducing the number of workstations from 12 to 5.

## A Successful Cultural Change

Although this project began as a quest for productivity and reduced lead times, this internal revolution—driven by six years of *kaizen* thinking—brought something much larger: a true cultural change. This is a company where workers had been working for almost 40 years with a strong organi -

zational culture and were not used to change. Still, they accepted the changes in their production processes quite well and absorbed the philosophy of *kaizen.*

Lean allowed Jorge Pinto to instill leadership in his management team and made it possible to identify those who were up to the task and those who were not. As we see in so many cases, CEO involvement was critically important. "Only with the commitment of top management is it possible to make such a project move forward," says Pinto. "Otherwise, it dies. If we had not restructured the company in time, and if we hadn't started these projects to improve productivity, I do not know if Caetano Bus would be here today, to tell the truth."

# Kenyan Flour Producer Uses *Kaizen* to Increase Capacity, Improve Efficiency

Maximizing capacity and reducing inventory are critical metrics for Unga Limited, one of the largest milling operations in Kenya. Through a comprehensive series of *kaizen* projects, the company has made substantial improvements in both these areas without investing in new equipment and has made visible progress toward a positive and caring work culture.

"Unga has succeeded in implementing *kaizen* practices across diverse business units, manufacturing processes, locations, and management teams," says Vinod Grover, director of the Kaizen Institute Kenya, whose consulting team assisted with the transformation. "This across-the-board success of the *kaizen* journey not only is truly remarkable, but it also has a clearly assignable cause—leadership."

Founded in 1908 by settler Lord Delamere, Unga grew to become Kenya's largest milling company, complementing its flour-milling operations with animal feed plants that recycle the flour-milling by-products. Faced with tough global market conditions, the company ran into financial trouble in the 1990s and sought outside partnerships to help restore profitability.

In early 2000, the U.S.-based Seaboard Corporation took a significant stake in the company and, in addition to financial backing, has provided access to milling expertise and trading infrastructure that allow Unga to better participate in regional markets. At the same time, Nick Hutchinson, a well-traveled Kenya-born agricultural associated industry business veteran, joined the company as CEO.

In the company that Hutchinson inherited, cash flow was tight, and a major capital investment to replace aging equipment seemed inevitable. It also was clear that attitudes needed to change within the company if it was to be successful.

Hutchinson had been impressed with what he learned about *kaizen* during previous trips to Japan, but he had been advised that *kaizen* probably would not work in an African context. This view changed when in 2006 he and two of his senior managers attended a seminar hosted by the Kaizen Institute and the Kenya Association of Manufacturers. "When we realized that other companies in Kenya had had some really good results with *kaizen*, we decided we'd give it a go," says Hutchinson.

Hutchinson engaged the Kaizen Institute to create roadmaps of the company's current and desired future states. As initial benchmarks, he chose two plants—a flour-milling facility and an animal feed production facility. "I really wanted to see what was different between the businesses and to see the overall business from both perspectives," says Hutchinson.

Equipment downtime owing to aging of equipment was the most visible problem area. "We were having a lot of breakdowns and emergency downtime," says Hutchinson. "Everybody was saying that because our plants are quite old, the only solution was to buy a whole lot of new equipment."

The cost of inventory was another concern. "We were borrowing money that I didn't think we needed to borrow," says Hutchinson. "After we did the roadmaps, I fully realized how much we had tied up in inventory."

Next came a series of week-long *gemba kaizen* workshops where cross-functional teams learned the basic *kaizen* tools and began to work collaboratively on solutions. Employees from all levels of the organization were brought into the process. The initial sessions had strong and enthusiastic participation, but Hutchinson and his senior managers realized that more had to be done to ensure that the sessions were being followed up with concrete action. "After about six months, we realized that the teams weren't necessarily being led by the right people," says Hutchinson, "so we adjusted the leadership. We made some changes to ensure that top management was more engaged."

After the first year, the company decided to expand the initiative to the remaining three plants in Kenya and to its plant in Uganda. In addition, the company worked with the Kaizen Institute to develop what the company calls its *PaTaMu model for continual improvement. PaTaMu* stands for "*Pamoja Tuangamize Muda*" ("Together let's eliminate waste").

## Taking on the Key Issues

In response to the equipment downtime issue, the teams developed an autonomous maintenance program modeled on the Toyota approach to total productive maintenance (TPM). The process provides a framework in which operators maintain the equipment themselves, eliminating costly waits for maintenance personnel. Unga has taken on the first three steps of this process—cleaning, eliminating contamination, and creating a checklist for cleaning and lubricating—and standards are strictly enforced. "We are now being quite ruthless," says Hutchinson. "If the machine isn't being sustained at step three, then we bring it back down to step two."

The result has been a dramatic reduction in downtime and, consequently, a substantial increase in capacity. "We've more or less doubled our capacity at one plant," says Hutchinson, "and haven't bought a single piece of equipment. Two years ago I would've written that up for a massive capital expenditure project."

Other teams took on the issues of inventory and flow. Plants were relying on "safety stock" to compensate for their frequent equipment breakdowns and for special orders. "We introduced the *kanban* system between production and the finished product warehouse," says Hutchinson. The *kanban* system now provides visual information on levels of inventory throughout the plant, reducing the need for large safety stock. Plants also have been able to significantly reduce inventories of packaging materials and engineering spares.

Supply-chain partners are next. "We still haven't backlinked to our suppliers, and also, we haven't linked our finished product to our distributors," says Hutchinson. "We're just starting to work with that now. We wanted to make sure that our internal processes and procedures were in place first."

## Emergence of a New Culture

Perhaps the most widely visible aspect of Unga's transformation has been an extensive 5S deployment that has extended from the plant floor to the executive offices. "The basic housekeeping, organizing, and all of that, we call the 'five Ks' in Kenya because it's five Swahili words that are translated with Ks," says Hutchinson. "We basically put together an initiative where every single person including myself is on a 5K team, and we have a competition that runs for 12 months at a time where we make efforts to improve on our housekeeping."

The program has created an atmosphere of inclusiveness and is breaking down barriers. "It's been a lot of fun," says Hutchinson. "It's about getting senior-level management engaged, actually doing things. I'm on a team—I'm not the leader of the team. I do what I'm told—I go to meetings when I'm told to go to meetings, and I clean my office up when I'm supposed to clean it up."

Shekhar Deshpande, senior consultant with the Kaizen Institute, feels this has been the deciding factor. "Unga has become a real success story because it has used all three vectors," says Deshpande, "Lean and *kaizen* tools, employee engagement, and most important, top-management commitment to a degree that is missing in other companies."

Today, the company is seeing a new level of employee initiative. "We have a program now where employees can give continual improvement ideas," says Hutchinson, "and we're just starting to see folks contributing ideas for new *kaizen* things we can do. So it's very different from where it was, but I would also say that like every *kaizen* journey, there's still a long, long way to go."

The *kaizen* journey, "PaTaMu—Everyone, Everywhere, Everyday," has become a reality at Unga. "Every year we're looking for a new idea, some new initiative, something new to get people excited about," says Hutchinson. "Once you've harvested the low-hanging fruit, it gets to be really hard work. But the opportunities are there—there are endless opportunities."

# *Kaizen* as the Foundation for Innovation at Medlog

Many times workers already know what needs to be done to make a company better. When this awareness exists, senior management must provide the tools and support necessary for these ideas to reach their full potential. When these measures are implemented properly, this creates an environment where innovation can flourish and big changes are possible.

Such was the case with Medlog, a Portuguese health care company that provides logistics and marketing services to the pharmaceutical industry. It was founded in 1975 when a group of pharmacy owners launched Cooprofar's Cooperative, which now owns more than 1,250 pharmacies.

Medlog's *kaizen* journey began when administrators realized the need for a structured process that was responsive to the improvement suggestions that work teams were already making. During the holidays, for example, an employee had developed several apron models with special pockets to help resolve some of the difficulties he was experiencing with his job. Many workers had ideas like this that could help the company but did not have the proper channel to implement their ambitions.

"The ideas and willingness to work were there, but there was no common thread linking all these suggestions," says Raquel Miranda, Medlog's administrative advisor. "With the support of the Kaizen Institute, we now have a structured process in place for this purpose. This has resulted in an increase in motivation and a strengthening of our company culture. People were eager to contribute ideas, as well as implement them."

## *Kaizen* Builds Motivation

In order to create a truly *kaizen*-minded company, warehouse employees—especially those who worked in the *gemba*—had to adopt the philosophy of continuous improvement in their day-to-day lives. Medlog, which had a total of five warehouses, began its first *kaizen* workshops in the northern

Portugal municipalities of Gondomar and Guarda, followed by initiatives in other cities.

The purpose of the workshops was to rearrange the layout of the warehouses' work areas using 5S and visual management. Maps were posted, activity indicators were described, and actions were planned and executed. At the same time, the *kaizen* teams began to create work standards so that any improvements deemed to create better results could be absorbed and perpetuated by all. All these changes set the stage for an environment that could be improved on a daily basis.

A daily *kaizen*, a process whereby all employees attended short daily meetings to discuss problems and solutions, also was implemented and was one of the largest contributors to the commitment and motivation of the work teams. "The most difficult process is not to implement new measures but to sustain them," says Miranda. "And what ensures maintenance is the involvement of all employees."

During these morning meetings, workers analyzed indicators, their workload, and the cleanliness of the warehouse, which, in turn, helped them to understand the importance of 5S. Daily mini-audits also were established, and these showed the teams that the organization was concerned with the performance of the implemented measures and also reinforced the involvement of all employees and their desire for daily improvement.

## The Success of 5S and Visual Management

Workers had already come up with several improvements, such as housekeeping, before they had ever heard of *kaizen*, but all of them were too busy to actually initiate those solutions. The moment management began to support them in their improvement ambitions, time was made available in a structured manner so that the entire team could collaborate on solutions.

In the Gondomar warehouse—the first location to be targeted by the new methodology—the process began with the organization of a multidisciplinary group of 36 people from four different departments and a team of consultants from the Kaizen Institute. The first few workshops introduced the tools necessary to make improvements. All training sessions were focused on logistics and used practical examples to illustrate positive and effective ways to eliminate waste and focus on adding value. In this way, employees were able to gain a sense of how to transcend theory and actually apply what they were learning.

The warehouse then was divided into several work areas, and the implementation of 5S began. "Workers' involvement was so effective that we began to include recommended improvements in addition to the 5S initiative," says Miranda. "The desire to provide input was huge, and we ended up accepting some of the employees' ideas." Many of the ideas were simple, such as converting a stand for bar code readers into a cart by placing wheels on it or changing stock-picking procedures, but the improved efficiencies of these changes began to add up.

## Visible Improvements

Team satisfaction began to rise as work areas became cleaner and more organized. Employees now had a more pleasant work environment and got less tired when performing their tasks. Furthermore, people were no longer doing unnecessary walking within the warehouse. "The company can now do more with the same people," says Miranda. This change was particularly valuable given the continuous growth in volume that Medlog was seeing.

As a result of the warehouses' reorganization and the processes' redefinition, productivity in the Medlog warehouses increased considerably, ranging from 7 percent in some areas to 25 percent in others, such as the manual picking system (MPS) storage area. The company managed to gain an extra 41.5 hours of productivity per day, the equivalent to five full-time employees and 60,000 euros per year.

In the shipping area, there was an overall productivity gain of approximately 25 percent, and the revised layout avoided the use of hundreds of feet of conveyor belt, thus reducing energy consumption by about 25 percent. The maintenance department is also expected to reduce average downtime due to breakdowns from one hour to 30 minutes. Furthermore, the number of invoices with complaints in them was reduced from 1.3 to 0.8 percent.

## From Incremental Improvements to Innovation

The establishment of a *kaizen* culture in the organization made it possible to successfully implement a fundamental change in the way products were distributed to hospitals. In 2009, Medlog began a large-scale innovative project that would use *kaizen* tools to help improve the distribution of medicines, medical devices, and other products on a national scale. The initiative was in response to a need for hospitals to improve the quality of their services, as well as control their costs.

The project was called *Sig-Log* (Integrated Management and Logistics), and it gave rise to a new business unit called *Logistic Health Solutions* (LHS), which delivered medication, medical devices, and other health care products to hospitals.

The project was supported by a consortium of national and international partners from several different industries, including the Institute of Mechanical Engineering and Industrial Management (INEGI), three public and private referral hospitals, Creative Systems, Knapp, and the Kaizen Institute.

LHS was different from traditional delivery systems. Normally, a hospital orders products from a supplier, who makes the delivery to a warehouse. From there, the hospital's own logistics team distributes the products to their various health units according to demand.

With the new Medlog service, the goods are delivered directly to the health units. This leaner approach to distribution eliminates many sources of *muda*, such as storage space in hospitals, duplication of shipping and packaging, and involvement of skilled health care workers in tasks that do not add value.

The project was implemented in stages, culminating in a pilot project at the Trofa Health Group, which was made up of three hospitals. It was there that the entire system was tested and optimized, along with a number of supporting technology systems. These included an information technology (IT) system supporting logistics software, a robotic solution for the handling of pharmaceutical shipping containers, a radiofrequency identification (RFID) system, and GPRS tracking.

## *Kaizen* Culture Enabling Innovation

The project yielded a number of direct benefits to the customer organizations. Trofa Health Group, for example, reported

- Freeing up of 400 square feet of storage space
- 25 percent savings in labor to handle pharmaceutical products
- 50 percent reduction in shipping costs

In the final phase, the program was rolled out to the other hospitals, and similar gains were reported. At the 1,124-bed São João Hospital (HSJ), the figures were

- ▲ 60 to 65 percent reduction in space requirements
- ▲ 40 to 42 percent savings in labor to handle pharmaceutical products

The work at Medlog shows that once a strong base of *kaizen* culture and *kaizen* methods has been established, large-scale innovation can be implemented successfully. This is in contrast to many organizations that try to implement a "big idea" when the people in the *gemba* have no experience with change and do not have the tools to improve their systems. By putting "first things first," Medlog ultimately was able to help its customers reduce waste and become more successful.

# Growing with *Kaizen* at Supremia

Rapid growth can force many changes in a company, and if the situation is not handled properly, these changes can be very disruptive to work processes and can cause the company to get into trouble. Major change, however, also creates opportunities for major improvements.

After only 10 years of operation, Supremia Grup, based in the city of Alba Iulia, became the largest producer of food ingredients in Romania, with annual revenues of 24.3 million euros. Because of this phenomenal growth, management knew that it needed to adopt the kinds of tools and methods required to run a large company. When the operation was moved to a larger and more modern facility, the company took advantage of this opportunity to implement *kaizen* successfully.

The new facility gave the company many new advantages. The latest technology in grinding, sieving, blending, and sterilization of food ingredients was adopted, and the factory building was sized to accommodate a continuation of the growth the company had experienced.

The new facility, however, posed some challenges for existing work processes. Because it was seven times larger than the previous facility, this meant that, potentially, operators would have to cover much larger distances, and in addition, there could be increased costs in transporting stock and materials. The factory also was using a new generation of equipment, with which the workforce and management team had no experience.

The management team was very young, and many managers were new to the management role. With the workforce doubling to match the scale of the new facility, the company wanted to ensure that the team was equipped for the challenges ahead.

Management began to look for guidance on how to face these challenges. On October 25 and 26, 2010, the executive manager of the company, Ciprian Gradinariu, participated in a presentation of the Kaizen Management System organized by the district authority (Consiliul Judetean Alba) and the Kaizen

Institute Romania. Following the seminar, he presented a copy of *Gemba Kaizen* to the owner of the factory, Levente Hugo Bara.

Both agreed that this was the system they were looking for. "*Kaizen* is the concept we need," said Bara. "We will implement the Kaizen Management System in the new factory in order to be able to become stronger and to develop our business in the whole of Europe. Investing in a new factory is not enough to be successful, we need to strengthen our system and people as well."

After several meetings with representatives from the Kaizen Institute Romania, Bara decided to start implementation of the Kaizen Management System (SMK), a *kaizen*-based management system developed by the Kaizen Institute, in the new factory in Alba Iulia. The project started at the end of February 2011.

## Training Roadmap

The company began "developing the organization by developing the people," as the process is described in the principles of the Toyota management system. This was to take place over a three-year period.

Julien Bratu, general manager of the Kaizen Institute Romania, defined an extended concept "*Kaizen* by Harmony," based on Toyota principles. This divided training into the following quadrants:

| | |
|---|---|
| Organizational strategic *kaizen* | Organizational operational *kaizen* |
| Individual strategic *kaizen* | Individual operational *kaizen* |

The main objective of "*Kaizen* by Harmony" was to develop all quadrants simultaneously, leveling up the performances and creating a dynamic equilibrium among all elements.

The training was delivered through two programs—an external training program by the Kaizen Institute Romania and an internal training program developed by Ciprian Gradinariu, executive manager in the company and SMK implementation coordinator. The two programs were synchronized in order to maximize the effects on the organization and people.

## Organizational Strategic *Kaizen* Activities

The entire project began with establishment of a strategic vision for the company. The vision was called "Supremia 2015," and this was developed during a two-day *hoshin kanri* ("policy deployment") session. This provided both the direction and a strong feeling of motivation for the management team.

The "Supremia 2015" vision is "to become one of the top 10 European producers in the food ingredients industry by developing an internal climate generating individual and team performance and by implementing the Supremia Management System (SMS) for effectiveness and efficiency." According to the new vision, all management activities are conducted according to *kaizen* balanced scorecard (KBSC) objectives generating the improvement projects and the action plans.

The management team then began development of the comprehensive management system for the company, the SMS. Early steps included the definition of material and information flows in the factory. Performance metrics, including quality, cost, delivery, motivation (QDCM), and key performance indicators (KPIs) also were established along with strategic objectives.

## Organizational Operational *Kaizen* Activities

In any *kaizen* organization, strategic planning is only the beginning. The heart of the project is teaching and empowering the workers in the *gemba* to develop the *kaizen* culture that makes attainment of the strategic plan possible. There are, of course, many objectives here: Make work processes as efficient as possible, eliminate waste, ensure quality, create a safe and orderly environment for workers, and optimize the use of machinery, all using basic *kaizen* tools and techniques.

This process began with a series of workshops in all the production and storage areas, which included

- Identifying and eliminating the seven *mudas* ("wastes")
- Fundamentals of 5S
- Visual management principles and techniques
- Standard work processes

While these workshops involved basic training, they also were action-oriented, allowing the groups to act immediately on proposed improvements. At the same time, standardization of work began immediately, creating the opportunity to capture and replicate improvements, for example:

- Reduction of distances covered by operators
- Better placement of tools to avoid work interruption
- Improvements in safety and in work conditions in general

The main benefit of this stage was worker awareness of *muda* and how it can be eliminated, as well as stabilization of the *gemba* processes through standardization. In this first phase, a total of 161 work standards were created.

The teams then moved forward through a series of value stream mapping (VSM) workshops, where waste and inefficiency in a number of product value streams were identified, and future state maps were created to set the objectives for improvement. The major focus here was to reduce lead times wherever possible, which is a major concern in the food industry.

*Muda* in the process was very carefully observed, and the operators identified solutions in order to optimize the workflow. Layouts were redesigned, and some of the equipment was repositioned so that the flow of material became faster and easier. Some of the operations were eliminated, and some were combined without decreasing the quality of the products but reducing lead time and increasing productivity.

The exercise led to a number of measurable improvements. The following was achieved in an area where ingredients are prepared for production:

- Distance traveled by operators reduced from 51 to 23 meters (55 percent)
- Several operations eliminated
- Synchronization between operators optimized
- Lead time for ingredient preparation reduced from 5 to 3.5 minutes (30 percent) with a corresponding improvement in productivity of 30 percent

Because these improvements were applied to high-volume products, the gains were significant for the company. As a result, the new processes have become the standard for ingredient preparation in other areas.

## Additional Tools

In the areas using machinery, workshops using special techniques such as single-minute exchange of die (SMED) and *kobetsu kaizen* were implemented in order to improve equipment efficiency and effectiveness. In the area where ingredients are blended, for example, changeover time was reduced by 30 minutes (15 percent), which led to an improvement in lead times, and as well as a 2.5 percent increase in capacity. In another area where the product is finally assembled, overall efficiency improved by 10 percent.

An autoquality matrix (AQM) also was introduced to bring tighter quality control to some of the processes. This tool makes the defects of the products more visible, making it easier to detect the source of a defect or the process that generated it. Once these relationships are brought out into the open, operators became more responsible, and their concern regarding the quality of their work and of the products is higher.

The company also adopted standards for regular *kaizen* sessions, where teams learned to analyze daily activities and respond to incidents rapidly and efficiently. Managers learned to "speak based on facts," improving their analysis skills and honing their ability to communicate concisely using all important and relevant data.

As of this writing, all these techniques are being rolled out systematically so that they will become the standard for the entire company.

## Worker Engagement

One of the best measures of worker engagement is the collection of new ideas. Companies that have implemented *kaizen* typically receive many times more suggestions from workers than traditional companies.

Six months into the implementation of SMK, Supremia began to track the number of employee ideas and found that the average of one idea per employee was achieved over a three-month period, a very impressive number (see Figure CS-19).

## Individual Strategic and Operational *Kaizen* Activities

The management team also took part in five training courses designed to develop their leadership and personal performance. On the individual

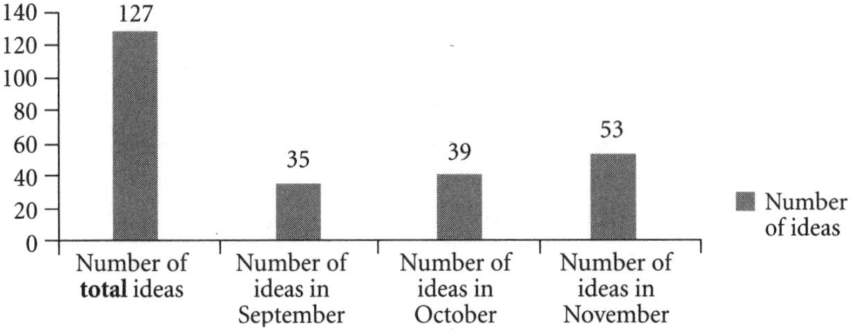


**Figure CS-19** Number of *kaizen* ideas from Supremia employees.

operational side, the managers are learning the importance of daily personal habits such as fitness and mental 5S. These are giving them more energy and a better vision of their daily objectives, preparing them for performance. Working daily with the plan-act-learn (PAL) approach, they organize their activities more efficiently, obtaining the expected good results, and moreover, they learn from everyday experiences.

On the individual strategic side, the managers began to acquire skills to help them define their own values and individual mission and to align their own values and mission with company vision. Each and every manager has his or her own personal development plan, followed by action plans, in order to achieve his or her objectives.

## Measuring the Results

At the end of the first year, Levente Hugo Bara, general manager, performed an annual audit with the assistance of Kaizen Institute consultants on the implementation process of the SMK and on company processes as well. The following progress was cited:

- A broad awareness of the new company vision with personal goals and objectives aligned with company goals and objectives
- Wide adoption of *kaizen* culture, with an attitude oriented toward continuous improvement
- A strong sense of order in the facility based on 5S principles, coupled with a community-minded 5S spirit
- Stabilization of work processes based on standardization
- Strong lines of communication established for quality issues based on the *kaizen* principle "speak based on facts!"
- Improvement in the professionalism of the management team
- Improvement in work conditions and job quality

Encouraged by these results, the company set forth 11 strategic projects to be fulfilled by 2015, which will make Supremia one of the top 10 food ingredient producers in Europe. With the Supremia Management System and a *kaizen* culture firmly established in the organization, management is confident that the company is ready to make this vision a reality.

# Exceeding Customer Expectations at Walt Disney World

At Walt Disney World, *gemba kaizen* spirit is alive and well. *Gemba* employees at Disney are placed at the top of the organization. Walt Disney once said, "You can dream, create, design, and build the most wonderful place in the world, but it requires people to make the dream a reality." This case of Walt Disney World shows how Disney management's faithful adherence to housekeeping and standardization contributed to the success of the business.

Fifty-seven years after the opening of Disneyland in California in 1955 and 41 years after Walt Disney World was opened in Florida in 1971, the frontline cast members are still the most important in the company. At Walt Disney World, employees in the park are called *cast members*, and customers are called *guests*. The guests' satisfaction is Walt Disney World's top priority, and housekeeping and standardization are the two major means to this end. Many visitors to the Walt Disney World Resort make repeat visits because they are so impressed with its clean and safe environment.

## A Flawless Performance

On careful observation, guests will find that waste disposal containers are installed everywhere in the park. On my recent visit, I could count six such containers from the spot where I was standing. Walt Disney believed that no guest should have to walk more than 25 steps to discard litter. The containers are designed to blend unobtrusively into their surroundings. During the procession of Mickey Mouse and his gang on Main Street in the afternoon, I found many guests leaning against the containers or sitting next to them; some were even sitting on top of them munching snacks.

At regular intervals, the waste bins inside the containers are replaced in a swift, efficient manner. A cart carrying several empty bins is brought to the site, and the bin inside the waste container is replaced with an empty bin. A cast member host makes a circuit of the park every 10 or 15 minutes with an elongated pan and a broom, picking up trash on the streets, under the benches, and in the shrubbery. Any Walt Disney World cast member walking through the park who happens to find litter is expected to pick it up. "If Mr. Michael Eisner were there, he would be doing the same," I was told. The collected trash is speedily conveyed to an underground station and sent to the processing plant through vacuum tubes. Thus guests are spared the sight and odor of garbage.

Another reason cited by guests for wishing to come back to Walt Disney World is the friendly and well-groomed Disney cast members. Walt Disney's dream was to provide services that not only satisfy the guests but also "exceed their expectations consistently." Walt Disney World is a place where guests are brought to the stage. The cast members are supposed to play their roles on the stage to entertain the guests. The cast must pay attention to safety and cleanliness and wear proper costumes at all times. Just as in film or onstage, imperfection (in this case, litter on the street, unpleasant odors, and the like) is not allowed. Therefore, every task, every movement of the cast, every building, every facility, every event, and every attraction must become a means of richly satisfying the guests. To do this, every newly hired cast member, including part-timers, must go through a two-day orientation program that instructs trainees in Disney's philosophy, the company's history, and the details of the job.

Cast members consist of full-time, part-time, and seasonal workers, and their jobs fall into about 1,500 different categories. Every job has its own job description and standard operating procedures (SOPs), and the 37,000 people working in the park are expected to follow the standards. If no such standards were provided and each of 37,000 cast members were to start working in his or her own way, management soon would find that there was no way to manage the cast members' behavior and the business and therefore no way to ensure the satisfaction of the guests.

Each new cast member receives the following list of guidelines for serving the guests:

1. Make eye contact and smile.
2. Greet and welcome each and every guest.
3. Seek out contact with guests.
4. Provide immediate service recovery.

5. Display appropriate body language at all times.
6. Preserve the magical guest experience.
7. Thank each and every guest.

The cast members selling tickets at the entrance are told that their job is not to sell the tickets but to communicate with the guests. As the first Walt Disney World cast members to meet guests, the ticket sellers are taught to make eye contact, smile, and greet the guests. These cast members are supposed to be well informed about the day's events.

A cast member selling balloons to children is expected to kneel so as to place himself or herself at the same eye level as the child—and employ body language that demonstrates friendliness and intimacy. A cast member who finds a guest taking a snapshot of other guests is expected to volunteer to take the photo for the group.

The housekeeping hosts or hostesses also have their own job descriptions and SOPs. They are reminded that their primary role is as stage players who entertain guests; the sweeping task is a secondary responsibility. Rather than stoop inelegantly to pick up litter, these cast members are expected to use a long-handled pan and broom or a long stick with a scoop on the tip to retrieve the trash and place it in the pan gracefully. Management must provide special training for performances of this kind. Often guests do not notice the housekeeping cast members as such because they mingle with the crowd so naturally.

## Giving Cast Members Discretionary Powers

Walt Disney would say that everything we do now is imperfect, and we therefore must constantly strive to do a better job; the moment we believe that we have reached perfection, we stop improving. Cast members are empowered to take initiatives whenever necessary to exceed guests' expectations.

For example, when a newlywed couple arrived at a Disney hotel, a receptionist cast member noticed that the bride was feeling ill. As soon as the guests were shown to their room, there was a knock on the door, and hot chicken soup was delivered. The cast member was able to make this gesture because she had been given discretionary powers that allowed her to order the soup. The guests were so pleased and grateful that they later wrote a letter of praise to management.

While cleaning a Disney hotel room being used by guests with children, a housekeeping cast member came up with the idea of arranging a menagerie of stuffed animals on the table in the room to look as if they had been having a party while the children were away. Imagine how enchanted the children were on returning to the room.

## An Eye for Standards

Each cast member is provided, during his or her job interview, with a booklet called "The Disney Look" that stipulates the importance of appearance; before a job offer is made, the cast member must agree to comply with the dress and grooming policies described in the booklet. "The Disney Look" specifies rules to be followed on such items as

- Aftershave, perfume, and deodorant
- Costumes
- Hair coloring
- Pins and decorations
- Sunglasses
- Tattoos
- Hairstyle
- Mustaches, beards, and sideburns
- Fingernails
- Jewelry
- Shoes and hosiery
- Makeup
- Skirt length

Some examples of acceptable and unacceptable practices defined in the booklet include

- *Sideburns (for men).* Sideburns should be neatly trimmed and may be permitted to extend to the bottom of the earlobe, following their natural contour. Flares or mutton chops are unacceptable.
- *Jewelry (for women).* Rings such as class rings and wedding rings, earrings, and conservative business-style wristwatches are permitted. Necklaces, bracelets, and ankle bracelets are unacceptable. Earrings must be a simple, matched pair in gold, silver, or a color that blends

> with the costume. A single earring in each ear is acceptable. Earrings can be clip-on or pierced and must be worn at the bottom of the earlobe. Their diameter must not exceed one inch.

The booklet also describes the procedures for supervisory cast members to follow in disciplining cast members for infractions of the appearance policies. For instance, the booklet stipulates that should a cast member need to be reminded of the policies, the supervising cast member should do this coaching in private.

Because of the popularity of the Disney approach to human resources development, Disney University professional programs are offered at Walt Disney World, enabling participants to actually view examples in the theme park on field trips and learn about Walt Disney World's strategies for people management, quality service, leadership, creativity, orientation, and standardization firsthand.

# *Kaizen* Experience at Alpargatas

The largest manufacturer of textiles and sports shoes in Argentina, Alpargatas is a joint-venture partner of Nike USA. The company's sport shoe division produces different lines at four plants with annual sales of $200 million. The Tuchman plant, where the *kaizen* effort was undertaken, is dedicated to producing Nike shoes (2 million pairs per year).

This case illustrates two aspects of *kaizen*. First, the company's *kaizen* team chose one of the most serious quality problems in the *gemba* as the focus for improvement and discovered that, in addressing the quality issue, the team also found the best way to reduce costs. Second, the team members strictly followed the eight steps of *kaizen* (known collectively as the *kaizen* story), as suggested by *kaizen* consultants, and found that following these eight steps helped them to achieve their target.

The *kaizen* story is a standardized format to record *kaizen* activities conducted by small groups such as quality circles. The same standardized format is employed to report *kaizen* activities conducted by staff and managers. The *kaizen* story includes the following steps:

*Step 1: Selecting a theme.* This step addresses the reason why a particular target has been chosen for improvement. Targets are often determined in line with management policies. Their selection is also based on the priority, importance, urgency, or economics of the circumstances.

*Step 2: Defining the goal.*

*Step 3: Understanding the current status.* The members of a *kaizen* team must understand and review current conditions before starting the process. Going to the *gemba* and following the five *gemba* principles is one way to do this. Collecting data is another.

*Step 4: Collecting and analyzing data to find the root cause.*

*Step 5: Establishing and implementing corrective countermeasures and actions.*

*Step 6: Evaluating.*

*Step 7: Establishing or revising standards to prevent recurrence.*

*Step 8: Reviewing the process and starting work on the next steps.*

The *kaizen* story follows the plan-do-check-act (PDCA) cycle. Steps 1 through 5 relate to *P* (plan), step 6 to *D* (do), step 7 to *C* (check), and step 8 to *A* (act). The story format helps anyone to solve problems based on data analysis and enhances visualization of the problem-solving process. It also provides a way to keep a record of *kaizen* activities. *Kaizen* stories based on data analysis use various problem-solving tools to help participants understand the *kaizen* process.

*Kaizen* was first applied at Alpargatas in June 1994 by a pilot team made up of production, industrial engineering, and technical staff. Two operators also were assigned to the team to work on *kaizen* projects on a full-time basis. The team's target areas for improvement were related to raising the product quality of Nike shoes to meet the company's stringent quality standards.

The project posed two challenges. First, the issue of craftsmanship had to be addressed because producing shoes involved many manual operations. Second, the failure of many previous quality-improvement efforts had produced a high level of employee skepticism, which had to be overcome.

The team, which was assigned to work on the project on a full-time basis for three months, met formally once a day, with informal meetings throughout the day as the flow of work demanded. The *kaizen* consultant joined the team for three full days per week and, at the beginning, led and coordinated the whole process. After a few weeks, the company *kaizen* coordinator began to lead the group, whereas the consultant guided the team in the use of the *kaizen* story and *gemba* approach. During the three months, the team worked on solving two main problems: excessive glue and heel quality. The remainder of this case study focuses on their work in the latter area.

*Step 1: Subject definition: Assembly quality at the heel.* Assembly quality at the heel is one of the most important determinations of footwear quality. During the most recent quality audit, an American shoe consultant has pointed to heel assembly quality as the most urgent problem to be solved. To accomplish the desired improvements, the *kaizen* team chose minifactory number 1, which conducted cutting, stitching, and back-part molding operations.

*Step 2: Goal definition.* See Table CS-4.

*Step 3: Current situation.* See Figure CS-20.

*Step 4: Cause analysis.* See Figure CS-21. When the analysis began, few members of the pilot team expected that their work eventually would involve activities in other departments, such as upstream processes (stitching, cutting, and counterskiving), as well as maintenance, product development, and design. The analysis showed that the adhesive material that glued the counter to the heel actually deteriorated the counter material, causing inconsistency in the quality of the bond between the counter and the heel.

*Step 5: Corrective actions.* See Table CS-5.

*Step 6: Evaluation.* See Table CS-6 and Figure CS-22.

- ▲ To implement new methods, it was sometimes necessary to modify operator working positions, build new tables, modify existing tables, and develop additional devices and tools.
- ▲ Throughout the project, the supervisor of the sector—a member of the pilot group—was consulted. He took part in the *kaizen* process, followed up on the learning curves, and supported the implementation.

**Table CS-4** Goal definition.

| Indicator | Defects % Current Value | Defects % Target Value | Improvement % by 7/18/94 |
|---|---|---|---|
| 1. Assembly margin | 37 | 7.5 | 80 |
| 2. Counter position | 19 | 4.0 | 80 |
| 3. Counter centering | 27 | 5.5 | 80 |
| 4. Flatness | 33 | 6.5 | 80 |
| 5. Opening | 54 | 11.0 | 80 |
| 6. Heel centering | 23 | 5.0 | 80 |
| 7. Neck position | 47 | 9.5 | 80 |
| 8. Average defects % at back-part molding exit | 34 | 7.9 | 80 |

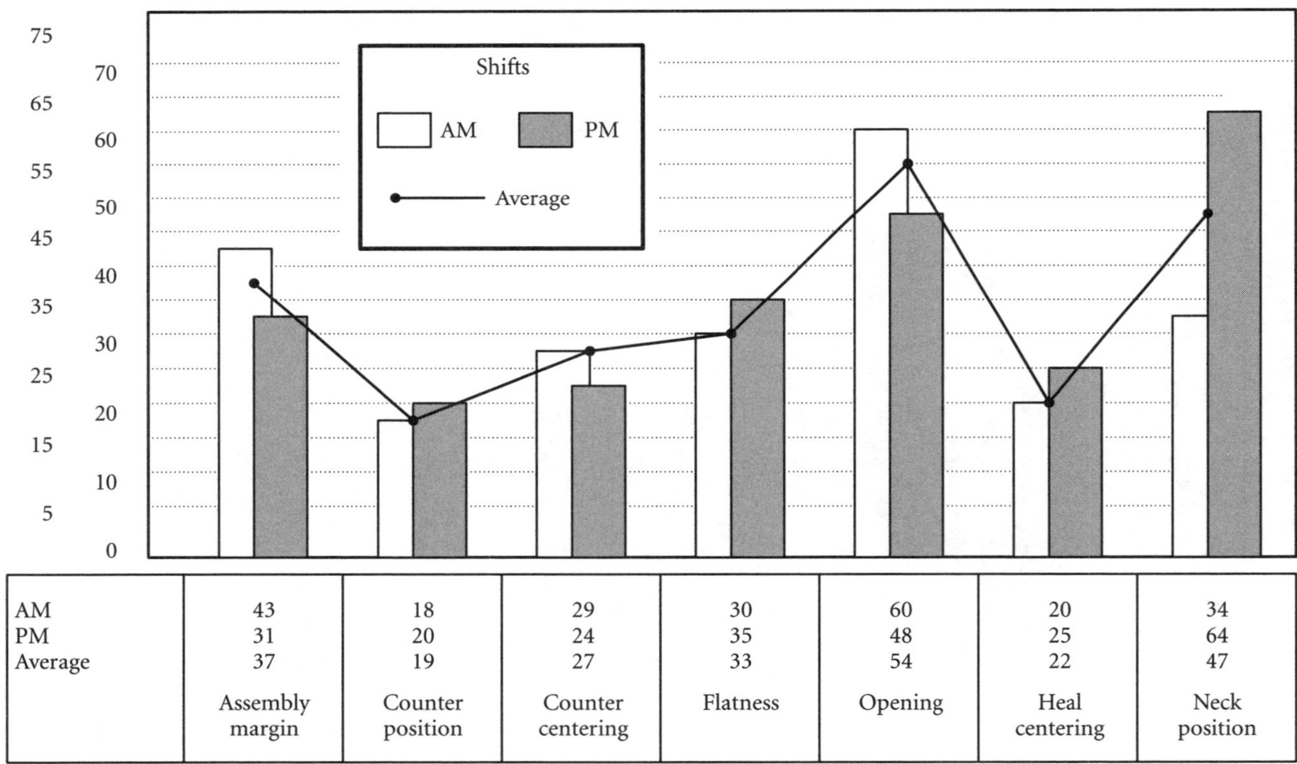


| | Assembly margin | Counter position | Counter centering | Flatness | Opening | Heal centering | Neck position |
|---|---|---|---|---|---|---|---|
| AM | 43 | 18 | 29 | 30 | 60 | 20 | 34 |
| PM | 31 | 20 | 24 | 35 | 48 | 25 | 64 |
| Average | 37 | 19 | 27 | 33 | 54 | 22 | 47 |

**Figure CS-20** Current situation.

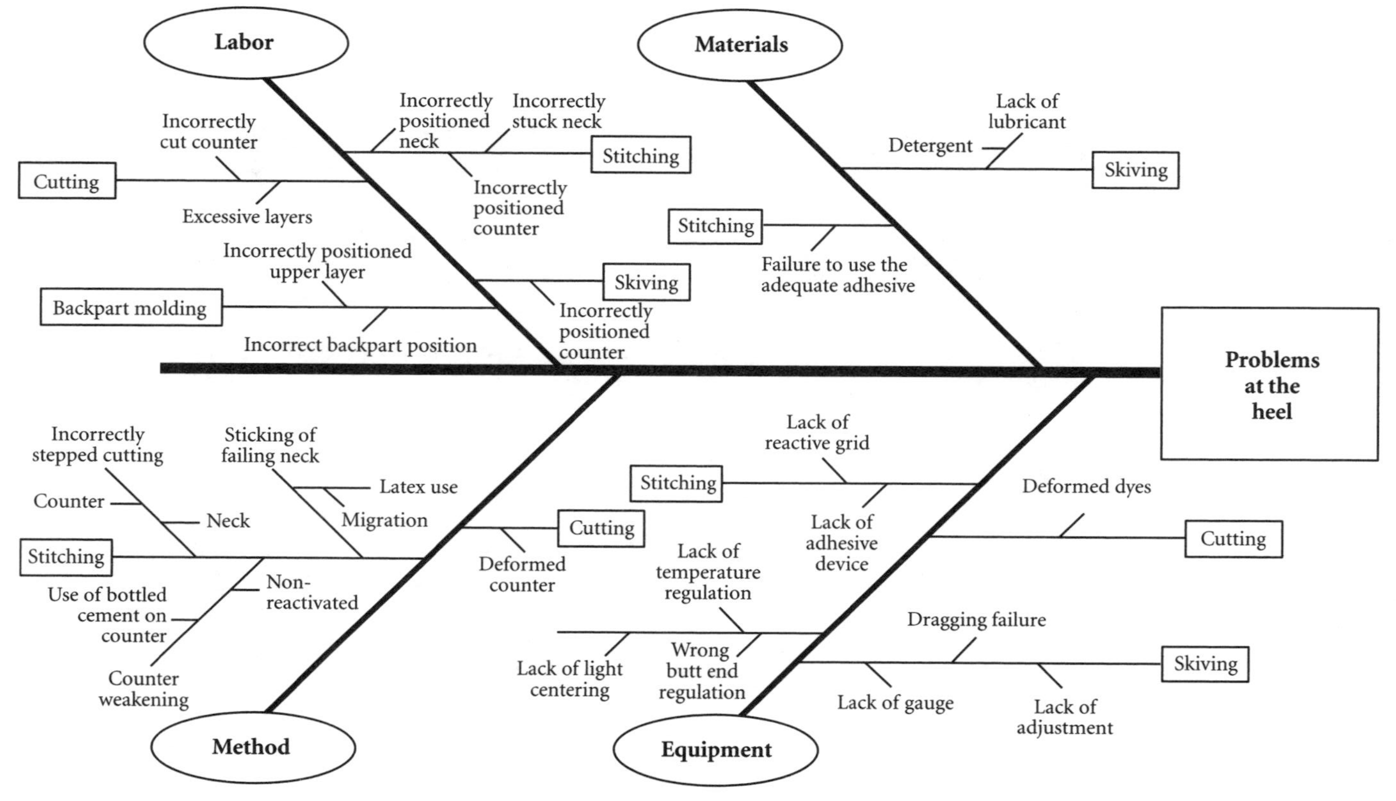


**Figure CS-21** Cause analysis.

**Table CS-5** Corrective actions.

| Problem | Cause | Action |
|---|---|---|
| Inadequate skiving | Lack of dragging at skiver<br>No gauge for skiving width<br>Lack of detergent | Verify sharpness, speed-advance relation, belt tension<br>Verify original gauge<br>Establish minimum stock level |
| Inadequate cutting | Overlapped cutting<br>Cutting greater number of layers than stipulated<br>Deformed die | Operation control – instruction manual<br>Operation control – instruction manual<br>Modify die height<br>Redesign not to allow sharp corners<br>Change to forged material |
| Inadequately positioned counter-stitching | Doesn't meet 6 mm standard | Train workers<br>Develop standard gauge |
| Inadequately positioned neck | Doesn't meet 9 mm standard | Train workers<br>Develop standard gauge |
| Inadequate blast counter | Heating temperature not meeting standard (maximum 80°C) | Use adequate heating device<br>Discontinue use of additional glue<br>Reactivate glue on counter |
| Glue wasted at neck dumping | Inadequate mixture of glue in use | Define appropriate glue |
| Inadequately positioned heel in back-part molder | Heel is not centered<br>Doesn't butt at back-part molding | Place centering light<br>Regulate butt to between 12 and 15 mm<br>Adjust centering faces to standards |
| Inadequate definition of heel border<br>Inadequately positioned heel | Lack of temperature regulation<br>Wrong reference points<br>Wrong use of reference points | Verify three times per shift<br>Redesign parts<br>Train workers |

**Table CS-6** Evaluation.

| Indicators | Initial Value | Target Value | July Value |
|---|---|---|---|
| Assembly margin | 37% | 7.5% | 7.6% |
| Counter position | 19% | 4.0% | 5.6% |
| Counter centering | 27% | 5.5% | 2.0% |
| Flatness | 33% | 6.5% | 4.0% |
| Opening | 54% | 11.0% | 9.5% |
| Heel centering | 23% | 5.0% | 1.0% |
| Neck position | 47% | 9.5% | 8.1% |
| Average defects | 34% | 7.0% | 5.4% |

*Average values for July 1994.*

- The supervisor helped to maintain close communication between the *gemba* workforce and the *kaizen* group, which allowed the workers to adapt to the new methods.
- The group prepared the instruction sheets as the basis for worker training. This brought about consistent operations in both shifts.

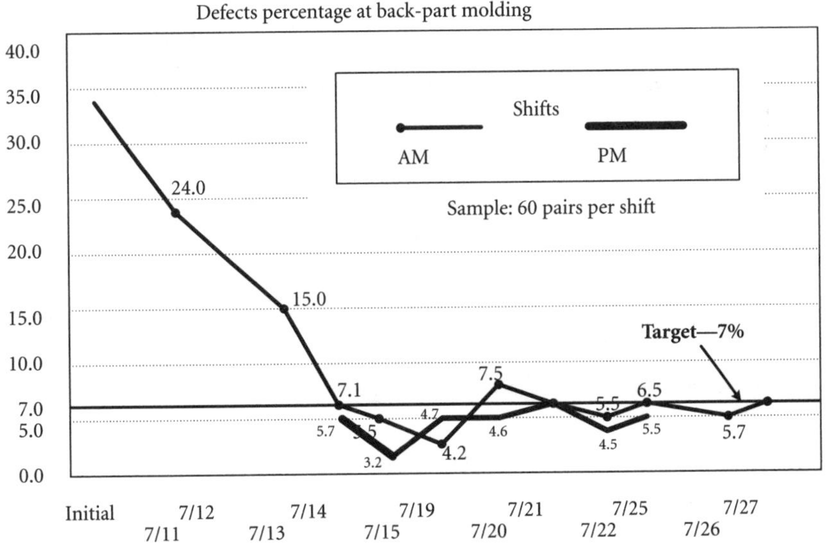


**Figure CS-22** Evaluation.

- ▲ A checklist including points of adjustment was placed at the back-part molder machine, which shapes the shoe heel. This enabled workers to adjust the machine when deviations were observed.

*Step 7: Addressing problems and preventing recurrence.* The team installed a control chart for indicators, and workers prepared instruction manuals. The team then introduced checklists for product quality and failures, and management disseminated the new standards throughout other affected sectors.

*Step 8: Follow-up.* Transfer these experiences to the remaining areas of the plant, and contact other glue suppliers.

## General Observations and Reflections

Team members methodically followed the eight-step *kaizen* cycle and found that these steps helped them to undertake the problem-solving process in the right sequence. Team members also learned that using such tools as fishbone diagrams and Pareto diagrams helped them to work on the project in a systematic and orderly manner and to find solutions more readily. Furthermore, the eight steps helped them to detect opportunities for future *kaizen.*

- ▲ The new work method paved the way for a subsequent project: implementing the one-piece flow of just-in-time production in the neck-dumping operations.
- ▲ This project identified many additional checkpoints.
- ▲ Team members found that having data greatly facilitated communication between supervisors and workers.
- ▲ The new method minimized glue waste.
- ▲ Along with quality improvement, productivity improved in the neck-dumping and counter-placing operations.
- ▲ Standardizing operators' tasks provided a standard for doing the job and facilitated training.
- ▲ Initially, the outside shoe consultant suggested that the back-part molding machine currently in use was obsolete and unfit for use. Completion of the project revealed that that machine was reliable with adjustments and maintenance.

## Savings

The *kaizen* activity in the pilot area saved 34,000 Argentine pesos per year. Applying the same procedures to other areas engaged in footwear heel assembly forecasts total savings of 225,000 pesos per year. (The Argentine peso maintained a one-to-one exchange ratio with the U.S. dollar at that time.)

## Senior Management Support

The success of this project owes much to the support of senior management in several ways:

- Holding the initial training course and meeting
- Participating in group work meetings and getting involved in the details of the discussion
- Participating in formal presentations of the work done by the team and encouraging its members to keep up their good work

# Transforming a Corporate Culture: Excel's Organization for Employee Empowerment

Excel Industries, Inc., is a supplier of $600 million worth of goods to the ground transportation industry. Excel Industries supplies a broad customer base of automotive original equipment manufacturers (OEMs), such as heavy truck, mass transit, and recreational vehicle manufacturers. The company is the leading supplier of window and door systems to the automotive OEMs. Excel has approximately 4,000 employees and operates 10 manufacturing facilities. Twenty-one percent of Excel's factory labor force is unionized.

This case addresses the issue of building a good, solid foundation for the house of *gemba*. It shows how Excel tackled the task of changing the corporate culture by clarifying the roles of managers vis-à-vis employees, providing training for employee empowerment, and building various infrastructures to carry out those tasks.

## Meeting the *Kaizen* Challenge

Excel Industries initially embarked on the *kaizen* process in March 1992 with the assistance of the Kaizen Institute of America. The motivation for implementing *kaizen* was straightforward. Without a disciplined process to achieve continuous improvement, Excel's ability to remain an ongoing, independent entity was in jeopardy. Heightened global competition to meet customer demand for continuous improvement in quality, cost, and delivery demanded a response.

To address customer needs, Excel formed a corporate steering committee in March 1993. This committee is cross-functional: Attending members included the company's president, three vice presidents of strategic business units, the vice president of human resources, three general managers of operations, the director of manufacturing operations, the director of corporate purchasing, and the vice president of value management.

After a year of *kaizen*, the committee saw impressive results in its 15 *gemba* workshops. The results included productivity gains of 57 percent, a 73 percent reduction of work-in-process (WIP), cycle time reduction of 78 percent, and floor space reduction of 44 percent. The *kaizen* committee recognized that *gemba* workshops were unlocking the human potential of Excel's employees. Excel wanted to find a way by which the potential and enthusiasm generated by participation on a *gemba* team during a workshop could be carried over to daily life within the company. The challenge was how to institutionalize the *gemba* workshop culture. The committee also wanted to ensure that Excel could sustain the *kaizen* process over the long term. To make sure this happened, the committee planned to use corporate and outside *kaizen* consulting resources on an ongoing basis to guide the process.

Following the plan-do-check-act (PDCA) process, Excel set out to benchmark companies with experience in the *kaizen* process in order to gain insight into whether these companies captured the potential and enthusiasm sparked by *gemba* workshops—and if so, how they did it (see Figure CS-23). The benchmark study revealed two factors as being key to sustaining *kaizen*: (1) strong management support for redefining responsibilities and (2) empowered employees. Managers in companies whose *kaizen* efforts were successful found it necessary to change their corporate cultures from top-down-driven to supportive.

Excel's new challenge was to define the steps or processes that would be required to help support a change in culture. The *kaizen* steering committee addressed this challenge by redefining the roles and responsibilities for the corporation (see Figure CS-24). The next step in changing the corporate culture was to redefine roles and responsibilities. The supportive process would require extensive training and education to help managers and other employees better understand their redefined roles and responsibilities. Senior management needed to know why change was necessary and to lead that change. Middle management needed to define what needed to be changed and support the change in culture. Employees needed to define how change should be implemented and accept responsibility for implementing the same.

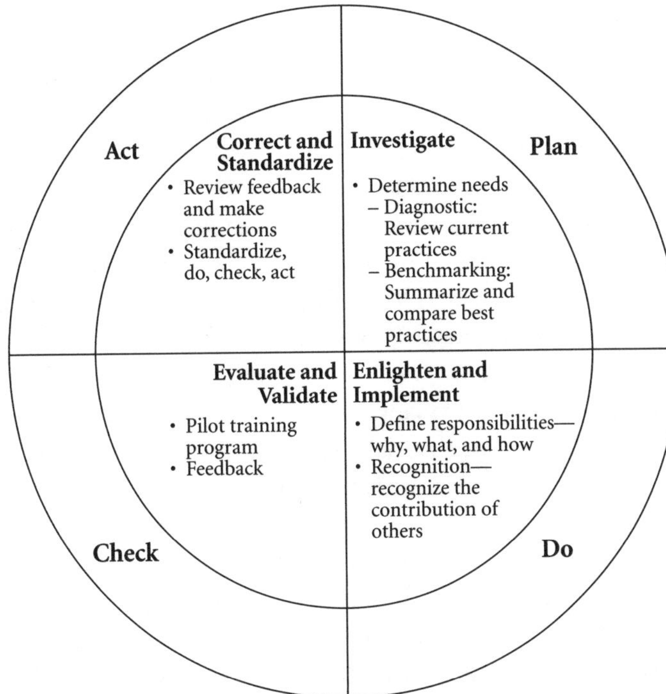


**Figure CS-23** The PDCA process.

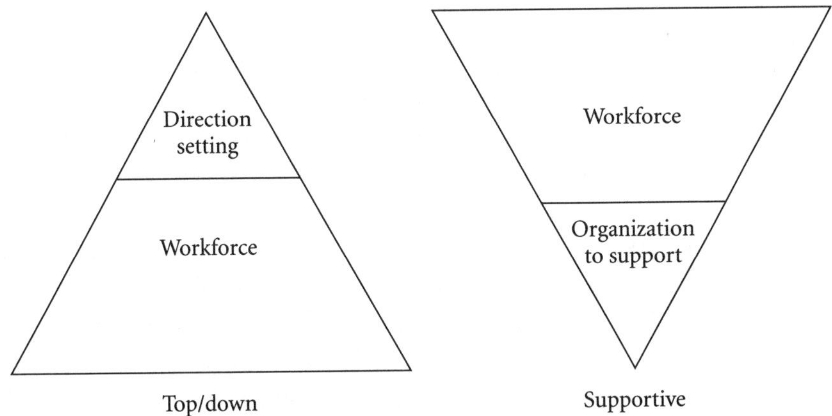


**Figure CS-24** Changing the corporate culture roles and responsibilities.

## Employee Empowerment Training

Excel Industries brought together a dedicated group of professionals from the Kaizen Institute of America and Trinity Performance Systems, plus a team of Excel employees spanning several areas, to form the Excel 2000 Empowerment Team. It assembled a cross-functional empowerment team with professionals from every area, including administration, engineering, manufacturing, quality, and human resources. Two individuals from each of Excel's nine plants were asked to participate. Excel's *kaizen* steering committee team members were empowered to establish a training objective (empowerment), develop a budget for empowerment training for the next five years, draft a training curriculum, and serve as an advisory board for the format and presentation of materials. The empowerment team met monthly for 10 months, and each meeting lasted three to four days.

Excel knew that its customers had reduced and would continue to reduce their supply base over the next few years. Customers with 1,500 suppliers today will have only 700 suppliers. All suppliers use tools such as statistical process control (SPC). Some use value engineering and *kaizen*, but few use empowerment. Excel intends to be a leading supplier in the future and believes that empowerment will enable it to accomplish this vision and attain a competitive advantage.

The understanding and definition of employee empowerment varies among organizations. At Excel, *employee empowerment* means that everyone has the authority and responsibility to improve their own work as long as they are part of the team, have the appropriate data, and follow a standardized improvement process. Employee empowerment is *not* synonymous with participative management, shared decision making, pushing decisions down, or "anything goes." Figure CS-25 illustrates the components of empowerment at Excel.

The Excel 2000 Empowerment Team's mission is to equip Excel Industries with an educational system that provides all teams with the knowledge, skills, and behavior to willingly accept ownership for the continuous improvement of their standardized processes. The Excel 2000 team developed the following training modules to accomplish this task:

A. Team kickoff (team, staff, and support groups)
   1. Define empowerment.
   2. Define teamwork.

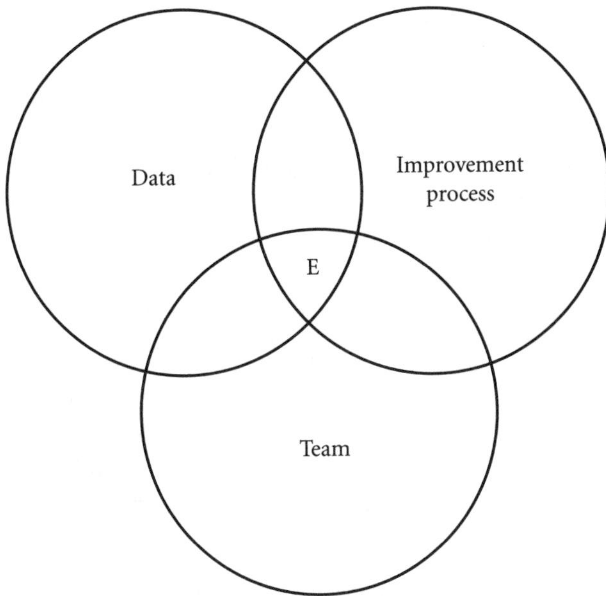


**Figure CS-25** The components of empowerment. *E* represents employees working in teams using data to prioritize and direct improvements throughout the *kaizen* process.

3. Speak with data.

B. Sponsor (staff)
   1. Define team goals.
   2. Determine the resources.
   3. Commit to support team.

C. Standardize work team—standardize-do-check-act (SDCA)
   1. Standardize the work process.
   2. Map the process.
   3. Develop the team.

D. Improve work processes—plan-do-check-act (PDCA)
   1. Define value added.
   2. Define waste (*muda*).
   3. Conduct a *muda* walk.
   4. Introduce data-gathering tools.
   5. Identify process improvement by speaking with data.

In addition, there are nine modules designed to train the supervisor/coach to manage the process, develop people and teams, and integrate systems.

All the preceding training promotes a shift within Excel toward a culture of incremental improvements. The company will accomplish its cultural change by establishing teams empowered to standardize, improve, solve problems, and be innovative in their work so that results better fit the customer's requirements. The SDCA and PDCA processes develop these teams both naturally and in a cross-functional manner. Excel has a totally committed and supportive management team, and this support and commitment is active, not passive. Empowerment teams are given a team sponsor. The sponsor for the company's first such team, the Excel 2000 Empowerment Team, is the plant general manager. Subsequent teams will be sponsored by members of the plant's managerial staff. Sponsors set team goals, allocate and ensure that resources are committed to support the process, mentor team coaches and future sponsors, and "walk the talk." Figure CS-26 depicts the process of empowering employees at Excel.

## The Need for Additional Training

As Excel's empowerment team was developing training materials for empowered shop floor work teams, it became evident that additional training modules would be needed to support the vision of changing the corporate culture. Also, the team found that training shop floor employees, first-line supervisors (coaches), and management staff personnel was just the first phase in the process. The team proposed a three-phase approach that would include every employee in the Excel organization. The team felt that additional training in support functions at both manufacturing sites and corporate headquarters would help to focus the efforts of all employees to support the empowered shop floor work cells and better serve the needs of Excel's customers.

The next step in the empowerment process will address phase 2: training related to manufacturing plant support functions. This training will focus on training these functional groups to support the empowerment team. Roles and responsibilities will be clearly defined. The Excel 2000 Empowerment Team will develop a needs analysis for the support groups. Gaps will be defined and training will be developed to better support the shop floor empowered teams and focus on customer needs. After phase 2, a third phase will address training related to corporate support functions.

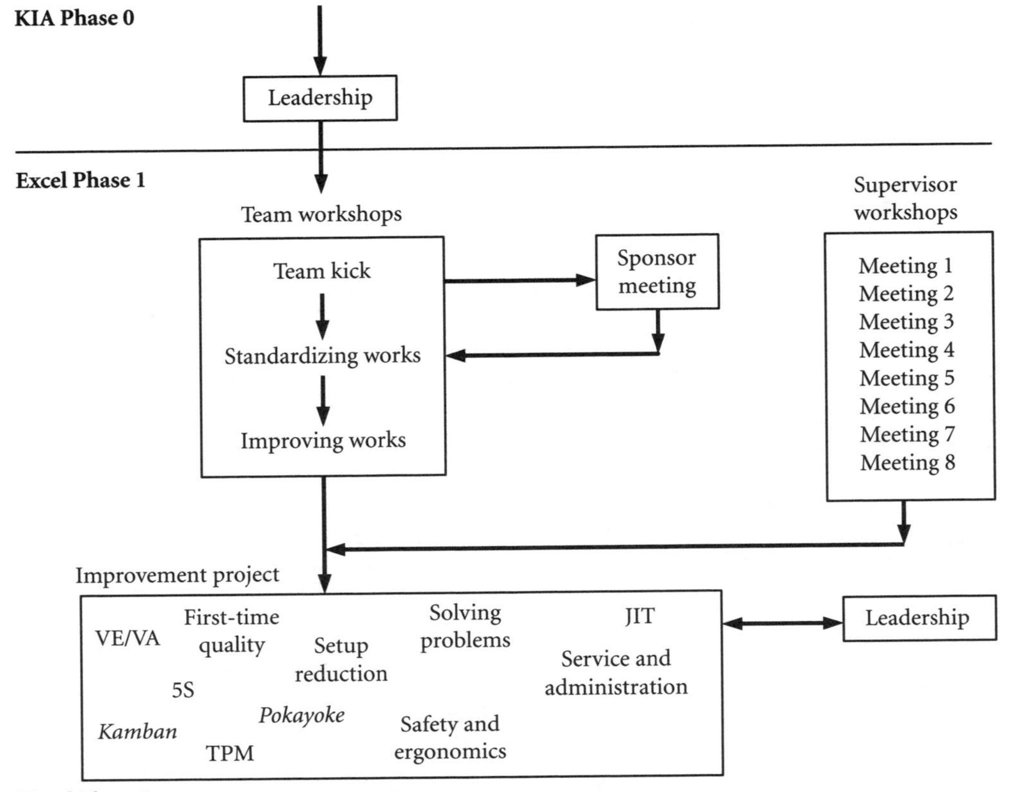


**Figure CS-26** The process of empowering employees.

This training will focus on technical and administrative groups to support the empowered manufacturing plants.

## Action and Accountability

Excel's senior management has clearly defined *why* a cultural change is necessary. Global competition is exerting pressure for continuous improvements in quality, cost, and delivery. Failure to comply with customer needs threatens Excel's ability to survive as an independent entity.

Middle management has clearly articulated *what* needs to change in the corporate culture. Through the efforts of the Excel 2000 Empowerment Team, middle management now possesses the process, skills, tools, and training necessary to unlock the human potential of Excel's employees.

Excel employees have gained the ability to define *how* to implement change and accept responsibility for change. Employees now have the authority and responsibility to improve their own work as part of a team working with data and following a standardized process.

The key to Excel's future success is directly related to the accountability among senior management, middle management, and all employees. The accountability of senior management will be measured by the ability to *lead* and *support* the process. Middle management will be assessed by its ability to *develop trust* by coaching and teaching the process and by providing support. Nonmanagerial employees' accountability will be measured by the extent to which they can *implement change* and *accept responsibility.*

# Quality in a Medical Context: Inoue Hospital

Inoue Hospital in Osaka, Japan, specializes in hemodialysis. It has 22 doctors and 420 staff. Its hemodialysis division has 127 beds for hospitalized patients and 180 beds for visiting patients. This is another case in which the collection of data (scare reports) has proved to be a crucial step toward improvement in the hospital environment.

## Scare Reports as a Quality Tool

In 1985, the hospital's director, Dr. Takashi Inoue, learned about scare reports being used in the manufacturing industry. The system requires that every time an operator in the *gemba* witnesses a potentially hazardous situation, he or she must submit a scare report, which is then used as a basis for correcting the conditions that allowed the situation to arise. Since the hospital was not immune to accidents, Inoue liked the idea of collecting data on scares to prevent an accident from actually taking place.

Often, scares happen as a result of somebody else's careless handling of the preceding tasks, and filing a scare report is tantamount to pointing the finger at someone else's mistakes. In introducing scare reports at the hospital, Inoue made it clear to everybody that the purpose of the report was to ensure the safety of the customers (patients), not to accuse colleagues who had made mistakes. Improvement of quality assurance was the main goal, and to do it, he said, everybody must be frank enough to admit mistakes. Otherwise, there would be no hope for improvement.

The hospital staff learned Heinrich's law on safety. Heinrich found that of every 330 industrial accidents, 300 are accidents causing no damage, 29 are accidents causing minor damage, and 1 is an accident of grave consequences (see Figure CS-27). In order to avoid that one serious accident, Heinrich argued, both the total number of minor accidents and the total number of accidents causing no damage should be reduced.

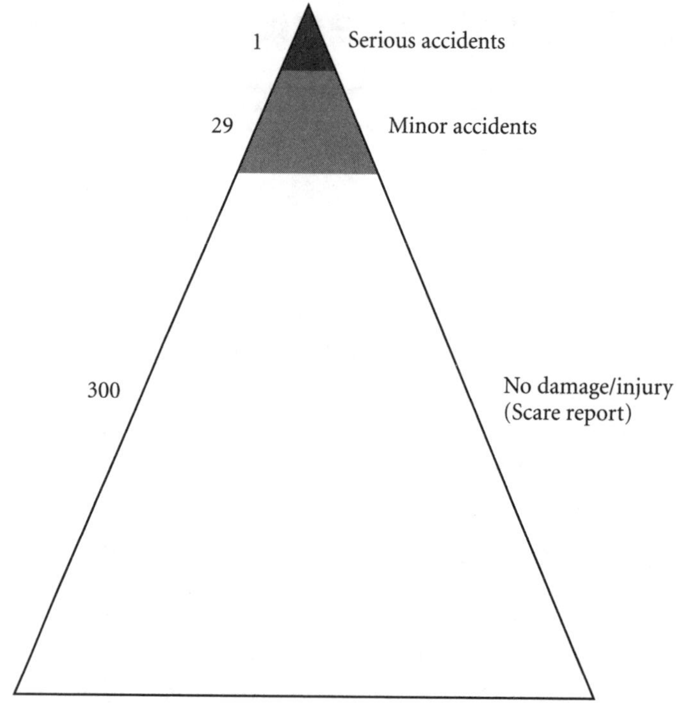


**Figure CS-27** An illustration of Heinrich's Law on safety.

Inoue Hospital classified its scare reports according to the categories in Heinrich's model, and standards were established. Scare reports at the hospital are now required in the following instances:

1. *Air.* If air has entered a patient's body during dialysis.
2. *Hemorrhage.* If any hemorrhage over 10 milliliters has occurred.
3. *Blood coagulation.* If the dialyzer circuit has had to be exchanged.
4. *Leakage.* If any rupture has taken place.
5. *Wrong medicine or wrong shots.* If any incident in which the wrong medicine or solution has entered a patient's body has occurred, even if it has not caused any harm.
6. *Wrong sequence in withdrawing the needles.* If the needle has been withdrawn completely from the body, even when hemorrhage has not resulted.
7. *Circuit malfunction.* If the dialyzer has needed to be replaced.

8. *Water release.* If water has been released in any amount 500 grams more or less than the stipulated amount or if it has taken 30 minutes longer than expected.

Scare reports must be submitted every day, and this tool has greatly enhanced the safety awareness of the nurses and paramedical staff. In the early days, the people responsible for causing the scare would ask, "Who reported it?" Sometimes doctors themselves were heard asking this question. Over time, though, everybody at the hospital has come to accept the scare report as a daily routine and a way to review and improve everyday work processes. Most problems arise from a failure to follow the correct procedures. Scare reports therefore help staff to review their own working procedures.

For instance, a nurse once tried to give a hemostatic shot to a patient. The patient said, "I don't usually get any shots." When the nurse checked the record, she found that the patient was right. This scare happened because this was the first time the nurse had worked on this particular patient, and she did not get all the pertinent information from the previous nurse. Normally, mistakenly giving a shot of hemostatic agent does not constitute a serious accident, but at Inoue Hospital, it must be classified as an accident. A scare report must be submitted, and a measure to prevent recurrence must be devised.

The reports are assembled at each nurses' station and submitted to management every day. Every month, management compiles the reports and sends a summary to the staff. Each department must implement countermeasures right away and report them. If the solution is more complicated and will require more time, the subject must be taken up by the hospital's quality circles as a joint task.

Other examples of scares reported at the hospital have included the following:

- At the time of starting dialysis, the artery chamber was found empty. This was due to the failure of priming.
- The heparin switch was not turned on; this problem was discovered on the second check. Fortunately, no blood coagulation had taken place.
- After the hemodialysis, the needle was pulled out in incorrect sequence, and a slight hemorrhage occurred.
- The wrong solution for continual injection was delivered, but this mistake was discovered before the solution was used.

The following table shows the incidence of dialysis problems at the hospital:

| Number of dialysis year | Procedures | Water release margin of error | Blood coagulation Accidents | Hemorrhages |
|---|---|---|---|---|
| 1991 | 63,522 | 88 (0.14%) | 21 (0.03%) | 27 (0.04%) |
| 1992 | 72,082 | 109 (0.15%) | 55 (0.08%) | 34 (0.05%) |
| 1993 | 73,240 | 147 (0.20%) | 75 (0.10%) | 14 (0.02%) |
| 1994 | 71,792 | 105 (0.15%) | 49 (0.07%) | 17 (0.02%) |

In 1993, the hospital had a total of 839 scare reports in the following categories:

| Number of year | Circles | Number of staff involved |
|---|---|---|
| 1983 | 10 | 127 |
| 1985 | 18 | 132 |
| 1990 | 23 | 282 |
| 1995 | 41 | 429 |

## Quality Circles

Another feature of *kaizen* activities at Inoue Hospital is active quality circles, which were begun in 1983. The following table shows how the number of quality circles at the hospital has grown over the years:

| Following categories: | |
|---|---|
| Wrong medicine or injection | 41% |
| Water release | 23% |
| Blood coagulation | 17% |
| Air inclusion | 10% |
| Hemorrhage | 9% |

The main subject areas addressed by quality circles are, in order of the number of projects, quality, efficiency, safety, and cost. A total of 189 projects have been completed since the first quality circle was organized.

Other topics addressed by the quality circles of the hospital have included the following:

- ▲ Improvement of clinical diagnosis forms
- ▲ Improvement of failsafe switch for blood pump to catch abnormalities in dialysis apparatus
- ▲ Elimination of dosage mistakes
- ▲ Reduction of waiting time for dialysis
- ▲ Elimination of switching mistakes in the air-detection apparatus
- ▲ Optimal inventory of drugs
- ▲ Reduction of x-ray film loss
- ▲ Reduction of mistakes in serving special dietary meals

In order to evaluate the hospital's effectiveness, management always works hard to collect information from the following sources:

- ▲ Patients' claims
- ▲ Patients' remarks at the time of leaving the hospital
- ▲ Review by a third party
- ▲ Countermeasures against an accident
- ▲ Cases in which death was involved
- ▲ Specific patients' symptoms
- ▲ "Hot mail" from patients to the hospital director

The hospital encourages its staff to experience medical treatment from the patient's perspective. In 1994, 16 nurses underwent hemodialysis, one administrative staff used a wheelchair, two secretaries tried laxatives, and one clerk underwent an examination by stomachic camera. The following remarks are from employees who underwent such experiences in 1995:

- ▲ *Nurse A:* "I underwent the experience of getting hemodialysis as a patient. I was expecting the pain of having the needle in the vein, but when I had to stretch my arm for three and a half hours, the muscles around my shoulder and elbow ached badly, and this was the hardest part. Since I couldn't use my right arm, it was not easy to have a cup of tea and lunch lying on the bed."
- ▲ *Nurse B:* "I am right-handed, and since the needle was attached to my right hand, I had to eat with my left hand. I was unable to eat the Chinese food and could only eat rice dumplings, and I was very hungry. I was anxious to know who would be attending me as my nurse and hoped that

it would be someone good at injecting the needle painlessly. Normally, when I am acting as a nurse and I hear the patient say something like that, I feel upset because I am always trying to do my best."

- *Nurse C:* "When I was lying down on the bed, I felt very uneasy. I suppose that patients feel the same way. I was very relieved when the nurse came to me and said, 'Are you okay?' When you take off your white uniform and lie down on the bed, you feel very feeble. From the bed, people standing by you look very tall, and the doctor looks really great! It's a different feeling from what you get as a nurse. Everyone looks great from the bed. So I think we shouldn't talk to them as if we are looking down upon the patients."
- *Nurse D:* "It seems that some patients hesitate to call a nurse even if they know the nurse well enough. It will be better if we can anticipate such needs and go to patients' beds without being called. I feel that they should not hesitate to call us but think we should take the initiative of calling them."
- *Administrative staff member:* "I have experienced sitting in the wheelchair. I found that the button on the side of the elevator was too high and inconvenient to push. We never had such a complaint from the patients, but when unattended, the patient will have to ask someone else to help."

In addition to gaining firsthand experience as patients, staff members are encouraged to experience working in areas outside their usual jobs. This helps them to better understand how business is conducted in other departments and assists them in building cross-functional teamwork.

# The Journey to *Kaizen* at Leyland Trucks

Leyland Trucks, Ltd., Britain's leading commercial vehicle maker, designs, develops, and produces a range of civilian and military trucks sold throughout the world. The company was formed in 1993 when its management team led a buyout of the Leyland Assembly Plant in Lancashire, England, and of its associated businesses, from DAF BV. In 1998, Leyland Trucks was acquired by PACCAR, which had acquired DAF Trucks in 1996, thus reuniting the two. Through the PACCAR Production System, *gemba kaizen* continues to be an important driver of competitive excellence at these companies.

The managing director, John Oliver, who joined Leyland Trucks before DAF BV went into receivership, not only went through the management buyout process but also led the company through *kaizen* and into a new lean production system to attain many dramatic improvements, including achieving the lowest cost of any European truck maker. Oliver has vividly described Leyland's experiences in his report submitted to me in 1995:

> In the 1980s, we invested an enormous amount in technology with little results. Now we emphasize building employee commitment, teamwork, slim organizational structure, and effective communication.
>
> The past five years have taught us that the most effective cost reduction programs start without any intent to reduce costs at all. We have frequently seen initiatives designed for quality, whether quality of process, product, or people, generate highly welcome financial benefits. It may be far-fetched to state that the more you ignore cost reduction, the bigger the eventual saving will be. However, many occasions show that to be exactly the case.

As in the Excel case study, this case addresses the issue of building a good, solid foundation for the house of *gemba*, but Leyland used another approach. Leyland's management directed its efforts toward making organizational changes to meet the new challenges, delayering the management structure, improving systems efficiencies, and introducing new measures to change the culture.

## Making Organizational Changes

To start the process of *kaizen* at Leyland Trucks, Oliver initiated three major changes to the company's organizational structure: (1) introducing business units, (2) cutting excess layers of bureaucracy, and (3) improving systems and procedures.

### *Business Units*

When John Oliver arrived at the Leyland assembly plant in mid-1989 as operations director, he found employee morale to be very low. Their attitude toward Leyland DAF could best be described as grudging acquiescence. Some were downright hostile.

Oliver and the other managers resolved to improve the quality of working life. Improve an employee's job satisfaction, went the argument, and you would improve his or her self-esteem and eventually his or her affinity with the company. Management realized that the key to improving the quality of life was building infrastructure appropriate to employee requirements—and *not* focusing on improving the design of individual jobs or assignments. "Our traditional hierarchical and functional structure was too remote, too slow, too impersonal, and too bureaucratic to meet employee needs," Oliver says. "If we were serious, then we had to find a new model."

Eventually, management came up with the concept of the *business unit*, a liaison team composed of employees representing each functional area (industrial engineers, planners, quality technicians, logistic specialists, etc.). The members of the business unit would work on the shop floor, next to the lines—in direct and regular contact with the regular shop floor employees. The business unit members would be trained in interpersonal skills, including team building. The objective was simple Management 101 pursuits. As a result of these efforts, quality defects dropped from 28 per vehicle in 1986 to 4 in 1995 (see Figure CS-28).

"I am convinced that had we set out with the specific objective of cost reduction, we would have failed," says Oliver. "The single-minded pursuit of quality of process and enhanced employee affiliation brought these huge savings."

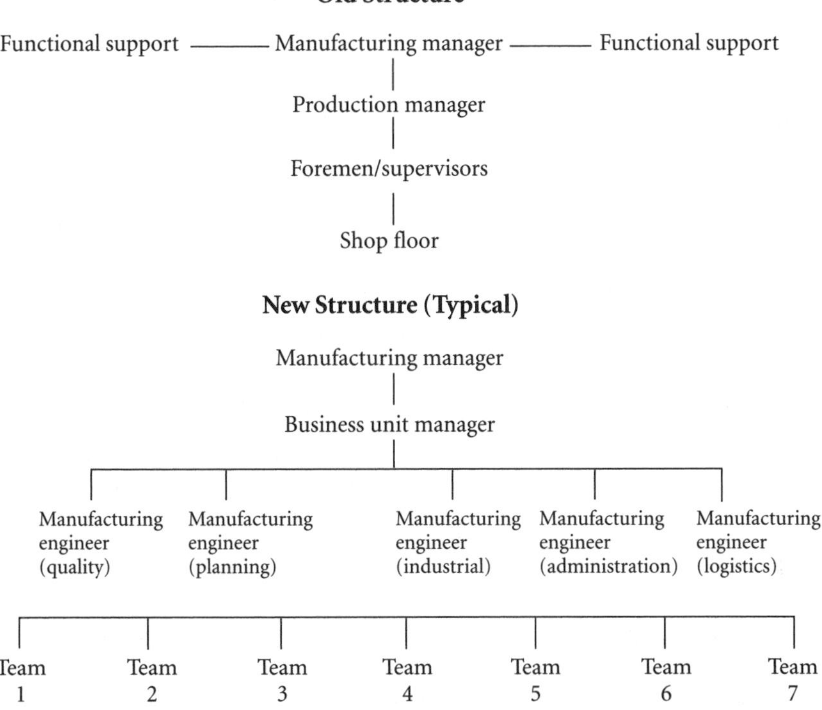


**Figure CS-28** (1) Chart illustrating the frustrations and cost of doing things "the old way" at Leyland; (2) organizational charts showing the old structure and new business unit concept.

## *Organizational Delayering*

Traditional management structure presented many obstacles to building an empowered organization. However, any move toward a flatter, wider pyramid challenged the status quo. Any organization resists radical change; there is never a "right" time for such change. Nevertheless, management realized that the traditional multilayered organization had to go.

The results, says Oliver, were well worth the effort: "Over two years, 42 percent of senior and middle management positions disappeared. Loosening the chains of the old hierarchical bureaucracy improved the added value of the team by 30 to 50 percent. Increasing individual spheres of influence—at first feared and resisted by anxious managers—very quickly led to improved self-confidence, self-esteem, and job satisfaction. Colleagues who had spent the previous 20 years doing largely the same thing day after day experienced a new resurgence of spirit, a new level of energy."

## *Building Effective Systems*

Management's third organizational change improved systems and procedures. On this subject, Oliver offers the following observations:

**Memo on signing for replacement tools and new tools.** When a tool broke on the assembly line, the operator would look for his or her supervisor to make out a replacement note. This note was then taken to a store, where it was exchanged for the new tool. The problem with the process was that the operator could not always find the supervisor, and this could cause the following problems: The operator's work was not complete; tools were being shared; tools were being stolen; and in some cases, the operator would request additional tools to ensure that he or she would not go through the same frustrations.

The results meant that everybody had a locker full of tools. To add to the problem, when the tooling budget costs got excessive, an instruction was passed to the managers that they had to sign for all replacement notes. Of course, this caused more frustration because if people had difficulty finding the supervisor, you can imagine how difficult it would be to find a manager.

So this was one of the first areas of empowerment that was developed down to the team.

Here's a story that made us realize the frustrations being caused:

One day when we were coming out of a meeting, an operator met me and asked me to sign his note. I took the note from him, looked at the tool being requested, and when signing, mentioned that the tool cost the company £10. He replied very aggressively, "You have got it wrong; the tool has cost you £19." I asked why, and he said, "Because I've been looking for you for one hour and 30 minutes at £6 per hour. Plus, I've not completed my operation." So, this gave us the incentive to empower the teams.

An initiative called *systems effectivity* addressed this issue of process clarity. Again, the objectives contained nothing about cost reduction, but pointed toward some form of quality. We first employed a self-audit, designed to enhance involvement. For up

> to five days, groups of employees kept personal diaries describing everything they did. The participants and their colleagues analyzed the data to distinguish added-value activities from those which were consequences of nonconformance elsewhere in the system. People enjoyed both the novelty of self-assessment and the challenge of analysis.
>
> Our expectations from the exercise were modest. We wanted a more transparent bureaucracy. We wanted to challenge the status quo. We got all that and more. Thirty percent of the paperwork from the operation disappeared—and along with that, a whole host of meetings, agendas, minutes, reports, and miscellaneous bumph [paperwork] which simply cluttered our daily lives and scarcely added any value whatsoever. In short, yet another example of almost inadvertent cost reduction on a very sizable scale.
>
> When we started it, the objectives were improved perceptions of "ownership," simplicity of systems, clarity and availability of information, greater customer satisfaction, and so on. But the final analysis showed a £1 million annual saving on the conventional cost of quality performance—a staggering 83 percent reduction.

## Changing the Culture

There is a Japanese saying to the effect that a statue of Buddha will amount to nothing if the person who carves it fails to put a soul into it. Even after Leyland introduced the "hardware" aspects of business units, delayering, and system effectiveness, management found these measures insufficient to take the company where it wanted to go. Additional efforts to involve and empower employees were needed. Of these efforts, which came to be known collectively as project "Bridging the Gap," Oliver writes:

> Small focus groups of representatives across a range of functions and levels addressed some of the more long-standing and vexatious questions. International visits to key competitors exposed representatives to benchmarking of work practices.
>
> All this took place while the above-mentioned activities were going on. In addition, a small experiment that introduced cell

working in our machining area got very positive internal publicity. The local management of the area recruited the operators to work and involved them in the design of everything from plant layout to process design. This departure from previous practice drew attention as all the prospective members of the new cost center worked together planning their new world of work.

The consequences of this experiment were startling. Quality levels improved sharply. Floor-to-floor times shrank by 80 percent. Inventories fell to previously unheard of levels. Efficiencies improved significantly. And equally important, absentee levels in the same group of people fell from an exceptionally high 8 percent to less than 2 percent.

We realized a dynamic at work which, if reproduced across the company, would transform its fortunes. The question was, *How?* Sadly, even after 18 months of successful business units, cellular manufacturing, organizational delayering, and so on, we had not achieved our fundamental requirements of winning the hearts and minds of the workforce at large. Things improved, but not sufficiently to remove the ancient barriers of mistrust and suspicion. Something had to weld the entire workforce into one coherent, mutually dependent and mutually supportive unit.

## *Encouraging Employee Initiative*

Realizing this, John Oliver decided to encourage *employees* to initiate changes in the company:

With the help of local consultants, we offered every employee the opportunity to express his or her *real* views and concerns in a structured manner. The key benefits of the change initiation programs were

- Independently generated information—the key word being *independent*.
- Quantifiable conclusions which gave an indication of relative importance.
- A representative, cross-sectional sample size so that all groups felt involved.

- Feedback facilitated by a consultant who was independent of the company to avoid bias and partiality.
- But most important, ownership of both the problem and the solution by the workforce. It was important that they knew and understood that it was their *collective* view.

Through this process, we were trying to (1) persuade the majority of the workforce that change is inevitable; (2) demonstrate that these changes are mutually beneficial; and (3) generate real ownership of the change process. To make it happen, the following six key stages were developed:

- *Stage One: Presentation and communication of the program.* It was important that the trade unions were involved.
- *Stage Two: Identification of needs through in-depth interviews and groups discussions.*
- *Stage Three: Identification of priority and satisfaction.* This involved asking the workforce to take a view of an ideal job for themselves in an ideal world, identify 4,550 needs, and rank them according to their aspirations.
- *Stage Four: Analysis.* This ended up in drawing a hierarchical picture of needs, priorities, and satisfaction from function to function and from department to department.
- *Stage Five: Feedback of data to employee groups.*
- *Stage Six: Diagnosis.* Various employee groups were organized to move into problem-solving mode.

Standard attitude surveys were often seen as solutions imposed on employees to problems defined by management, often with a hidden agenda suspected. In this process, change initiation, need, its priority, and the level of dissatisfaction were *all* defined by the workforce and therefore owned by them.

## *Recognizing a Job Well Done*

The first step toward recognizing employee efforts happened in the area of idea generation. Overcomplexity, overbureaucracy, and wholesale dissatisfaction had defeated previous suggestion plans. Clearly, a new approach was needed. As John Oliver writes:

A brainstorming exercise on the plans concluded that the main objective behind suggestion plans was to get people to make suggestions! That's not quite as silly as it may seem. The lesson we learned was that success should be measured by the number of times employees thought positively, constructively, and imaginatively about the company, not how much successful suggestions could save us.

We devised the "Every Little Counts" plan, where every employee who makes a suggestion and submits it on a formal entry form gets a £1 voucher for a national chain store. It did not matter whether the suggestion saved 1 pence or £100,000. As long as it was done constructively, the voucher was handed over with a "Thank you."

The ELC scheme has mushroomed since, with its own self-managing infrastructure. Financial tradeoffs are avoided to ensure that the controversies over previous models are not repeated. Contribution levels are high by traditional standards and by current British best practice.

While ELC offered an organizational response to greater involvement, it also tested the relationship between the managers/supervisors and the workforce during the working day. After years of impersonal, task-oriented management, we needed a much more interactive style that demonstrated awareness of the social dimension of the working world. Managers had to learn to greet people at the start of shift, to use first names, to get to know the whole person—not just the work characteristics. This was not easy. To change these habits of a lifetime was a different proposition, and they needed a helping hand.

Accordingly, the mouthful "informal individual recognition token" was created. For every 100 employees in a particular cost center or management area, the manager was given 25 tokens to hand out during the coming year for exceptional work, whether consistent or incidental. In year two, the token became a rather nice metallic tape measure. The tape measures, obviously useful in the world of work, became a source of some pride—a major step forward from the fear of being singled out. In year three, a rather large sports bag suitably embossed was chosen. Its size was important, to elevate the visibility of the recognition process. Recognition became a comfortable, mutually acceptable part of everyday life. In year four, the token is a company T-shirt. The only way anyone can lay their hands on this garment is to be singled out by a manager for exceptional performance. To wear one has become a matter of pride. Our key thrust within the area of recognition is, however, the acknowledgment of the contribution of the team. Every three months, nominations for the quarterly team awards are canvassed energetically. Care is taken again to avoid the financial tradeoff between recognition and reward. We have had enough experience to conclude that monetary gain demeans the process. Although teams who are commended receive only a simple certificate applauding their efforts, the level of pride is there for all to see. An informal but public ceremony is used to ceremonially acknowledge their contributions. The winning team has its name endorsed on a large shield. It is a very comfortable, very much accepted process which is now embedded into the culture.

## *Working as a Group*

Teamwork, says Oliver, has been a way of life at Leyland from the beginning:

> When we started our journey in 1989, we visualized a company whose ethos was grounded on the concept of teamwork. It could be functional or multifunctional, formal or informal, horizontal, vertical, or diagonal. What really counted was the spirit and the readiness to use the group as the core of the business.
>
> Involvement and participation have come to life. The autonomous working groups on the tracks with their peer key operators have demonstrated efficiencies never previously encountered. Multifunctional project groups, properly sponsored, trained, and facilitated, spring forth regularly and deliver remarkable results free of bureaucracy and senior management intervention. Ad hoc teams appear regularly and tackle issues in a structured fashion that overcomes functional barriers.
>
> Involvement and participation can be talked about and can be encouraged. But the acid test is in the execution. Group working is the ideal vehicle to practice what we preach.

## The Start of a Journey

Oliver concludes his remarks by saying:

> The foregoing describes just a few of the mechanisms designed to foster involvement and participation. However, I believe the fundamental task is to keep pushing the barriers back, to keep on introducing new ideas to maintain the momentum. Leyland is only at the start of its journey.

# Tightening Logistics at Matarazzo

Matarazzo of Molinos Rio de la Plata Company, a member of the Bunge & Born group in Argentina, is a manufacturer of fresh and dry pasta and other products. The company delivers its finished products to the group's distribution center, located five kilometers from the plant. This case shows that a great improvement in logistics management was registered by collecting data, observing operations in the *gemba*, and using a common-sense approach to solve problems.

Before the company embarked on its *kaizen* project, deliveries took place between 6 a.m. and 6 p.m. every day; total time for each delivery (including loading at the plant, transport on the road, unloading at the distribution center, and returning to the plant) averaged about three hours. Six or seven trucks made a total of 10 to 14 trips per day. Moreover, delivery personnel were usually required to work on Saturdays because they could not meet all the daily requirements between Monday and Friday.

Since the loading and unloading operations required the driver to step on the product in order to cover and uncover the pallets containing the product, goods often were damaged. Creating another bottleneck was the fact that the trucks had to be weighed four times—twice at the plant and twice at the distribution center (once when empty and once when full). To eliminate such logistical inefficiencies and optimize the operation, a *kaizen* group was organized.

Six months after the *kaizen* project began, the company found that it needed only two trucks and four to six trailers, and daily working time was reduced by 7 hours to 11 hours. The operation time per truck was reduced by 22 minutes (an 88 percent improvement), and the cost per load was reduced by 35 percent.

The company achieved these results through the following *kaizen* activities:

- In the new procedures, a truck picks up a loaded trailer at the plant, travels to the distribution center, leaves the loaded trailer there (to be unloaded), picks up an empty trailer, drives back to the plant, and leaves the empty trailer there. The driver then picks up a loaded trailer and returns to the distribution center.
- The distribution center has designated two unloading docks for Matarazzo's exclusive use.
- Trailers have been fitted with structural roofs and sliding curtains, eliminating the need for the drivers to step on the product.
- A display board has been installed to show the weight of each truck and trailer. It includes all possible combinations, so the loads only need to be weighed once, rather than twice, at each location.

The time saved by these improvements has made it possible to increase the daily number of trips per driver from one to five. And employees no longer have to work on Saturdays.

## The Next Step: *Kaizen* on the Supplier Side

Shortly after the new procedures were in place, employees from the logistics and operations departments formed a group that included quality-control and production people from each department. As the subject for *kaizen*, the group chose lead-time reduction of supplier service. Because of inadequate preparations at the plant to receive deliveries, the loading and unloading lot was perennially congested, resulting in an average service time of 3.5 hours. The *kaizen* team set out to bring the time down to under 2 hours. Toward this end, the group focused its work on the following items:

- Designing a schedule for receiving supplies in order to eliminate bottlenecks and allocate support personnel efficiently.
- Advising suppliers on how to improve delivery scheduling.
- Prioritizing suppliers by grouping them into critical and noncritical categories based on business requirements.
- Setting time targets for completing the unloading task and giving prompt service to suppliers who complied with those targets.
- Monitoring the time expended for each task and the suppliers' time inside and outside the plant.

Six months after the project began, and after 18 group meetings, supplier service time came down to an average of 70 minutes. Encouraged by this achievement, the group set a new goal: to reduce service time to less than an hour.

Stepping up their cooperation, several sections of the company produced a report on the time expended for each event (supplier arrival, waiting time outside the plant, unloading attention time, and time of departure from the plant). The suppliers welcomed this information because it enabled them to exercise precise control over their own transport, especially the contracted freights. The data also encouraged suppliers to abide by the agreed-on scheduling.

Two months later, supplier service time had dropped to 45 minutes. As soon as the second goal was achieved, the group devoted itself to standardizing tasks and collecting data in order to stabilize, control, and maintain the target values they had achieved. Finally, as an alternative to manual data gathering, with all the cross-checking that entailed, the group developed a database application that allowed instant access to information such as activities carried out, classification by input and supplier type, service time, comparison with programmed events, and so on. The database allowed the group to produce different kinds of reports, summaries, and graphs depending on its needs.

# Stamping Out *Muda* at Sunclipse

Based in Commerce City, California, Sunclipse sells and distributes industrial packaging and corrugated shipping containers. The company's president, Gene Shelton, says: "We are not in a high-tech business; we are, however, in a high-energy business. Ours is a low-capital business. Anybody with good connections with customers can start the distribution business as long as he or she has a telephone and a desk. They don't even need to own a truck to transport products, because they can lease the truck. We recently had a new entry into our market from Taiwan. That's why we must work harder, think smarter, and keep satisfying our customers better than anybody else." This case shows various ways, including *muda* elimination, introduced by Sunclipse to improve its competitive edge.

## Relying on Coworker Input

Over the years, the company has introduced various *kaizen* activities involving coworkers. (Sunclipse prefers the term *coworkers* to *employees* because of the latter's connotation of management-labor confrontation.) There are two major vehicles for coworker input for *kaizen* at Sunclipse. One is the opportunity-for-improvement (OFI) sheet (see Figure CS-29), whereby coworkers can write down any idea for improvement on an OFI form and submit it to their supervisors. If the supervisor is unable to solve the problem, the subject is brought before the problem-solving team. The other vehicle for coworker input is the customer satisfaction form, on which coworkers can report any customer complaint or other problem (see Figure CS-30).

## Continuing Improvements

Each Sunclipse division has a facilitator who devises various programs to ensure coworkers' support for continuous improvements. Each division also

# O.F.I.

OPPORTUNITY FOR IMPROVEMENT

DATE:__________ REPORTED BY:__________

The following situation is making it difficult for me to do my job right the first time:

OPTIONAL:

What has already been done:

What could be done:

LOG #:

Figure CS-29 An example of an opportunity-for-improvement sheet.

CUSTOMER SATISFACTION FORM

LOG #: ______

DATE: ______

CUSTOMER: ______

ADDRESS: ______

PHONE: ______ ACCT #: ______

CONTACT: ______ SALES REP #: ______

CUSTOMER CALLED FOR:

| | | |
|---|---|---|
| ☐ LATE DELIVERY | ☐ WRONG QUALITY | ☐ WRONG PRICE |
| ☐ CUST. RET. PRODUCT | ☐ TAX VS. RESALE | ☐ PROD. NOT TO SPECS. |
| ☐ ON HOLD TOO LONG | ☐ WRONG PRODUCT | ☐ NOT GIVEN FOLLOW-UP |
| ☐ NO RESPONSE BACK ON QUOTE/PROBLEM | ☐ BACKORDER | ☐ WRONG ADDRESS |
| ☐ SAMPLE DELAY | ☐ INVOICE NOT REC'D OR WRONG | ☐ MISSED APPOINTMENT |
| ☐ ATTITUDE (GOOD/BAD) | ☐ COMPLIMENT | ☐ OTHER (SPECIFY BELOW) |

EXPLICIT DESCRIPTION: ______

ACTION REQUESTED: ______

ISSUED BY: ______

ROOT CAUSE OF PROBLEM: ______

CORRECTIVE ACTION TO BE TAKEN: ______

COMPLETED BY: ______ DATE: ______

**Figure CS-30** An example of a customer satisfaction form.

has a quality improvement team (QIT), which meets once every two weeks to go over unresolved problems and discuss how to carry out programs.

## *Muda* Miles

At Sunclipse's St. Hart Division in California, which produces corrugated paper products (and which places particular emphasis on eliminating *muda* or non-value-adding activities), facilitator Pat Arnold introduced "Muda Miles" as a way to visually display improvements made by coworkers. A map of the United States is posted in a prominent place, and each time a worker's suggestion helps to reduce *muda*, the improvement is converted to a mileage value and plotted on the map. The object is to "travel" across the country, with St. Hart as the starting point of the journey. Alongside the map is a list of *kaizen* projects that have been implemented and their corresponding mileage.

## The Wish Tree

At the Orange Division of Sunclipse's Kent H. Landsberg operation, the major activities are administration, distribution, warehousing, and sales. When the division first introduced the suggestion system, coworkers were very interested and enthusiastic, and there was a deluge of new ideas. A few months later, however, facilitator Stacey Snyder found that the initial burst of enthusiasm had given way to inertia. Realizing that a more accessible program was needed, Snyder came up with "The Wish Tree." She developed a form with a simple structure that allowed coworkers free expression of their wishes.

An orange tree (in keeping with the division's name) was placed in Snyder's office. For every idea submitted, a white ribbon attached to the "I wish . . . " form was detached from the form and placed on the tree. When people began working on the idea, the white ribbon on the tree was replaced with an orange flower, and when the problem was solved, the flower was replaced with an orange. The quality team is now considering the next step for involving coworkers in *kaizen* projects.

Each supervisor at this division is asked to submit a monthly report to management on how *muda* was handled in his or her area during the previous month. At bimonthly meetings, supervisors read their reports and exchange information on the current status of *muda* elimination.

## Getting the Sales Force Involved in *Kaizen*

The Kent H. Landsberg operation of Sunclipse, tasked with selling not only Sunclipse products but also the products of other manufacturers, is pivotal to the company's success. Nevertheless, management had a difficult time getting salespeople to participate in continuous improvement.

The sales representatives maintained that since their job was to increase sales, they were already involved in continuous improvement. The reps offered every excuse for not attending problem-solving meetings: They were too busy; they had their hands full just getting orders into the system and products to customers; they were not, to the best of their knowledge, causing any quality-related problems in the company. At times, the salespeople seemed to believe that the earth revolved around them.

Finally, management decided to get back to basics—to start with the voice of the internal customer and use data to convince the sales force. Managers learned that a large number of voicemail messages from sales - people lacked one or more of the following pieces of information:

- Customer name
- Purchase order
- Buyer name
- Purchase quantity
- Method of transportation

At the next biweekly sales meeting, the 80 sales reps present were told that some 700 incomplete orders had been received from them in the previous month. Still, every salesperson believed that "somebody else must have done it." Each salesperson then was handed a sealed envelope containing a record of the incomplete orders he or she had issued. The sales reps opened their envelopes, and a long silence followed. For the first time, the sales reps realized that they, too, had to change. By the next month, the number of incomplete orders had dropped to 289.

Today, the information systems department is working with the sales force to develop a digital paging system, including a mechanism for taking remedial action immediately when the system fails. They have developed an online fax monitoring system. Again, corrective action will be taken the moment mistakes are found.

The coworkers in these departments also have developed a data-collection system that allows them to handle customer orders proactively. When a machine producing a particular product for a customer goes down, for example, the revised delivery date will be relayed to the sales rep through the digital paging system. The salesperson then can inform the customer of the change.

## Recognizing Employee Efforts: Q Bucks

Another feature of *kaizen* at Sunclipse is an employee recognition system called *Q bucks*. Q bucks may be awarded to coworkers who participate in the quality-improvement process and participate in one or more of the following:

- ▲ Completion of a quality-related education or training session
- ▲ Submission of an OFI leading to corrective action
- ▲ Improvement of a work process
- ▲ Solution to a problem by the problem-solving team
- ▲ Performance of a measurement
- ▲ Achievement of a departmental goal
- ▲ Membership in the company's quality council, a corrective action team, a QIT, or some other quality-related committee such as a "Quality-Is-Fun" committee.
- ▲ Other quality-related contributions at the discretion of the divisional QIT.

Sunclipse has a contract with a merchandise redemption company that allows coworkers to redeem the Q bucks they have earned for products or services of their choice. The products and services, which coworkers choose from a catalog, range from $5 worth of merchandise to a two-week Caribbean cruise.

Greg Brower, Sunclipse's vice president, director of training, and head of the companywide *kaizen* project, told me the following story: One Sunclipse truck driver had to deliver products to a warehouse each day at the end of his day shift. Because the warehouse was never ready to receive the products when he arrived, the driver had to wait in the truck for a long time. By the time he began unloading, it was well into the night shift. Thus the driver was receiving an average of 20 hours' overtime pay per week. Troubled by this waste of resources, the driver made a suggestion. If he could arrange for a nightshift worker at the warehouse to do the unloading, he could leave the truck at the warehouse unattended and go home.

This suggestion earned the driver $380 in Q bucks. He and his family enjoyed perusing the catalog before finally settling on a brand new color TV. "He was very happy, and we were delighted!" said Brower. "Once you get started in *kaizen*," Brower added, "it's difficult to stop. It just happens."

# Housekeeping, Self-Discipline, and Standards: Tokai Shinei Electronics

This case shows how quality can be improved dramatically when the two pillars of *gemba kaizen* activities—housekeeping and standardization—are introduced. Tokai Shinei, an electronics firm with slightly more than a hundred employees, started out as the sole supplier of printed circuit boards to one company. Initially, the company had no research and development (R&D) capabilities in-house and depended entirely on its customers to provide engineering drawings.

It was difficult for Tokai Shinei President Yoshihito Tanaka to find qualified workers in Shinei's hometown, a country town located about 150 kilometers north of Nagoya, Japan. The issue that haunted Tanaka was education. Since he was unable to hire employees with good educational backgrounds, he felt that the employees needed to be taught subjects such as statistical quality control and electronics. He asked a local high school teacher to lecture his employees on the principles of electricity. However, the classes proved too difficult for the employees, so Tanaka invited a middle school teacher to the company. This teacher, too, gave up after only a few lessons. Tanaka then invited a consultant on quality control to give a series of lectures. But this consultant soon ceased visiting the *gemba*, preferring instead to come to Tanaka's office for chats because nobody in the *gemba* could understand his lectures.

Thus Tanaka repeatedly found himself frustrated in his efforts to use outside resources for employee education. Suddenly one day it dawned on him that he had been expecting a third party to teach his people when such education should be the job of the president himself. He had neglected to share with employees his aspirations and visions for the company, as well as what he considered the company's problems. Realizing that he needed to take the initiative of teaching and sharing his ideas, he decided to hold a series of meetings with his employees.

In 1988, a mutual learning session was instituted, with Tanaka as the leader. The learning sessions were two-day programs held on the first weekend of every month. Employees took turns participating in the sessions, and every employee was required to do so at least once a year.

Saturdays were devoted to discussing issues of common concern. Tanaka saw that when employees were discussing issues of direct interest to themselves, they became much more excited and involved than they had during their lessons on electricity and quality control. Employees were involved, developed a sense of responsibility about the problems they were discussing, and came up with many possible solutions. On Sundays, the employees cooked and ate together at the nearby picnic ground, activities that greatly enhanced camaraderie.

The plant has been rebuilt since these sessions began, and a classroom specifically for the learning sessions has been added. Guest lecturers are invited to share their ideas, and the sessions are open to members of the community at large.

Currently, discussion subjects at the learning sessions include management plans, equipment purchase, recruitment, and bonuses. The company's financial reports and monthly performance figures are also reported. Other subjects addressed at the sessions include recreation, safety, communication, and financial management.

Hidesaburo Kagiyama, president of a local automobile parts supplier with a very successful nationwide network, was invited to speak at one session. Kagiyama has a unique management philosophy—management should start with housekeeping and end with housekeeping. He is a particularly firm believer in cleaning toilets—and does so himself every day.

Tanaka was so impressed with Kagiyama's lecture that he decided to put his ideas into practice right away. The next morning, he arose early and went to clean up the grounds of the shrine in his neighborhood. The shrine has a candy store on its premises, and the ground is carpeted with wrappers discarded by children. After cleaning up this litter, Tanaka set about cleaning the public toilets, which were so dirty the township had decided to close them. Tanaka went to the town hall and persuaded the local authorities to reopen the facilities, promising that he himself would clean them every morning.

Tanaka had heard Kagiyama say that housekeeping causes people to change their behavior, and to his surprise, he found this to be true. He enlisted his children's help in cleaning the toilets. The children would say, "Today the toilets were really dirty. It was wonderful!" What they meant was that they were gratified by the knowledge that people were using the toilets they had cleaned and that they were happy to make them clean again. Tanaka realized that pleasing other people is the starting point of pleasing oneself.

Tanaka found that once the candy wrappers had been picked up and the toilets were being cleaned, children stopped throwing litter, and people began trying to keep the toilets from getting dirty. Tanaka learned from this experience that self-discipline does not arise spontaneously but rather as a result of participation in some beneficial activity such as cleaning the environment.

Aside from employee education, another issue that had been haunting Tanaka for a long time was employee self-discipline. In the company's early days, management had difficulty hiring workers who could be expected to carry out their appointed tasks. An operator was once found smoking when he should have been working. When reprimanded by a supervisor, he became so angry that he began striking the machine with a hammer. Tanaka realized that education in technologies and skills was utterly useless if there were fundamental problems in human relations and self-discipline. Tanaka came to believe that self-discipline should be the starting point of all activities taking place in his company. He came up with three activities to serve as the pillars of self-discipline: housekeeping, greeting each other, and etiquette.

Once Tanaka introduced employees to these three pillars of self-discipline, he was amazed to see how greatly they improved human relations, enhanced employee awareness of other quality issues, reduced equipment breakdown, and changed employees' attitude toward customers. Community relations also improved. In other words, an awareness revolution was taking place among employees. As yet another benefit of self-discipline, the reject rate dropped by half.

Tanaka launched a full-scale housekeeping project, and now the working day at Shinei Electronics starts at 7:30 a.m., when all the employees roll up their sleeves and join in cleaning the factory floor, offices, hallways, toilets, and even the cars in the parking lot and the roads within a one-kilometer radius of the company. Everyone concentrates on housekeeping for 15 minutes before regular work begins.

When I visited the plant in July 1995, the first thing I noticed was the parking lot, which was so immaculate that it looked like an automobile dealership's lot. The sales personnel in particular wanted their cars to look neat because they use them when calling on customers. I also found the *gemba* spotless. Although chemicals such as sulfuric acid are used in Shin - ei's operations, not a drop of liquid could be found on the floor. Before the cleanup, operators worked in boots and aprons because the floor was covered with chemical liquids. Now the employees wear slippers and normal working clothes.

Employees offered the following comments on this, the first housekeeping experience in which they had been directly involved:

- "By working with others in cleaning up the premises, I was able to communicate with people whom I had never had a chance to talk with before and have come to feel much closer to them."
- "In the beginning, I just took pride in finding my own place neater than others'. But now, whenever I find other areas dirtier, I volunteer to pitch in and help. I used to think that what I was doing was best, but now I am ashamed I was so naive. I have grown as a human being as a result of cleaning. Cleaning is indeed a marvelous thing."
- "I have learned that in order to improve myself, I must help others to improve. I have come to believe that whatever I can do to help others, I should, though it is not always easy. I think I have become more patient."
- "When sales and production personnel cleaned together, we were able to communicate and understand each others' troubles."
- "I have become much more attached to and affectionate toward tangible items such as machines and buildings and readily notice abnormalities, such as which spot on the machine gets dirtier sooner than other spots."
- "This experience has made possible joint work among sales, engineering, and production, which used to regard each other as adversaries before."
- "I expect that these positive results that come out of cleaning together not only help our work but also benefit our family life."

Even after these housekeeping and other activities to enhance self-discipline had taken root in his company, Tanaka felt that something was missing. In late 1994, Tanaka told a *kaizen* consultant that, aside from quality, one of his major problems was that employees started work very slowly in the morning, getting busier as the day went on and becoming busiest toward the end of the day. He said the same was true for monthly production—that is, production started slowly at the beginning of each month and picked up at the end of the month to meet customers' orders.

The consultant's advice was as follows: "You have invested a sizable amount of money in your equipment. You hire a given number of people. Both equipment and people should be available to work at full capacity at all times. The uneven distribution of workload must be costing the company a lot of money. The reason for the uneven workload distribution lies in some inappropriate systems or work procedures in the company. So why don't you address this problem? The biggest problem is that you have accepted such uneven distribution as something unavoidable and never questioned the situation. The first thing you need to do is to go through an awareness revolution."

"For instance," the consultant continued, "why does work start so slowly in the morning? It must be because the machines are slow to start due to inadequate setup preparations. Why can't you change the work procedure so that machine setup is completed before the end of the day? In other words, the existing standards and work procedures must be reviewed. In particular, if workers' operations are not standardized, there is no way to establish proper line balancing."

Tanaka decided to carry out the consultant's advice and declared that a review of the existing standards would take place right away. The company had many work standards in place, but the standards had been prepared by engineering staff; *gemba* workers were expected to follow them unquestioningly. Often engineers prepared standards without checking beforehand how they would affect the *gemba*.

On Saturday and Sunday of the same week in which Tanaka had met with the consultant, all the employees were summoned for a review of the standards. (Employees were used to attending weekend discussion sessions.) The employees showed up at the *gemba* bringing with them existing standards (work sequence sheets) together with past records of problems. In order to review the work sequence, methods, and tools used for a given task, employees formed teams of three or more. A veteran operator performed a task according to the usual procedure while other operators looked on. Referring to the standard sheet, the onlookers corrected the veteran's actions when necessary. A second operator then tried to follow the sequence of work as demonstrated by the first operator. If the second operator encountered difficulty, employees discussed how the procedures might be made easier and revised the standard accordingly.

At each process, there are several key points that must be observed for technical reasons; these points were incorporated into the new standards. Thus the new standards specified the point that had to be observed at all times. Another feature of the new standards was that parameters previously left to individual discretion were quantified to the greatest extent possible. Processes also were simplified so that operators had only to push buttons on the machines.

These standards review meetings carried over to Sunday and involved managers, engineers, and veteran operators. The two-day session enabled employees to identify existing operational problems. The workers learned that making problems visible is the starting point of *kaizen*. They also found that although initial standards were written by engineers or line management, the nature of some tasks had changed considerably over the years, as had operators' understanding of the work procedures. Furthermore, operators often changed hands. The standards review showed employees that work speed differed from product to product, as well as from person to person. They found that adopting a uniform speed greatly increased efficiency and improved line balancing.

In the following weeks, the employees began implementing the new standards. Three months later, they held a two-hour standards review session during normal work hours. This time, part-time employees also were involved. The review sessions helped to reduce careless mistakes, and operators became much more confident in their jobs. The sessions also promoted the "awareness revolution" among employees.

The engineers, who had once assumed that their role was to teach and guide employees in the *gemba*, now work with those employees in establishing standards that are practical. Following are summaries of comments from operators who participated in the standards review sessions:

▲ "Today, I wrote a work sequence standard. I have been working here for 10 years, and up to now, I have relied on my personal experience and hunches to do my job. It was not easy for me to write down what I do in my job. There were some *kanji* (Chinese) characters I couldn't write. I could not put into words what I was doing. I felt so helpless that I got a headache."

- "As I look at my daily work, I find that I have practically no work to do in the morning. Then, at about 4 p.m., there is an onrush of work. So we need to distribute the workload evenly. Since I am engaged in inspection, I can only stand to work until 5 p.m., since it is very tiring to inspect tiny pieces. Please arrange the workload in such a way that I can return home on time. Thank you for giving me a chance to review my own work."
- "I feel that I have been doing my job in such a way that I am the only person who knows how to do it. As a result of today's session, I learned that if I do my job according to a set procedure, someone else can do it even when I am absent."
- "No matter what kind of a job we do, I believe the most important thing is our attitude. I realized the importance of morale in doing my job."
- "I used to think that I knew what I was doing. But once I started writing it down, I was surprised to find many items that have slipped out of my mind or items I have newly recognized. I was surprised to find that some coworkers did not know enough *kanji* characters to write their comments. We helped each other write and found it a wonderful opportunity for communication."
- "All the participants forgot about the time and put their full efforts into the task. It was a wonderful learning experience."
- "We labeled the machine switches so that anybody can operate the machines. For those who don't know how to operate the machines outside their job area, the work sequence sheet and the switch labels were very helpful, and I believe even newly hired employees can easily use these machines."

A supervisor offered the following comment:

Today's theme was how to write a work standard to eliminate *muda*, *mura*, and *muri* (waste, irregularity, and strain). I realized that, until now, I had let the operators do the job the way they wanted to do it. Every time operators changed hands, there were deviations in product quality, and the key parameters were not observed. When the operator in charge was absent, nobody else

> could do the job. For these reasons, I realized how important it was to prepare work standards. I also realized how difficult it was to communicate the right procedure to our own people. From now on, I will stick to the work standard as the basic rule of work, and each time there is a problem, I should look for the root cause; I should check whether it arose because the worker did not follow the standard or because the standard was inadequate; and also whether the standard included important control points. Thus the work standard should be the starting point of *kaizen*.

Six months after the first weekend standards review session at Tokai Shinei, the reject rate had dropped to one-quarter its previous level. Overtime also had gone down. More important, although sales had dropped during this period, profits had improved because some work formerly performed by veteran employees had been transferred to part-time employees. Many night-shift jobs also were transferred to part-timers. What enabled this transfer of labor? The standardization of work procedures. Tokai Shinei had achieved all these improvements without investing in any new equipment and without hiring any new employees.

# Solving Quality Problems in the *Gemba*: Safety at Tres Cruces

Most problems in the *gemba* can be solved if (1) the five *gemba* principles are followed or (2) data are systematically collected and analyzed. Some problems can be readily identified and solved if one takes the trouble of going to the *gemba* right away, stays there for five minutes, and keeps asking "Why?" until reaching the root causes of the problem. In such cases, observation is the key, and solutions can be reached on the spot in real time. Most problems in the *gemba* can be solved in this way. However, other types of problems require collection of data in the *gemba* and take some time to solve.

This case study describes how safety problems were solved at Tres Cruces Cold Storage Plant, a company in Argentina that manufactures such products as skinless sausages, hams, and salamis. Between January 1993 and May 1994, 27 accidents occurred at the company, costing it 78 worker-days. The company organized a group made up of a supervisor and three workers at the raw materials receiving depot. They had to design a safety project to reduce accidents while meat was being unloaded and transported. (The company was handling about 100 tons of meat per day.)

The group started its project by collecting information on the current status of accidents. Since no systematic means of collecting data existed at that time, only post-1993 data could be found. The group determined that 52 percent of accidents resulted in skin bruises, 33 percent in cutting injuries, and 15 percent in other types of injuries.

To gain a better understanding of the situation, group members held brainstorming sessions aimed at defining the causes of the most frequent accidents. They designed a "scare report" for operators to submit during the following four weeks every time they were frightened by near accidents, creating a database for analysis.

Such scare reports are often used to report close calls in Japanese *gemba* (see Figures CS-31 and CS-32 for typical examples). The number and types of scare reports filed during the four weeks at Tres Cruces are shown in Figure CS-33. Based on these findings, the *kaizen* group was able to identify major accidents and their frequencies and to plot them on a Pareto chart (Figure CS-34).

Once the group became familiar with the nature and frequency of accidents, it was able to analyze possible causes. The team developed a cause-and-effect diagram, shown in Figure CS-35.

**Scare Report**

Name: ____________________

Supervisor: ________________

1. When:

   Month__________ Date __________ Hour ________ Minute ________

   Where:

   What happened:

2. *Kaizen* Ideas

   If you have good ideas, please write them down.

———————————◆———————————

1. This is how I dealt with the problem. Date __________________

   This is how I am going to deal with the problem. Date ___________________

2. I cannot deal with the problem for the following reason. Date _______________

**Figure CS-31** A typical scare report form used in a Japanese *gemba*.

**Safety Classification (Safety, Transport, Quality, Energy, Resources, TPM, Production, Others)**

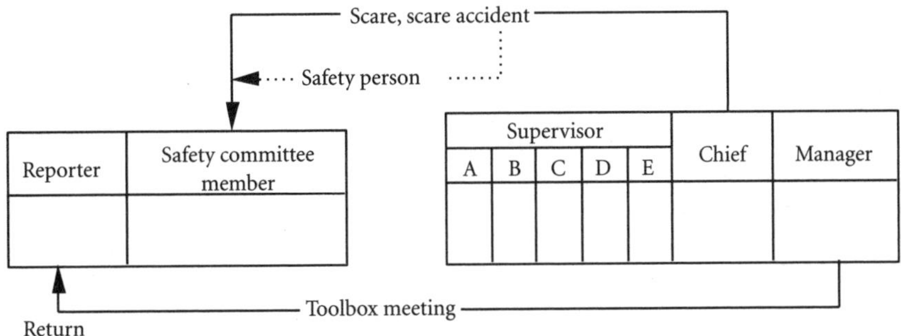


**Scare Report**
**Toolbox Meeting Report**

| | |
|---|---|
| When | Date: ______________<br>Before work   During work   After work   During break |
| Where | |
| Who and what | |
| What happened? | (Major indication of trouble)<br>1. Almost got fingers caught   2. Almost got stuck<br>3. Almost got hit   4. Almost got cut<br>5. Almost got burnt   6. Other |
| | - - - - - - - - - -<br>- - - - - - - - - -<br>- - - - - - - - - - |
| Why? | - - - - - - - - - -<br>- - - - - - - - - - |
| Superior's opinions and instructions | - - - - - - - - - -<br>- - - - - - - - - -<br>- - - - - - - - - -<br>- - - - - - - - - - |
| | Countermeasure adopted (People, *gembutsu*, or both)   Countermeasures not adopted yet (To be implemented by     ) |
| Participants | Person in charge:   Process:<br>Leaders, etc.:   Name: |

**Figure CS-32** A typical scare report form used in a Japanese *gemba*.

| **Scare Report** | | | | | | |
|---|---|---|---|---|---|---|
| **Scare** / **Period** | **First week** | **Second week** | **Third week** | **Fourth week** | **Partial** | **%** |
| Pork carcasses fall | /// | //// | // | 𝍸 | 14 | 9 |
| Cutting injuries | 𝍸 // | 𝍸 //// | 𝍸 𝍸 | 𝍸 𝍸 | 36 | 24 |
| Beef carcasses fall | 𝍸 𝍸 / | 𝍸 //// | 𝍸 𝍸 / | 𝍸 𝍸 / | 42 | 27 |
| Frozen raw material falls | /// | 𝍸 | 𝍸 | 𝍸 | 18 | 12 |
| Slide on truck floor | /// | //// | // | 𝍸 | 14 | 9 |
| Beef carcasses hit | // | /// | /// | / | 9 | 6 |
| Beef carcasses displaced | //// | /// | // | /// | 12 | 8 |
| Other | /// | / | // | // | 8 | 5 |
| Total | | | | | 153 | |

**Figure CS-33** The number and types of scare reports filed at Tres Cruces over a four-week period.

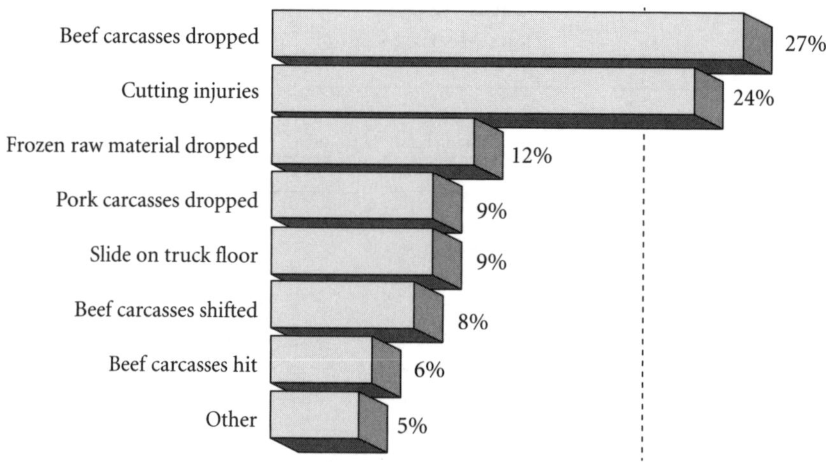


**Figure CS-34** Pareto diagram identifying the types of major accidents at Tres Cruces over a four-week period according to frequency.

## *Kaizen* Actions

As a result of the findings, the *kaizen* team at Tres Cruces took the following 10 actions:

1. Repair the electric hoist.
2. Ask supplier to quarter carcasses before shipment instead of doing so inside the truck.
3. Ask supplier to send pork after cutting the heads off the carcasses.

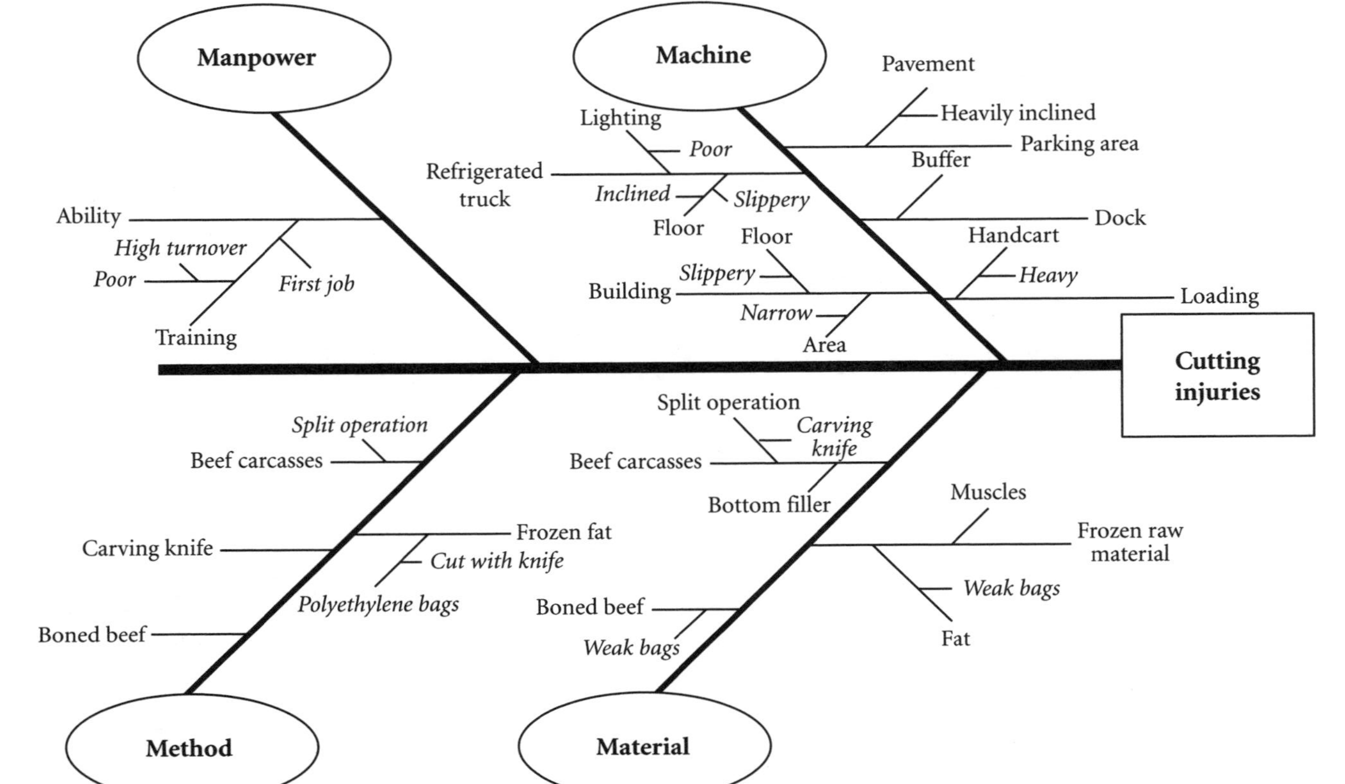


**Figure CS-35** Fishbone cause-and-effect diagram showing the nature of and relationship between accidents at Tres Cruces.

4. Shut the pork chamber's door while unloading beef carcasses.
5. Give operators safety devices such as shock-absorbing safety helmets.
6. Replace electric hoist's hook.
7. Attach protective cover to the unloading dock.
8. Improve method of cleaning floors.
9. Eliminate carcass-cleaning operations inside the refrigerated truck.
10. Use a portable conveyor belt to unload fat, muscle, and boned beef from the refrigerated truck.

The drop in the number of scare reports submitted since May 1994 has been a good indicator of the success of this project (see Figure CS-36).

## Standardization

As part of the *kaizen* effort, the following items or procedures at Tres Cruces were standardized:

- The scare report
- Quartering of carcasses by the supplier
- The procedure for cleaning the sector's floor
- The use of the conveyor belt

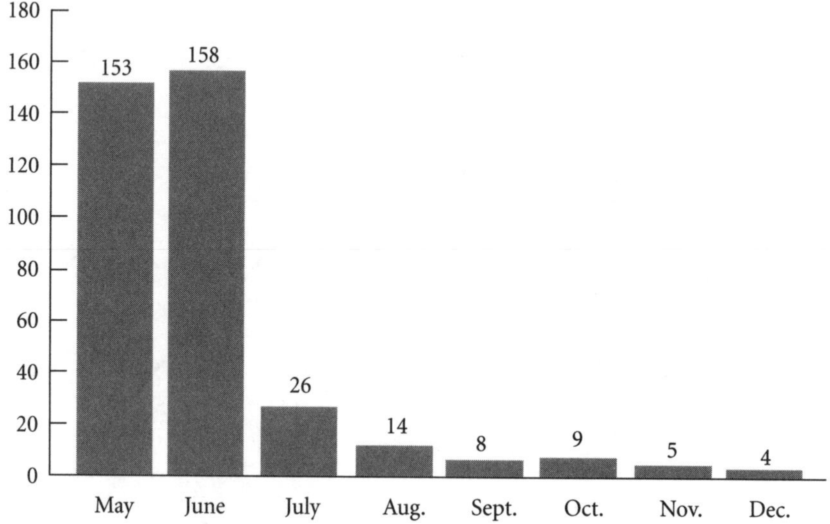


**Figure CS-36** This chart illustrates the dramatic drop in scare reports at Tres Cruces.

## Subsequent Steps

The *kaizen* project was followed up by the following actions:

- ▲ Study of the feasibility of an overhead conveyor system
- ▲ Study of the feasibility of unloading frozen fat directly into the cold-storage chamber
- ▲ Modification of the layout in order to improve reception of beef at the chamber

## Finding the Answers Within the *Gemba*

Managers tend to look to outside sources for solutions. For instance, when faced with safety problems such as those at Tres Cruces, management tends to turn to an outside safety expert for remedies.

However, the managers at Tres Cruces were able to solve their company's safety problems almost entirely on their own by following *gemba-gembutsu* principles and collecting data. I firmly believe that managers can find the answers to most of the problems facing them—and in fact already have solutions at their fingertips—if only they take the trouble to collect the necessary data by involving *gemba* people and asking "Why?" until they reach the root causes of the problem. They can then come up with countermeasures for each cause and implement them.

As the Tres Cruces example demonstrates so clearly, once management becomes serious about making improvements, starts collecting data, and commits itself to continual follow-up, the employees themselves gain an enhanced recognition of the problem and become enthusiastic about finding solutions and doing a better job. At Tres Cruces, this was evidenced by the sudden drop in the number of scare reports.

*Kaizen* is contagious. The improvement registered by the team at Tres Cruces' raw material receiving depot, which reduced accidents by 79 percent in 1994, had an immediate impact on another group, the meat deboning department, which reduced accidents by 60 percent during the first half of 1995.

In the course of these *kaizen* activities, people at Tres Cruces gained many valuable insights:

- ▲ Priority should be assigned in selecting *kaizen* projects; the receiving depot and the deboning department had the worst records in the plant and the highest occurrences of accidents.

- Employees worked hard on the project continuously throughout the year and realized that continuity was one of the conditions for their success.
- Employees realized that lack of data and the unreliability of existing data were the major barriers to embarking on *kaizen.* All accidents that had occurred in the previous year had to be checked, one by one, and a system to collect data by tracking every future accident had to be developed.
- All workers were involved, trained, and motivated to work on the project.
- The team started to work on the problems closest to its heart (accidents), creating expectations and concerns that fortunately were resolved early in the game.
- Seeing how seriously management dealt with industrial safety instilled a sense of trust in workers.
- Management realized the importance of scare reports and taught employees how to use them to preempt problems.
- Having employees fully involved is very important in building initiatives; at Tres Cruces, workers were involved in naming major scares.
- Based on the findings at Tres Cruces, a new form was prepared for workers to refer to whenever they experienced a scare.
- The reports were checked weekly, and the main causes of scares and accidents were identified using Pareto diagrams.
- Accident-free periods of record length (167 days) were registered twice, once in 1994 and once in 1995.
- Better working conditions, accident reduction, and various other improvements registered during this period resulted in productivity improvements.

# Glossary

**AQL.** Acceptable quality level is a practice between customers and suppliers that allows suppliers to deliver a certain percentage of rejects by paying penalties.

**Ask why five times.** See *Five whys.*

**Check *gembutsu*.** Examining tangible objects in the *gemba* when attempting to determine the root cause of problems.

**Conformance.** An affirmative indication or judgment that a product or service has met the requirements of a relevant specification, contract, or regulation.

**Control chart.** A chart with upper and lower control limits on which values of some statistical measures for a series of samples or subgroups are plotted. The chart frequently shows a central line to help detect a trend of plotted values toward either control limit.

**Cost.** When used in the context of quality, cost, and delivery (QCD), the word cost usually refers to cost management, *not* cost cutting. *Cost management* refers to managing various resources properly and eliminating all sorts of *muda* in such a way that the overall cost goes down.

**Cross-functional management.** An interdepartmental management activity to realize QCD.

**Cycle time.** The actual time taken by an operator to process a piece of product.

**Delivery.** When used in the context of QCD, the word *delivery* refers to meeting both the delivery and the volume requirements of the customer.

**Don't accept it, don't make it, don't send it.** A commonsense slogan to be implemented in the *gemba* that puts into practice the belief that quality is the first priority in any program of QCD; for example, don't accept inferior quality from the previous process, don't make rejects in one's product, and if a reject has been produced, don't knowingly send it to the next process.

**Failure tree analysis.** Failure tree analysis is used to analyze and avoid safety and reliability problems in advance by identifying cause-and-effect relationships and probability of problems by using the tree diagram.

**Five golden rules of *gemba* management.** A set of the most practical reminders in implementing *kaizen* in the *gemba*: (1) Go to the *gemba* when problems arise, (2) check *gembutsu*, (3) take temporary countermeasures on the spot, (4) find and eliminate the root cause, and (5) standardize to prevent recurrence.

**Five Ms (5M).** A method for managing resources in the *gemba*—specifically those known as *5M*—manpower, machine, material, method, and measurement.

**Five Ss (5S).** A checklist for good housekeeping to achieve greater order, efficiency, and discipline in the workplace. It is derived from the Japanese words *seiri*, *seiton*, *seiso*, *seiketsu*, and *shituke* and adapted to the English equivalents of *sort*, *straighten*, *scrub*, *systematize*, and *standardize*. In some companies it is adopted as the 5C campaign: *clear out*, *configure*, *clean & check*, *conform*, and *custom & practice*.

**Five whys.** A method of root-cause analysis used in problem solving in which the question "Why?" is asked repeatedly until the root cause is understood.

**Flow production.** One of the basic pillars of the just-in-time production system. In flow production, machines are arranged in the order of processing so that the workpiece flows between processes without interruptions and stagnation.

**FMEA.** Failure mode and effect analysis is an analytical tool used to predict and eliminate in advance any potential design defect in a new product by analyzing the effects of failure modes of component parts on the final product performance. FMEA is also used for design review activities of a new production facility (called *process FMEA*).

**FTA.** See *Failure tree analysis.*

***Gemba.*** A Japanese word meaning "real place"—now adapted in management terminology to mean the "workplace"—or that place where value is added. In manufacturing, it usually refers to the shop floor.

***Gembutsu.*** The tangible objects found at the *gemba*, such as workpieces, rejects, jigs and tools, and machines.

**Go to the *gemba*.** The first principle of *gemba kaizen*. This is a reminder that whenever abnormality occurs, or whenever a manager wishes to know the current state of operations, he or she should go to the *gemba* right away because gemba is a source of all information.

**Heinrich's Law.** A principle related to the occurrence ratio of accidents with injuries. Heinrich expressed the ratio as follows:

Serious injury : minor injury : no injury = 1 : 29 : 300

This equation expresses that when you see 1 person who was seriously injured by an accident, the same accident might have hurt 29 persons slightly. At the same time, there might have been 300 people who luckily were not injured but experienced the same accident.

***Hiyari* KYT (*kikenyochi* training).** *Hiyari* KYT is the practice of anticipating danger in advance and taking steps to avoid it.

***Hiyari* report (scare report).** *Hiyari* report (the scare report) is a written form from a worker to a supervisor that reports a condition that is unsafe and could lead to quality problems and/or accidents.

***Ishikawa* (fishbone) diagram.** A diagram originally developed by Professor Kaoru Ishikawa to show causes (process) and the effect (result). The diagram is used to determine the real cause(s) and is one of the seven basic tools of problem solving.

**ISO 9000 Series Standards.** A set of international standards on quality management and quality assurance developed to help companies document the quality system elements to be implemented to ensure the conformance of a product to specifications.

***Jidoka* (autonomation).** A device that stops a machine whenever a defective product is produced. This device is essential in introducing just-in-time (JIT).

***Jishuken*.** In the early 1960s, *jishuken* (autonomous JIT study team) was started to implement JIT activities in the *gemba* among the Toyota Group of companies.

**JIT (just-in-time).** A system designed to achieve the best possible quality, cost, and delivery of products and services by eliminating all kinds of *muda* in a company's internal processes and delivering products just-in-time to meet customers' requirements. Originally developed by Toyota Motor Company, it is also called by such names as the *Toyota Production System*, the *lean production system*, and the *kanban system*.

**JK (*jishu kanri*).** *Jishu kanri* means "autonomous management" in Japanese and refers to workers' participation in *kaizen* activities as a part of their daily activities under the guidance of a line manager; it is different from quality circle activities, which are voluntary and are carried out by the workers' own volition.

***Junjo*.** Logistics system that prepares and delivers materials to the line or point of use in the sequence of use. The Japanese word for "sequence" is *junjo*.

***Kaizen* concepts.** Major concepts that must be understood and practiced in implementing *kaizen*.

- *Kaizen* and management
- Process versus result
- Following the plan-do-check-act (PDCA)/standardize-do-check-act (SDCA)
- Putting quality first
- Speaking with data
- Treating the next process as the customer

***Kaizen* story.** A standardized problem-solving procedure to be used at each level of an organization. A *kaizen* story has eight steps: (1) select a project, (2) understand current situations and set objectives, (3) analyze data to identify root causes, (4) establish countermeasures, (5) implement countermeasures, (6) confirm the effect, (7) standardize, and (8) review the preceding process and work on the next steps.

***Kaizen* systems.** Major systems that must be established to attain a world-class status.

- Total quality control (total quality management)
- Just-in-time production system
- Total productive maintenance
- Policy deployment
- Suggestion system
- Small-group activities

**Kanban.** A communication tool in the JIT system whenever batch production is involved. A *kanban*, which means a "sign board" in Japanese, is attached to a given number of parts or products in the production line, instructing the delivery of a given quantity. When the parts all have been used, the *kanban* is returned to its origin, where it becomes an order to produce more.

**Kosu.** Manufacturing operations can be divided between machining hours and personnel hours. *Kosu* refers to the specific personnel hours it takes to process one unit of a product in a given process and is calculated by multiplying the number of workers involved in a process by the actual time it takes to complete the process and dividing that by the units produced. It is used as a measure of operators' productivity. *Kosu* reduction is one of the key measures of productivity improvement in the *gemba.*

**Morning market.** A daily routine at the *gemba* that involves examining rejects (*gembutsu*) made the previous day before the work begins so that countermeasures can be adopted as soon as possible based on *gemba-gembutsu* principles. This meeting involving the *gemba* people (and not staff) is held first thing in the morning.

**Muda.** The Japanese word meaning "waste," which, when applied to management of the workplace, refers to a wide range of non-value-adding activities. In the *gemba*, there are only two types of activities: value-adding and non-value-adding. In *gemba kaizen*, efforts are directed first to eliminate all types of non-value-adding activities. Elimination of *muda* in the following areas can contribute to significant improvements in QCD: overproduction, inventory, rejects, motion, processing, waiting, transport, and time.
*Muda* elimination epitomizes the low-cost, commonsense approach to improvement.

**Mura.** Japanese word meaning "irregularity" or "variability."

***Muri.*** Japanese word meaning "strain" and "overburden."

**One-piece flow.** Only one workpiece is allowed to flow from process to process to minimize *muda* in a JIT production system.

**Pareto chart.** A graphic tool for ranking causes from the most significant to the least significant. It is based on the Pareto principle, first defined by J. M. Juran. This 80-20 principle suggests that 80 percent of effects come from 20 percent of the possible causes. The Pareto chart is one of the seven basic tools of problem solving.

**PDCA (plan-do-check-act).** The basic steps to be followed in making continual improvement (*kaizen*).

**Pull production.** One of the basic requirements of a JIT production system. The previous process produces only as many products as are consumed by the following process.

**Push production.** The opposite of pull production. The previous process produces as much as it can without regard to the actual requirements of the next process and sends them to the next process whether there is a need or not.

**QA Best-Line Certification.** An in-house certification system to certify a world-class level of quality assurance performance of a particular process.

**QC circles.** See *Quality circles.*

**QCD (quality, cost, and delivery).** Quality, cost, and delivery are regarded as an ultimate goal of management. When management is successful in achieving QCD, both customer satisfaction and corporate success follow.

**QCDMS.** In the *gemba,* often *M* (morale) and *S* (safety) are added to QCD as targets to be achieved.

**QFD (quality function deployment).** A management approach to identify customer requirements first and then work back through the stages of design, engineering, production, sales, and after-service of products.

**QS 9000.** A U.S. version of ISO 9000 series imposed by the "Big Three" automotive companies to their suppliers compared with the general description of requirements by ISO 9000. QS 9000 specifies additional requirements, in particular, the need for continuous improvement of the standard and corrective actions.

**Quality.** In the context of QCD, *quality* refers to the quality of products or services delivered to the customer. In this instance, *quality* refers to conformance to specifications and customer requirements. In a broader sense, *quality* refers to the quality of work in designing, producing, delivering, and after-servicing the products or services.

**Quality circles.** Quality improvement or self-improvement study groups composed of a small number of employees (10 or fewer). Quality circles were originated in Japan and are called *quality control (QC) circles*. The QC circle voluntarily performs improvement activities within the workplace, carrying out its work continuously as a part of a companywide program of mutual education, quality control, self-development, and productivity improvement.

**Scare report.** See *Hiyari* report.

**SDCA (standardize-do-check-act).** The basic steps to be followed to maintain the current status.

**Simultaneous realization of QCD.** Top management must make certain that all levels of the company work to achieve QCD. The ultimate goal is to realize QCD simultaneously, but first of all, priority must be established among the three, quality always being the first.

**Small-group activity.** Shop floor group activity to solve problems that appear at their own workplace. Groups are usually formed by 5 to 10 shop floor operators. Their activities are mostly similar to those of quality circles. However, small-group activities are implemented not only for such activities as quality improvement, cost reduction, total productive maintenance, and productivity improvement but also for recreational and other social activities.

**Standardization.** Standardization is one of the three foundations of *gemba kaizen* activities and means documentation of the best way to do a job.

**Standardized work.** An optimal combination of worker, machine, and material. The three elements of standardized work are *takt* time, work sequence, and standard work-in-process.

**Standards.** A best way to do a job, namely, a set of policies, rules, directives, and procedures established by management for all major operations that serve as guidelines that enable all employees to perform their jobs to ensure good results.

**Statistical process control (SPC).** The application of statistical techniques to control a process. Often the term *statistical quality control* is used interchangeably.

**Statistical quality control (SQC).** The application of statistical techniques to control quality. Often used interchangeably with *statistical process control* but includes acceptance sampling as well as statistical process control.

**Storeroom.** The place where work-in-process and supplies are stored in the *gemba.* A storeroom is different from the normal warehouse because only *standardized* inventory is kept in the storeroom.

**Suggestion system.** In Japan, the suggestion system is a highly integrated part of individual-oriented *kaizen.* The Japanese-style suggestion system emphasizes morale-boosting benefits and positive employee participation over the economic and financial incentives that are stressed in a Western-style system.

***Takt* time.** The theoretical time at which a producer must produce a piece of product ordered by a customer. *Takt* time is calculated by dividing the net available production time by the number of units required during that time.

**Three Ks (3K).** The Japanese words referring to conventional perception of the *gemba: kiken* ("dangerous"), *kitanai* ("dirty"), and *kitsui* ("stressful")—in direct contrast to the idea of the *gemba* being the place where real value is added and the source of ideas for achieving QCD.

**Three Ms (3M).** *Muda* ("waste"), *mura* ("variation"), and *muri* ("over - burden"). These three words are used as *kaizen* checkpoints to help workers and management identify the areas for improvement.

**Three Ms (3M) in the *gemba*.** The three major resources to be managed in the *gemba*: manpower, material, and machine. The three M's are sometimes referred to as 5M with the addition of *methods* and *measurement*.

**Total productive maintenance (TPM).** Total productive maintenance aims at maximizing equipment effectiveness throughout the entire life of the equipment. TPM involves everyone in all departments and at all levels; it motivates people for plant maintenance through small-group and autonomous activities and involves such basic elements as developing a maintenance system, education in basic housekeeping, problem-solving skills, and activities to achieve zero breakdowns and an accident-free *gemba*. Autonomous maintenance by workers is one of the important elements of TPM. 5S is an entry step of TPM.

**Total quality control (TQC).** Organized *kaizen* activities on quality involving everyone in a company—managers and workers—in a totally integrated effort toward *kaizen* at every level. It is assumed that these activities ultimately lead to increased customer satisfaction and success of the business. In Japan, the term *total quality management* (TQM) is getting increasingly popular in usage and now is taking the place of TQC.

**Total quality management (TQM).** See *Total quality control (TQC)*.

**TQC.** See *Total quality control (TQC)*.

**Two-day *gemba kaizen*.** *Gemba kaizen* practices at Nissan Motor Company and its suppliers. A particular process is selected, and a group of internal *kaizen* consultants, engineers, and line managers spend two days in the *gemba* using JIT and other related checklists to attain the target.

**Value analysis (VA).** A method for cost reduction introduced by L. D. Miles at General Electric in 1947. It aims at reducing material and component costs at the upstream stages of designing and design reviews and involves cross-functional collaborations of product design, production engineering, quality assurance, and manufacturing. VA is also employed for competitive benchmarking.

**Value engineering (VE).** A method and practice for cost reduction developed by the U.S. Department of Defense in 1954. In Japan, both VA and VE are used almost for the same purposes (see *Value analysis*).

**Visual management.** An effective management method to provide information and *gembutsu* in a clearly visible manner to both workers and managers so that the current state of operations and the target for *kaizen* are understood by everybody. It also helps people to identify abnormality promptly.

***Yokoten.*** The horizontal expansion of successful results from *kaizen* in one area by sharing the learning with people in other areas.

# INDEX

# Worldwide Contact Information for Kaizen Institute

## Americas

### Kaizen Institute USA

13000 Beverly Park Rd, Ste. B
Mukilteo, WA 98275
Tel: +1-425-356-3150
Email: us@kaizen.com
www.us.kaizen.com

### Kaizen Institute México

Av. Chapultepec 408
Int. 3 Colinas del Parque
78260 México
Tel: +52 444 1518585
Email: mx@kaizen.com
www.mx.kaizen.com

### Kaizen Institute Brazil

Al. dos Jurupis, 452 – Torre A – 2º.
Andar 04088-001
São Paulo – SP Brazil
Tel: +55 (11) 5052 6681
Email: br@kaizen.com
www.br.kaizen.com

### Kaizen Institute Chile

Av. Providencia 1998 of. 203
Providencia, Santiago, Chile
Tel: +52 (0) 2-231-1450
Email: cl@kaizen.com
www.cl.kaizen.com

## Asia Pacific

### Kaizen Institute Japan

Glenpark Hanzomon, #310
2-12-1 Kojimachi
Chiyoda-ku, Tokyo 102-0083 Japan
Tel: +81 (0) 3 6909 8320
Email: jp@kaizen.com
www.jp.kaizen.com

### Kaizen Institute China

1027 Chang Ning Road
Suite 2206
Shanghai, China
Tel: +86 (0) 21 6248 2365
Email: cn@kaizen.com
www.cn.kaizen.com

**Kaizen Institute Singapore**
20 Cecil Street
#14-01 Equity Plaza
Singapore 049705
Tel: +65 (0) 6305 2410
Email: sg@kaizen.com
www.sg.kaizen.com

**Kaizen Institute New Zealand**
15a Vestey Drive Mt Wellington
Auckland 1060 New Zealand
Tel: +65 (09) 588 5184
Email: nz@kaizen.com
www.nz.kaizen.com

**Kaizen Institute India**
Office No. 1A, Second Floor
Sunshree Woods Commercial Complex
NIBM Road
Kondhwa 411 048
Pune, India
Tel: +91 92255 27911
Email: in@kaizen.com
www.in.kaizen.com

**Kaizen Institute Malaysia**
19-1, Jalan Putra Mahkota 7/4A
Putra Heights, 47650
Subang Jaya, Malaysia
Tel: +60 (0)3 5191 5112
Email my@kaizen.com
www.my.kaizen.com

## Europe, Middle East and Africa

**Kaizen Institute Austria**
Michael-Walz-Gasse 37
5020 Salzburg
Austria
Tel: +43 (0)662-423 095-0
Email: at@kaizen.com
www.at.kaizen.com

**Kaizen Institute Germany**
Werner-Reimers-Strasse 2-4
D-61352 Bad Homburg Germany
Tel: +49 (0) 6172 888 55 0
Email: de@kaizen.com
www.de.kaizen.com

**Kaizen Institute France**
Techn'Hom 3
15 Rue Sophie Germain
F-90000 Belfort France
Tel: +33 145356644
Email: fr@kaizen.com
www.fr.kaizen.com

**Kaizen Institute United Kingdom**
Regus House
Herald Way
Pegasus Business Park
Castle Donington DE74 2TZ UK
Tel: +44 (0) 1332 6381 14
Email: uk@kaizen.com
www.uk.kaizen.com

**Kaizen Institute Netherlands**
Bruistensingel 208
5232 AD 's-Hertogenbosch
The Netherlands
Tel: +31 (0)73 700 3440
Email: nl@kaizen.com
www.nl.kaizen.com

**Kaizen Institute Spain**
Ribera del Loira, 46 Edificio 2
28042 Madrid, Spain
Tel: +34 91 503 00 19
Email: es@kaizen.com
www.es.kaizen.com

**Kaizen Institute Switzerland**
Bahnhofplatz
Zug 6300 Switzerland
Tel: +41 (0) 41 725 42 80
Email: ch@kaizen.com
www.ch.kaizen.com

**Kaizen Institute Portugal**
Rua Manuel Alves Moreira, 207
4405-520 V.N.
Gaia, Portugal
Tel: +351 22 372 2886
Email: pt@kaizen.com
www.pt.kaizen.com

**Kaizen Institute Italy**
Piazza dell'Unità, 12 40128
Bologna, Italy
Tel: +39 051 587 67 44
Email: it@kaizen.com
www.it.kaizen.com

**Kaizen Institute Belgium**
Uitbreidingsstraat 84/3
2600 Berchem
België / Belgique
Tel: +32 3 218 2143
Email: be@kaizen.com
www.be.kaizen.com

**Kaizen Institute Czech Republic**
Vinohradská 93
CZ-120 00 Praha 2 Czech Republic
Tel: +420 736 620 849
Email: cz@kaizen.com
www.cz.kaizen.com

**Kaizen Institute Hungary**
Danubius utca 5. IX/5.
1138 Budapest, Hungary
Tel: +36 (1) 878 0703
Email: hu@kaizen.com
www.hu.kaizen.com

**Kaizen Institute Poland**
ul. Koreańska 13
52-121 Wrocław, Poland
Tel: +48 (0)71 335 22 75
Email: pl@kaizen.com
www.pl.kaizen.com

**Kaizen Institute Romania**
Ştirbei Vodă 162
010121 Sector 1, Bucureşti
Romania
Tel: +40 (0) 21-6372169
Email: ro@kaizen.com
www.ro.kaizen.com

**Kaizen Institute Russia**
105005, Россия, Москва,
Денисовский пер., 26
Tel: + 7 (495) 785 14 15

105120, Россия, Москва,
Костомаровский пер., 3, стр. 1
Tel: + 7 (495) 225 88 04
Email: ru@kaizen.com
www.ru.kaizen.com

**Kaizen Institute Kenya**
c/o KAM – Kenya Association of Manufacturers
3 Mwanzi Road
Opp Nakumatt Westgate Westlands
Nairobi, Kenya
Tel: +25 472 220 13 68
Email: ke@kaizen.com
www.ke.kaizen.com

**Gemba Academy online training:**
www.GembaAcademy.com

**Kaizen Institute Blog:**
www.gembapantarei.com

**Executive Master's Degree Program:**
www.kaizen.com/Master

To find more information for all KI locations and services, please visit www.KAIZEN.com

**Kaizen Institute Global Operations**
Bahnhofplatz
Zug 6300 Switzerland
Tel: +41 (0) 41 725 42 80
Fax: +41 (0) 41 725 42 89
Email: kicg@kaizen.com
www.kaizen.com